I0465294

Cover design by Joel Uber Interior design by Author

Published in the United States of America

ISBN: 9781981022335

Fiction, Science / Physics / Quantum Theory

Crystals of the Covenant

PROLOGUE
(June 6, 1202 A.D.)

"Oh, Children of the Earth my name is Pakraban Sinhad. I am a simple mortal of meaningless existence. I live in the valley of beggars in the mountains of Dumas, Syria. It is the year 1202 A.D. I have found a scroll in the caves of Abu Hannah, and I am compelled to write its warning for the world to read. I make it clear to you that this is a warning for future generations from an ancient entity called YuMuYa.

YuMuYa warns of the future of humankind. He warns that if the future generations of the Earth do not change their ways, then, this warning shall come to pass and it shall prevail and become reality. I do not claim to be a Prophet, but what I have read in this most secret of scrolls compels me to translate it and give to all of you to read so that you may know that you have been warned."

The Vatican

(June 6, 2040 AD.)

The first Black Pontiff reclines pensively in his favorite armchair in the new Vatican's white room. The brilliance of the sun's rays pay homage to the blazing gold-leafed armrest of his chair sending a spectrum of colors that filter through the large room. The Pope adjusts his eyes as he turns toward the small window. His gaze focuses on the crowds filling Saint Peter's Square below. The pleasing aroma of incense suddenly fills the room and he turns toward the entrance to see if someone has entered. There is no other in the room. On a cold marble top coffee-table beside him a cup of hot black coffee steams untouched.

He thinks he hears a voice. It is a soft whisper almost inaudible and directed as if internally in his brain.

"Paul," it says.

He quickly glances around the room again to see where the voice comes from but could see no one.

"Do not look upon me", the voice says, this time in clear tones.

The Pope is confused. He feels his arm trembling but not from fear.

"Take your pen and write what I tell you," the voice says.

The Pope hesitates.

It is Latin, he thinks. Whoever or whatever is speaking to him is speaking in clear ancient Latin!

He quickly picks up a white notepad bearing the insignia of the Holy See and prepares to write. He can sense a mysterious presence lurking behind him. From his peripheral vision he sees a man-like figure dressed in a long black robe. The man moves in front of him in full sight. A slight cold shiver runs through the Pope's veins as he realizes that the figure in front of him is no ordinary person. His face is covered in a long white veil that wraps around a head and falls gently over the chest. His robes are of a priestly nature.

"Who are you?" the Pope asks in a trembling voice.

"Do not question my presence, write what I tell you!" the apparition commands in soft musical tones. The Pope looks at the apparition and relaxes. The strange man towers more than six foot six. He cannot see a face or features that will reveal anything he can learn about it. The veil bears a strange golden emblem across the chest. The emblem is the image of a bald eagle with a humanlike head sporting a crown of jewels. A knife sticks through the center of the eagle's breast but its bulging eyes betray no pain. In its powerful claws is the dead body of a huge black serpent bleeding from deep wounds inflicted by the eagle's claws. The pope has seen this emblem before. As he prepares to write and recalls that the emblem is the original insignia of the Nicaea Council!

Through the veil, he could barely see a small brown and handsome face graced with a long white flowing beard. There is a glow about the face and the Pope bears no doubt that this is no ordinary person. The Pope assumes they are features of a Middle-Eastern priest or monk. The apparition stands silently in front of the Pontiff then slowly as if floating it moves beside his chair and rests a slender clean brown hand on the soft upper curve of the leather armrest while the other hand holds a long wooden staff. It is a comforting gesture. The Pope relaxes and focuses through the window on the crowd of worshipers in the massive square below. He can no fully longer see the figure but can sense his presence beside him.

The Pontiff continues to write as the strange man dictates in the musical notes of ancient Latin. For hours he takes notes listening intently to the words of the man he cannot see. At the end of the dictation the robed figure bids the Pontiff farewell and disappears into the depths of the Vatican. The Pontiff did not have the privilege of seeing his face. Tears roll down the Pontiff's eyes as he softly repeats the words,

"A Guardian, O Holy God, I have just encountered the rare apparition of a Guardian!"

For days thereafter, the apparition comes and instructs the Pope on what to write. More than one hundred volumes have been written, bound and archived. They are to be the kept in secret silence until an appointed day comes. When that day comes none knows but Pope Paul the seventh himself.

CHAPTER 1
(January 14, 2011 A.D.)

A flash of light breaks the darkness in the room and something tells me that I am in real trouble. It is dark and ominous and I cannot decipher where I am. My mind is still floating in a place that I do not quite recognize. I feel pain searing through my head and I instinctively reach out to touch the spot where the pain comes from but my hands cannot reach the spot. My hands are tied to the head post of a bed and I can barely move my sore head to see my surroundings. I try to open my eyes but cannot.

"He is alive" a voice states. A flash light again points at my face and I feel a hand trying to open my eye lids to see my insides. I am in horrific pain and the light burns deep into my head. Where the hell am I and who are these people? Slowly I force my eyes to open and I start turning my head around to see who they are. A tall heavy set man in a dark suit looks back into my eyes from across the room. He smiles and starts moving toward me. I turn away from him instinctively and look the other direction. Another man is standing by an open window looking outside as if expecting something to happen.

"Seth" the first man calls my name.

I tried to talk but could not my head hurt too much.

"Give him another dose" the first man says in a loud thick Southern accent.

"It will kill him" the other revolts.

"We cannot wait any longer we need the information now!' the first man states with some authority.

I feel the needle stab into my arm as the drug penetrates my veins. I feel an overwhelming feeling of peace and calm and for a moment the world seemed to be just right. I am relaxed and inhibited. I opened my eyes and feel a rush of light enter my brain. The whole room becomes clear and receptive to my thoughts. I feel energy surging through my veins. The pain is gone and all I feel is heightened sense of my being and my environment. The man in the dark suit opens the curtains to let in light and the room fills with a brilliant glow of colors that seem to me like flowing musical notes. I do not feel any resistance or resentment for where or what I am. I am just me, my being and my world.

"Your name is Seth Malaki," the man in the dark suit states as if I should know.

"Who are you" I ask looking around the room to ascertain that this is all real.

"My name is Mendeleev Polcheski, " the man in the dark suit states.

"We are from the CIA and I am here to interrogate you," he states blankly.

"Why?" I ask not really caring why but more so because they are disturbing my new peaceful pain free world.

"You were researching a secret scroll called the YuMuYa scroll," he says pointedly, "do you remember that?"

A faint memory comes into my mind as thoughts start rushing through my head as if they are being pulled from the vacuum of nothingness. Suddenly I know it all again. It is all there in my head.

"We have given you a dose of Memorin, a drug that is used to enhance memory. We need some important information for National Security," he says.

"Why are my chained down?" I ask.

"For your own security," the other man states as he walks toward my bed and looks into my eyes. I hint of recognition starts to flow through my mind. Then it all comes to me in a rush, it is Senator James Owen of South Dakota. We went to the same high school in Coral Springs Florida and we were best friends back then. After graduation he moved with his parents to South Dakota and he is now a Senator. I met him again during the 1999 Pittsburgh conference on analytic instruments and we rekindled our relationship and became very close friends. I have met him again several times since and we corresponded with follow up discussions on Space instrumentation and funding for NASA. He was always a featured guest speaker at NASA training centers. What is he doing here?

"You are delirious and in real danger. We had to restrain you to make sure you do not go off and endanger yourself," Senator Owens states reassuring me again. I feel a little more relaxed as he puts a warm hand on my head. At least he is a familiar face and I consider him to be my friend.

"Your fever is almost gone Seth and you will soon be Ok" he comforts me.

"What happened at the Space Station?" the man in the dark suit asks.

"Who is he?" I ask.

"Mendeleev is National Security. He is here to help us get some information that is locked in your head!"

"The Space Station?" I ask as a thousand thought come rushing it. I suddenly recall that I was in Space, I am an Astronaut!

"You and two Cosmonauts were launched to the Space Station in November in a Russian Soyuz TMA-22 spacecraft. You were supposed to be the first of a joint American-Russian mission after the decommissioning of the Space shuttles." The Senator explains.

"So how did I get here?"

"You tell us. All we know is from the recordings of the Station and they were so strange we could not tell fact from fiction. That is why we need first-hand information from you," Senator Owens states.

"And the other Cosmonauts?" I asked.

"They station is empty. They evacuated with the Russian return craft but they vanished in the Ocean. There is no trace of them, you are the only survivor."

I tried to recall what happened in the Space Station but my memory comes to a blockade. It is as if some metal fence prevents me from entering that area of my mind. I remember the evacuation, but then, I do not recall coming in with Berkoff and Mulenski the other Cosmonauts.

"How did you find me and what about the other Astronauts in the Space Station?" I ask.

"Your craft docked on the station but you did not enter the station. The entry lock was jammed. You and the two Russians remained in Soyuz for two days exchanging information with the Station. Then something happened that forced you to separate from the Space station. Your craft returned to Earth but fell into the Pacific Ocean about 100 miles off-course. We searched the area and found you floating on your automatic inflation suit. You are paralyzed from the chest down and so you could not move or activate your radios. Luckily a Japanese Cruiser found you and recovered you. No trace of the space craft or the other two Russians has been found." Mendeleev explains.

My thoughts reel to the moment when we arrived at the station. I was responsible for navigating the rendezvous with the Space station while Berkoff was in communications. Mulenski was preparing the air lock sequence. I try to pull in the memory of what happened but again my mind is pulling a blank. Mendeleev removes an IPod from his pocket and looks at me again with his enquiring eyes. He turns it on and points it at me as if that would make a better reception. I hear strange sounds and then my voice. I was shouting out commands to Berkoff, asking him to compensate for the spinning craft. He was shouting back commands at me while Mulenski was crying and swearing in Russian. It sounded like a desperate attempt to save our souls and we were not succeeding. There is a strange background squealing coming in. I could not recognize its origin. Mendeleev stopped the IPod and looked at me again. I am supposed to find some explanation for all that noise. I forced my mind to recall the squealing sound and it did.

"I remember putting on my safety suit at this point" I shouted to Senator Owens as if he will understand. I am beginning to feel guilty for my Russian colleagues.

"When the squealing started?" He asks.

"Yes. I activated my safety suit and the squealing started as if it was interfering with the craft's electronics!"

"Do you recall why you had to eject the craft from the lock?" Mendeleev asks.

"I remember that when we first attached to the lock, it froze and would not open. Then the power connection to attach to the station was not correctly done. The connector jammed and we could not attach it to the station. The pressure equalizer started working properly but then it would not stop at the designated pressure. It exceeded the station pressure and we were getting really anxious. We knew we had to bring the difference down otherwise the craft would be blown off the station as if it were a bullet." I explain.

"You were there for three days after that incident. You must recall that something happened during this time that we need to know about as soon as possible. Try to recall the first day," Mendeleev states concerned.

"We could not get the air-lock opened. The pressure was building up and we were very worried that we would die if it continued to build. I was already experience bends at some point. I recall the pain shooting through my ears."

"Your communication logs show that you were concerned about an incoming signal. You and Berkoff were exchanging calculations about orbits. The sounds that were recorded were not too clear."

Again, Mendeleev started the IPod. It rang out the squealing sound again, but this time there is a conversation between me and Berkoff. We were talking about Voyager! I recalled that we were doing calculations about the orbit of Voyager in deep space. Something about images in a viewer we were receiving. The recording suddenly stopped. My memory is faint of what went on but I do recall that we were talking about the YuMuYa scroll.

"The YuMuYa scroll!" I utter.

"What about the YuMuYa scroll?" Mendeleev asks.

"Seth when you were a young man, you worked in Africa with a mining company. Do you recall that?" Senator Owens asks.

"Yes, I do recall. I was a kid earning a living. But what could that have in common with the Space Station?" I ask.

"We do not know. All we know is that you were muttering about this scroll as if it was the object of the mission." Mendeleev states.

I am confused. I look at Senator Owens and tried to reach out to touch him. My hands move but they are constrained.

"Get him out of those restrains," Senator Owens pleads with Mendeleev.

Mendeleev thinks about it for a minute and then reaches out to get a set of keys in his pocket. He opens the cuffs and releases my hands. I look at my feet and point to them. Mendeleev releases them too. I try to move my body but I cannot. I can feel the commands going down my nerves to my feet but they do not move. I cannot move my feet! I relax my body and try again but to avail.

"You cannot move your feet Seth and you may never be able to." Senator Owens is terse and abrupt almost wicked.

"Tell us about the YuMuYa scroll" Mendeleev states, I hear a strong demand and a command in his voice. My brain goes back to Africa. The lush equatorial forests with the sweet smell of palm trees and Mango come back. The fresh air mixed with the musty smell of ant colonies and animal dung. A pleasant memory of past days, when I used to work in the mines, the diamond mines; the earth. Something tells me that I should not reveal some information, but what information? I do not know.

"Before you became the first civilian American sent to space with the Russians, you were a Mining Engineer in the small nation of Sierra Leone in the jungles of West Africa." Senator Owens states as if that should spark a memory. It does. It sparks a memory of a bully; my father, Tanius. My mind floats to the stairs leading from my father's back yard porch to the garden behind our house. I am sitting on the top of the concrete stairs looking at the flowers blooming a few feet from me. They are roses, red as roses can be. I thought I was alone. I hear the birds and the other sounds that make you remember what nature is like. I can smell the flowers from where I am sitting. The edge of the cold concrete porch is painted with green patches of algae and I am using a small branch to scrap off patches to create images. My mind is immersed in my creations and I am oblivious of the world around me. I start sucking my molars as I enter into my world completely lost in my reverie. I always thought that sucking my molars will take me to a new world, a world that only I have the power to enter. I feel myself vanishing from the real world and entering a new dimension, the dimension of love. It is funny how other people cannot feel or travel through this dimension and I am able to. My thoughts run to my lovely mother, Zina. She is my only consolation in this new world that I enter. She lives in the dimension of love. It is a real place where I can meet her and play and feel free. Sometimes I travel through this dimension to my cousin Mavian, who loves me very much. I would go far into the hills where she lives and enter her house. It is a small beautiful house perched on the side of a hill. She would pick me up and toss me in the air with pure joy. There is no age limit in this dimension. I can travel as a baby, a teenager or an adult.

My reverie is broken by a loud crack as the stick breaks. I hear my dad calling out my name in some other dimension; it is a dimension that I dislike. It makes me come back to the point zero where I started in the dimension of love. It is the dimension of hate or anger. My thoughts are again broken by my father's voice calling my name. I am propelled through a space of confusion only to land back on the stairwell where I started. As I come back to the real space and time I look at the sketches I drew on the algae bloom albeit subconsciously. It is a pattern of twenty seven circles forming a cube around the corner of the moldy stair case. I always do that when I travel into the other worlds and I never did know why until recently. I hear the whisper of my brother's voice in the background.

"Blood diamonds war," Senator Owens states.

"It always takes him to a place that he does not like!" He continues as if I am not in the room.   They sound so alike, my brother and my father. They always bring me back to the real world in a hurry. I start thinking about the war. It is not what they believed it was. It was a war that had a lot more deeper meaning than they could realize.

"Yes. Some locals believed that the war started when an invading group of warriors from Northern areas of the country entered the mining fields in search of a mythical diamond called "The Eye of M'Dulu"". I state blandly.

When I worked in the mines, I learnt that the Eye of M'Dulu is supposed to be the greatest diamond ever created by the ancient forces which the locals call Yumanya. That is what they are really after before I die. They want to know where to find the Eye of M'Dulu. I do not think that I should tell my brother or Mendeleev about the Eye of M'Dulu. Like everyone else, they will just get excited about the wealth it can bring them. They should not really know much.

"Is that what you were discussing with Berkoff and Mulenski?" Mendeleev asks.

"I do not know. I do not know why I would but perhaps memories of a fearful episode must have come into my mind during the stress," I explained not quite believing myself.

"In the recording you said that you discovered a scroll in Africa, and that the scroll predicted the coming of Voyager. Was this real, or was this just some delirious episode that was due to the stress?" Mendeleev asks.

"I do not know. I recall that we were all very stressed and worried and that the Space station was out of reach even though we were docked right there on its portal." I explain.

"Three days Seth. You spent three days out of touch with NASA and out of touch with the Station. Your communications were faulty and something was disturbing the signals we were getting. What was it?" Mendeleev asks.

I look at Senator Owens as if he could help me recall. The thoughts of Africa and the freedom of the wide open plains of grass come back into my head. I start to suck on my molars again. This time I am going to travel through the dimension of Hope. It is a dimension that makes what you want to believe comes true. The more you believe the deeper in you go. I have to go in deep to enter the past. It is a place where I want to go just now since I do not want to be back here in this room. They continued asking questions while my mind fades away into another world. Slowly, the face of Militant fills the screen in my head. He is smiling as if to encourage me to recall. I am back at the camp site in the mines. It suddenly comes back as if I am there-now. It is more than thirty years ago. I am drinking cold drinking beer with other guys and we are listening to Brazilian miner stories. It is a cold evening and for some reason I am thinking it is Saturday and I have to do something urgent before noon comes. It is almost as if I have to go to the bank before it closes for the weekend. But there are no banks in the African jungle. We are all sitting in a large circle around a large burn-fire filled with glowing amber logs. I recall cutting some of these from the forest behind the house. It is one of the chores that I had assigned to me for Saturday. One of the Brazilian miners, Militant Vegas is talking about a strange incident which he claims to have experienced first-hand while working with the Tuaregs in Central Africa. Militant is the type of guy you listen to. He carries an air of no nonsense about him. He is explain that he was working on oil drilling operations in Central Africa when they stumbled upon the grave of a Pigmy that appeared to be centuries old. Militant's long black hair is flowing in the wind as if confirming what he is saying to us. He should cut that hair it makes him look like a very ugly woman with a big nose. I try to pay attention to his story but the old Brazilian miner, his name is Bayano, has a very old looking girl-friend that keeps smiling at me as if I am an object of her desire. I hate that. I look away and focus on Militant again. According to Militant unlike the traditional burial rights of indigenous tribes the remains were mummified and beside the still intact bones they found an ancient scroll that he referred to as the YuMuYa scroll.

Militant is the type of guy that never finished a sentence before he jumps on to something else. He is always ahead of himself. While telling us the story Militant suddenly stands up and walks back into the house. He emerges moments later with a slim black briefcase made from alligator skin. Militant clears the coffee table in front of us as we gather around it to look at what he wants to show us. He wipes the table clean and removes the beer bottles and places them on the ground. Careful to avoid any damage, he places the brief case on the table and opens it to remove a carefully rolled cloth loosely tied with twine. He unfolds the package and places it on the table. Beneath an aluminum sheet foil is what appears to be an ancient scroll made from dried papyrus leaves. He removes a bunch of papers from his briefcase and then carefully sorts out some photos of what appears to be images of a scroll. He allows us to examine the scroll and the photographs carefully guiding the process to make sure we do not damage the scroll. The torn brown Papyrus cloth the scroll is written on had the scars of time. The Egyptian hieroglyphic writing is faded but legible and the rim of the cloth appears to have been chewed upon by animals. There is a translation of the scroll on a separate parchment that appears to be in very good condition. Militant states us that the scroll was written by an ancient entity he called YuMuYa. He explains that the scroll was a special document that foretold about the end times of the world. He points to pictures of the body of a pigmy. The body he explains was found mummified in a grave with the scroll well wrapped and protected on the mummy's chest.

Militant explains that at the time of the discovery a band of Tuaregs who were present during the discovery appeared to be very disconcerted by the opening of the ancient grave. They hastily left the mining site and reported the discovery to the Dogon tribal head who ruled that region of the forest. Some of the locals that had become close to Militant told him that graves are sacred to the Dogon and that by opening the burial site, demons from the stars will be summoned to punish the camp and the people around the camp. The site was getting very chaotic as word spread that a special grave has been assaulted by the foreign miners. Soon, Militant explains that they were approached by a large band of locals who took over the grave site and started escorting them out of the area into the thick of the jungle. They were commanded to leave the area immediately and stop all mining activities. They complied and left for the city of Bandiagara. A couple of weeks later a Tuareg miner who worked with Militant approached him in his hotel room and told him that he had some special documents that concern the grave site. The government investigator who was sent to examine and preserve the grave site was not particularly impressed with the legends of the Dogon tribe. He saw an opportunity to make some money. The man, Militant explains had confiscated the scroll and taken a lot of pictures of the mummy and gravesite. He gave them to the Tuareg to sell to the foreign miners. Militant told us that he then demanded that the scroll be translated before he pays for it. The official complied and Militant bought the scroll and photos.

My reveries are interrupted by a loud voice that did not belong to the camp site. It is coming from outside my head! It is a familiar voice that somehow has penetrated my world and is able to reach me where I am.

"These were the real scroll and photos?" the voice asks. I think for a moment about the voice. It is Mendeleev! I find myself torn between two worlds. One is where I am in the mines listening to Militant. It is where my soul is and I think it is as real a world as the other world in space and time. How did they get through? I wonder. I find myself answering a question.

"Yes. Militant is half-drunk we have had too many beers. The scroll appears to be genuine and so do the photos of the mummified pigmy."

I must have been speaking out loud. Militant stopped talking. He is looking at me as if he can hear me talk to the other side. Bayano, the old man smiles and starts laughing.

"I knew he should not be drinking so much beer, he is still a kid!"

I know he is talking about me and I am puzzled by the fact that I am interacting with the dimensions of Space, Time and Hope all at the same moment. I am in the past and future in one piece. I look around the camp to see if all is fine. I see the eyes of the other guys in the camp looking at me and laughing. Obviously they think I am drunk and talking to myself. Militant continues to talk. The voices fade away and I am back in the camp site. This time it is Militant and Bayano that catch my attention. The fire is duller and the flames are subsiding into a faint amber glow. Bayano Solvay is mocking Militant incessantly, teasing him about the scroll. "Every time Militant is drunk he starts talking about the scroll and you start talking to strangers", Bayano laughs.

"Can you describe the scroll, is it real?" Mendeleev asks again bring me through to the real world, or at least what I think is the real world.

I do not answer. It is getting dark and I decide to leave the camp fire and enter the house. I walk to my room and lock the door. In moments, I am sound asleep. I awake to a bright morning sun penetrating the window of my room. I am back in the hospital room or where Mendeleev and Senator Owens had brought me yesterday. There is no one in the room but myself. I want to close the curtains, but could not move. The sun's rays are too bright and my eyes are watering from its intensity. I cover my face with the blanket. That works, but then I cannot easily breathe and I have to remove the blanket from my face. I manage to place a pillow between my eyes and the sun and that solves the problem. I look around the room it is large with plain white walls. To the right of my bed is a single door that is half open and I can easily see if someone enters. I look beyond the door and see a far wall with a plaque hanging on it. Beyond the door must be either a corridor or hallway. I try to read the writing on the plaque. It looks like one of those certificates you find in a doctor's office. I can read the name Dr. Joseph Cooper. I must be in a hospital of some sort yet I do not sense the familiar smells I associate with a hospital. My head is not hurting much this morning, but I can sense that I will be restless all day. I feel anxiety and desperation at the loss of my mobility and that is not a good thing because I can be taken to those dimensions easily. I see nurses walking past my door and I somewhat hope that one of them will show up and comfort me. I hope Senator Owens and Mendeleev do not show up today. I am neither ready to talk about the past nor enter some dimensional gateway especially when Anxiety and Desperation are looming. My reverie is broken when a nurse walks in as the clock strikes 10. She closes the door behind her with bang. She is carrying a silver tray of food which she places beside my bed on a bed tray. She looks at me as if she can read my soul.

"You must be hungry" she says.

Her blue eyes look like two opals blazing with clarity and her blonde hair is neatly tucked under a blue head wrap. She looks sterile and clean and I can tell that she is one of those nurses that have pride in the profession. I do not respond to her comment. She sticks the bed tray over my chest and places a bottle of water beside the food bowl. The aroma of spinach and tomatoes fills my nostrils and I realize that I am very hungry.

"Your brother requested a special meal for you, he says you love spinach and rice," she says looking at me with a big smile.

"Yes, I do, thank you very much," I reply hoping that would start a conversation with her. I sort of like the way she looks at me with those big blue eyes.

"My name is Diana," she volunteers.

"Diana. That is a lively name," I reply smiling at her.

"You are an astronaut!" she exclaims "my father was an astronaut!"

I wait to see what more she will say but her smile is constant and I know that she is hoping I will say something back.

"Yes I am."

"You were in Space then?"

"Yes, I was on the last mission."

"I always pray when they lift-off. Ever since Columbia, I have had a special prayer for all the missions," she says. I could sense pride and humility.

I think about what she says for a second, "every mission?" Someone who prays for every mission must be really close to NASA. It takes a special interest for someone to do that.

"Where am I?" I ask.

"You must be hungry," she replies smiling, "you must eat your food before it gets too cold."

"My father was an astronaut. He was on Apollo 12," she volunteers again.

"Conrad or Bean?" I ask.

"Neither. Gordon," she answers, "he never walked the moon."

"November 14, 1969, the day I was born," I volunteer.

"The day of the launch! Now how is that for coincidence?" she smiles at me.

"Fete" I replied smiling back.

"The voices," she says, "did you really hear the voices?" she asks.

"What voices?"

She looks around the room suspiciously and comes closer to my bed. She cups her hands around her mouth and whispers.

"My father told me about the lights."

"Mmm, they must be aliens on the moon!" I joke.

She does not seem amused.

"What do you think about space?" I ask hoping to see if her interest would escalate.

"Ever since Challenger I thought space was not meant for humans to tamper with."

"Why?"

"Remember what McAuliffe said that she wouldn't really be the only teacher on her flight and that she is only one body, but there's going to be 10 souls that she is taking with her? Did she know that she was going to take souls with her?"

"She must have had a strange intuitive notion of reality. She was a real hero who did not fear the consequences of her mission. She only saw the privileges of the mission and so she cared very much," I try to console Diana.

"But why? Why do things like that happen?"

I decide that perhaps I should change the conversation since it is not in my best interest to continue on a path of thinking that involves emotional journeys.

"Where am I? What is this place?" I ask again. She realizes that I am not interested in a conversation about space and so she moves away from the bed and starts toward the door without looking back at me. I uncover the tray and start to consume my food. It is a delicious blend of rice and spinach with pieces of chicken and potatoes all mixed in with the greens. I eat voraciously cleaning my dish totally empty. I feel really good now and I so want to go out into the fresh daylight that spills through the room window. Outside the window I can see a delicious blue sky broken by blooming Japanese Cherries in the distance. I remember those cherries, the cherry blossoms. I must be in Washington D.C. that was where I last saw them. I relax back unto the bed looking at the empty ceiling tiles above. They have no designs of interest and no particular pattern I could play mind games with. The room is silent for a minute and then she returns again. This time, she does not pay attention to me. She removes the tray from my bed and heads toward the window. Before I could protest, she pulls the curtain shut and covers the room in darkness again. I plead for her to open the window, but she ignores me and heads out the door shutting it behind her. For an hour or so, I daydreamed about outside. I so want to go out and breathe the fresh air again, to feel the cool breeze on my face and enjoy fishing in a lake and all that. I so wish I can walk again. I must have fallen into a deep-deep sleep. I awaken to the sound of wheels rolling on tiles. My head is dizzy and all I can see are lights flashing past me high above. I hear voices calling out readings with urgency. They seem to be talking about me as if I am in trouble. I soon realize that I am on a hospital cart and I am being carted away urgently at high speed.

Moments later, I wake up in a different room surrounded by what appears to be doctors and nurses. There are needles all over my body and my mouth is gagged by what appears to be an oxygen mask. I do not know what all the fuss is about because I feel Ok and do not seem to be in any trouble. The doctors are talking about my blood pressure and pulse and it seems as if I have had a cardiac arrest and they have stabilized me. Mendeleev is asking questions about the stability of my health and the doctor is insisting that Memorin will kill me but Mendeleev is insisting that they give it to me anyway!

Memorin, just what I need now. They argue for a while determining what to do with my being. They have total control of my being and there is nothing I can do about that. I sort of want Memorin to take over my schizophrenic world again. It calms my transitions to my synthetic world of emotional fields. Mendeleev leaves the room and the doctor follows him. They shut the door with a loud inconsiderate bang. Loud noises always scare me out of my reverie and I can feel the vibrating waves passing through my entire being. The room is dark again and a beam of light from the sun escapes through a slit in the curtain to neatly knife the wall on the far side. I see the dust particles floating through the light creating patterns of random motions. I feel myself sucking my molars again. When that happens, I know that my mind is about to enter into a dimension of some sort. I sense the twitching of my eye as I start falling into the dimension of Love. Love, it is a beautiful world with many meanings. I find myself sitting on a mat in the floor of a little mud hut. I know this little hut very well. It is place where my nanny Masiray used to take me when I lived in Africa with my dad. It is raining outside but the sun is also shinning bright. The smells of the earth stir up the dusty patio and rise up to my nostrils to awaken a sense of belonging in me. I see the rain piling up into a little stream in front of the hut through the open doorway. The mud floor is cold and comforting and I am sitting on a small raffia mat beside a raised portion of the mud floor that forms a bed. The familiar smell of burning wood fills the air. Tubular rays of sunlight come streaming at regular intervals through small holes in the patched tin roof. I watch the dust particles play their random games as they pass through the light and then vanish into blackness. It is a never ending stream of particles that seem to have a brief existence only when they pass through the beams of sunlight. Outside these beams they do not exist. Masiray is naked except for a small cloth around her waist. She is preparing something in a small pot by the wood fire in the middle of the hut. I am sucking my thumb and my nose is runny. I watch her move about the hut barefooted. She approaches me with a cloth in her hand and I instinctively move my face away from her hand. She holds my head still and vigorously wipes my runny nose with the cloth. She sits beside me and starts to sing a familiar song. After a while, she gets up and removes a pot from the fire. It is simmering hot and I can see the steam boiling off as she approaches me. I sense that it is bath time and I instinctively try to stand up and run away from her. I cannot walk. I stumble as gravity gets a better hold of my tiny frame. I am a child barely able to walk. She grabs hold of me and takes off my clothes and places them on the mat. She removes my nappy and lays me on the mat as I revoltingly start to cry out in revolt. Her hands are gentle as she massages and cleans my body. I stop crying as the warmth of the cloth sooths my body.

"This will make the cold go away" she says.

She covers me with a towel and then clothes me with a fresh pair of pajamas. She lifts me and places me on the soft feather mattress on the raised earthen bed. I am feeling very good, warm and loved as she lies beside me and starts singing my favorite song again. I doze off into the dimensions of Love once again.

CHAPTER 2
(January 19, 2011 A.D.)

Memorin is the type of drug that makes you forget your pains and sorrows. It is able to take you to the doorways of emotional dimensions that are sometimes hard to get to. It opens the door to emotional dimensions and leaves you standing there looking at these other worlds of reality. It does not push you through the door or tell you what to do next but it at least makes it easy for you to get there without being prompted by all the emotional drama you will need otherwise. People sometimes need some grand emotional event to take them to these dimensions and when they get there they are sometimes totally lost trying to find the right door that goes to the right emotional dimension. Memorin takes you to there and opens the correct door.

When I open my eyes again, I am in the hospital room with the nurse Diana standing by my bed. They must have given me more Memorin because time passed quickly and I am again looking at Nurse Diana.

"How are you feeling?' Diana asks.

I do not answer. Diana is standing by the bed with a radiant glow in her eyes. Memorin makes her look fabulous and beautiful. She measures my pulse and blood pressure and I could see that she is concerned and that she is looking at me with some confusion.

"They wanted me to give you 100ml. but I only gave you 10ml," she states as if I should appreciate and understand the difference.

"Thank you."

"Your friend Senator Owens left and is no longer here. Doctor Cooper has taken charge of your case," she states.

"My case? What is wrong with me?"

"I am only following orders," she answers looking deep into my eyes. I can feel her thoughts and they are of real concern.

"Where am I?" I ask again hoping that she can answer the question.

She smiles and says "I am not at liberty to say."

"So, are you my nurse?"

I do not really want to talk right now, but I ask to engage her and get some more information. Perhaps she will open up if I can somehow relate to her state of being and engage her in a conversation.

"Yes. I am a private RN."

"What hospital do you work for?"

She smiles she knows I am trying hard.

"Try again I am not at liberty to tell you. You are in good hands, at least in my hands." I could sense that she genuinely means what she says and I can sense real concern in her voice. I need to probe deeper.

"Where were you born?" I ask hoping to keep her talking as I probe deeper using my special abilities. The Memorin is starting to do its thing and I am slowly feeling relaxed. My brain is entering into a state that opens up the fields of reality that very few can feel or sense. I look at her and wonder if she knows I can sense her fields of concern. In my world there is something that others call an "aura" that I can feel and sense. I can feel and see it all around a person but only if that person is tuned to me and my frequency.

As she talks of her childhood in South Carolina, Diana's aura becomes a bluish glow and looks like a soft smoky field that starts to surround her being. It is becoming particularly intense around her head like a holy glow. Unlike what others call an aura, my auric visions are real. They are fields that I can actually see and feel around people. They are almost electromagnetic in nature but in fact they are emotional fields. When I was studying physics to become an astronaut, I encountered a theory proposed by a scientist whose name I do not recall. It is called the Principle of Causal Conspiracy. I read two books that described fields that are associated with different types of quantum information and the mind. These quantum fields are called the annihilator-creator fields or the Josephus Nambu fields. I know these fields very intimately. They impart a sense of self to individuals and provide us with a world that is private. This is how the mind becomes this private "self-thing" that gives us a sense of oneself.

As she talks I decide to enter her self-field to look. I generally do not do that because it is a private thing and trying to enter another's field can be very embarrassing. I feel a compulsion to look into her field and so I projected my field into hers. She is telling me about her childhood and about her father. I enter into a room with a bright light sitting on the far end of a table. Her father is standing beside the table and she is looking up at him demanding that he stops what he is doing so they can go to the mall to buy a Barbie. Barbie was the thing then for kids. Richard Gordon, her father looks just like he did in the large gallery of NASA photos of the Apollo 12 mission. He is playing with a blue laser and seems to be tuning a mirror. She tugs at his pants and his hand pulls the mirror off course. He turns angrily and smacks her across the cheek. The laser is burning a hole in the wooden door to the garage and a fire starts. Smoke fills the room as a can of penetrating oil lights up with a flame. He scrambles to turn off the laser but it is too late. He picks her up and rushes outside the garage and commands her to go stand by a tree at the far end of the yard. He rushes into the house and comes back with a fire extinguisher in his hands. Diana is crying as she runs back into the house through the front door. I follow him to the door of the garage and watch as he tries to put out the flames. The flames have risen beyond his control and the fire extinguisher is not able to eliminate the fire. A woman who appears to be Diana's mother rushes out of the house to see what is going on. I enter into the burning door of the garage and through the thick black smoke and raging flames. I see the flaming can. I pick it up and throw it through the door way into the grassy lawn. I look around for something to use to put the flames out. A water hose is sitting by the garage sink. I connect it to the faucet and start spraying water on the flames. I can hear the commotion outside the garage as people try to put out the flames. Soon the flames subside and all I can see is the smoke and steam that fills the garage. As it clears, I see her father and mother standing outside the garage door and wondering what is happening.

"The fire is out!" she cries, "Did you see that? The fire is out," her mother shouts in utter amazement.

"Who, or what?" Richard shouts in the confusion.

"The hose, the water hose, someone or something connected it to the faucet and put the fire out!" she says. Richard walks over to the burning can of penetrating oils and smoother it with his shoe. He is kneeling beside the can and looking at it in astonishment. His gaze travels through the trajectory the can must have taken through the garage door.

The water hose is still running. Diana is watching them in wonder and her blue eyes are alight with anticipation and fear.

"Daddy, he was in there," she says.

"Who, who was in there?" he asks as he holds her arm and questions her.

"My friend, he always comes and plays with me in the garage," she states innocently.

"Pinocchio?" Her mother asks.

"No, my friend Barney," she states innocently.

They look at her in confusion.

"Strange. Strange." Her dad shakes his head as he starts into the garage to turn off the hose.

I come out of her aura feeling guilty. I have invaded her being and know something about her that I should not. But then, did I actually do that in the past, or was it all a figment of my imagination? If I did travel to her past, then we must have been dimensionally connected somehow. I decide to use the information to my advantage.

"Barney." I state interrupting her memories of her childhood,

"I am Barney!" I repeat so she can hear me more clearly.

She stops talking and looks at me as if I have uttered an insult.

"You are Seth not Barney!" She exclaims almost angrily but more confused as her voice drags with emotions.

"I put out the fire in the garage!" I state hoping it will hit the mark.

"The fire?" she asks confused, "Who told you about the fire?" She asks moving away from my bed.

"I have been there all along. That is why they want to capture my being in this place." I explain.

"Are you psychic?" she asks.

"Sort of, I am a traveler. A dimensional traveler and they want to capture my mind so they can control the world."

"Who wants to capture your world? And… and how did you know about the fire in the garage? How did you know it was put out by Barney?"

"I was there with you when you pulled your father's pants and tugged the laser. I am your friend. Help me please," I state hoping that she will get emotionally involved and help me at least understand what Mendeleev and the others are really up to.

"How could you have been there? Barney is just a childhood creation, no one knows about that."

"Trust me; I am able to find out things like that. I am a traveler. I travel through emotional dimensions."

"You are able to read my thoughts and memories?"

"No, I am able to actually be with you in a different dimension. It is a trick I have learnt since I was a child."

"Some gift. I am insulted; no one should know things like that."

"Ok, I know, but it is not like I do it to harm, I am just able to be there when certain curtains of emotions are lifted to let me in," I explain hoping that she will understand.

As if she can read my thoughts, she says,

"Why can't you find out yourself from Mendeleev's mind?"

"I do not know. There seems to be a blockade that was deliberately setup. I can only penetrate through their emotions and they are always bland when I look at them."

"It's the pendant they carry right?"

"What pendant?" I am curious.

"They each have a pendant that looks like a scarab beetle. They wear it on a chain around their necks. All the staff from the other building wear it. I thought it was weird that men should wear those types of things."

"Do you have a pendant?" I ask wondering why she wasn't issued one.

"No. Only people from the other building have it. They told us it is a disguise for an electronic identification security device for entering and leaving the main building."

I look into her eyes and find them hypnotizing me with their beauty. It is almost as if I have known her all her life and I hope that somehow we can be real friends.

"I must leave, they will soon come," she says smiling as she turns to walk out the door.

"No wait!" I shout, hoping she would tell me more about the scarab. She turns and approaches my bed.

"What can I do for you now?"

"The scarab; can you get one for yourself?"

"I don't know" she says, "I do not work in the blue building only workers that enter that building get one." She pushes her hair from her face and looks at me enquiringly.

"If you tell them that I am bordering your mind, maybe they will give you one." I say hoping she will at least make the attempt.

"You already border my mind, so what's next? Are you going to find out about my personal problems too, about my everyday life?" she asks smiling.

"No, I promise not to intrude again. I am worried about their plans. They are trying to capture my mind, and that is what I must prevent from happening for the good of the world."

"Why do you say that? They are nice people. I have known Mendeleev for many years," she states.

"Have you ever been in the other building?"

"No, never," she answers resignedly.

"Mendeleev is a nice guy," she repeats, "his only problem is that he always makes funny jokes, you know, about blondes," Diana grins disapprovingly. I sense a bit of resentment, she is blonde.

"I must leave now, they will be here any minute and I have other patients to take care of."

I watch Diana walk out the door resignedly. I sense a feeling of belonging. I somewhat know that she is connected to who I am and who I have been. It was so easy to enter her aura, I almost feel as if she wanted to let me in without a struggle.

The room is silent again and I am back in my world of reverie. Now that the curtains are closed I feel isolated and alone. Senator Owens is gone and I feel unprotected. I do not know what Doctor Cooper and Mendeleev are up to. I could not open up a passageway into their inner world, their auric world. My thoughts go to the first time I saw the aura. I was a young boy skipping along a horse track in the woods behind my Aunt Lila's house. Aunt Lila was tending to her garden as usual. She was a rich woman having all the toys anyone would ever want including a helicopter. I find myself diving into my head, opening up a passage into the world of emotions that surround me when I think of her. I start sucking my molars and my mouth contorts to a smiley face. I am back in her yard looking at her. She hands me a hot cookie.

"Your mother made these. Have one before you go wandering off into the woods," she says in a thick South London accent, smiling.

I thank her and tackle the cookie. I am skipping off into the thick woods behind her massive house in suburban London. In the distance the grazing horses look serine and peaceful. As I approach them Muscle raises its head to look at me. I can almost see her smile as she approaches in my direction. Suddenly there is commotion. Muscle and the other horses are scattering away from my direction. There is something happening behind me. I turn to look and I see a massive dark cloud looming just a few feet away from where I am. That must be a funnel cloud. Before I can think, I feel myself being pulled into a loud vortex of noise and confusion. I am floating off about twenty feet in the air. I feel myself falling and my mind enters a blank. I awake on a bed in Aunt Lila's house. My mother and Aunt Lila are standing a few feet from the bed and my brother Edward is looking at me as if something is wrong. A man who looks like a doctor stoops over me with a periscope aimed at my right eye.

"This will not hurt," he says.

I cannot speak. It is as if my mouth is gagged with a piece of tape. I reach to feel my face. There is a large bandage over my mouth, I must be hurt. The flash of light from the periscope enters my eye and blinds me. I try to wince but cannot. The doctor's fingers are forcing my eye lid open. Then he did the other eye and again flashes the light into my eye. Moments later, my eyes readjust to the light of the room, but there is more to room than just natural light. I can see a golden glow over my mother's head and body. It is almost like a faint flame burning around her. I want to shout and tell her she is on fire, but I cannot, my mouth is gagged. I struggle to look at Aunt Lila she also has a faint bluish glow over her head. There is a red patch of light near her breast area. It is a different sort of light. I point to her breast and muffle something. The doctor approaches again with his light and focuses on my right eye.

"He is in a trance" I hear the thick Scottish accent.

I open my eyes and look again and this time it is Mendeleev and another doctor looking at me. I am back into the world of space and time.

"Is he going to be OK?" Mendeleev asks.

I find myself answering his question. I am no longer gagged but my face is straining every muscle from the calming effect of the Memorin.

"There is nothing wrong with me," I shout.

"Just checking," the doctor says looking and smiling at Mendeleev, "we are on."

Mendeleev looks at me and I gaze back. His thick white beard betrays his face. His curly mustache looks almost artificial and threatening. I do not know if he could see my disgust. They are ready to manipulate me one more time to enter into my past, to learn the secrets of the Sacred Scroll.

"Seth, I want you to go back to the days in the mines, to the camp fire with Militant."

I find myself compelled to travel back to that day in history. There is no holding back. I am again sitting beside Bayano catching up on his laughter. He is laughing at Militant and Militant is pointing a fist at him jokingly.

"Did Militant know about the Poro ceremony?"

I find myself answering his question.

"No Militant got his story from the Scroll and he had never heard of the African god Yumanya." I answer to Mendeleev's query at the expense of Bayano mocking my sanity again. They keep asking questions and I keep answering them as if compelled by Mendeleev's hypnotic spell on me.

My pain is slowly returning and this time I feel the dimension of Hope slipping away from me. The immobility of my legs is becoming desperation and I am beginning to feel the torment of not being able to walk again. My mind has become split. I can move a part of it on to different dimensions that I like while they query that part of me that is sitting by the camp fire back in the 80s. I recall what my surrogate dad a Catholic priest named father O'Toole used to tell me in church. "Faith" heals everything". I believed him then and I believe him now. As an Altar boy in St. Anthony's Catholic Church, I used to be a child of profound faith. I believed everything happened for a reason and I was a true believer. Now, my Faith is contaminated by Science and other philosophical thoughts, but I still know how to enter the dimension of Faith by sucking my molars. With Faith, I have to also focus my eyes on something far away and clench my teeth as if I am biting onto something. When I do that I enter into the dimension of Faith and I could go back to experiences that make me faithful. The dimension of Faith is a very funny dimension. You cannot travel too far in it without encountering the strange landscape of holes in the ground. I call it that because it actually has large holes in the ground that you can fall through into other dimensions called Despair, Anger and Hate. I have entered all these dimensions in the past and I hate them. The deeper you go into them the more you accelerate and embed yourself in them. It is difficult to get out of these dimensions. I started seeing the holes in the ground when my grandmother told me a story about Faith. She said two of her sons were planning to travel to a faraway land. They came to say goodbye. She gave them long ugly shoes to wear for the journey, but the younger son being obstinate decided he did not need the shoes. The older son took the shoes and wore them. As they journeyed, the younger son suddenly fell into a deep hole and vanished. The older son, she said turned around and looked down only to see holes in the ground everywhere. What kept him alive was his faith in the shoes that were too long to fall through the holes.

"Faith," she had said "is what keeps you alive when there are holes of despair and hate everywhere in your life".

I have faith and I hope that my legs will come back. At least I can walk and run about in the other dimensions I travel to. My eyes open to the bright room again and I can see Mendeleev and the doctor standing my bed discussing something in quiet voices. The Doctor walks up to the side of the bed.

"Are you OK?" He asks if he really cares. Mendeleev takes the opportunity to ask more questions.

"What is the Poro society?"

The answers are automatic.

"The Poro Society is a local African society that is very powerful among the miners. They are feared and revered by the locals so much so that anyone who is a member of that society is treated like a lord in the mines," I answer. I hope that is enough, the pain in my arms and head is increasing and I can feel my heart pounding like a drum.

"The pain!" I shout hoping they will give me more of that Memorin, but then I realize that Memorin will only make me go into the dimensions they want me to go to and then I will talk and reveal more. Talking more is not what I want to do now. I would rather go into a dimension where I can walk, play and do things I want to do.

The question Mendeleev asks about the Poro spark new memories in my head. As far as I am able to tell the Poro had special powers akin to magical powers. They could pass swords through their bellies with no apparent injuries and they could swallow poisonous snakes and do other amazing things that normal humans are not supposed to be able to do. They are in tune with the natural fields of the universe, the fields that I am able to travel through. I am Poro. The YuMuYa scroll describes a future world in which a massive mythical crystal plays a major role in some sort of "Armageddon". When I was in the mines, this was a very exciting topic to my young mind and so I started my research into the ancient scroll and its relation to the African folklore. Pakraban Sinhad was a pygmy. He was probably born in Syria around the year 1170 A.D. When he found the "Scroll of the Eye", as it was later referred to in folklore, he was about 32 years old. It was believed that he became enlightened or acquired great knowledge by just making contact with the ancient document. After discovering the Scroll, he was compelled to travel over three thousand miles on horse-back into the dark jungles of Central Africa where he lived for the rest of his life. There, somewhere in the jungles of Central Africa, probably in the dense equatorial forests, he started a cult that worshiped the god of the Eye, Yumanya.

I later learnt that there is an African folklore that tells the story of a teacher that came from a faraway world into the jungles of Africa. This teacher also a pigmy was named Padaragan Sinjha. This name was also similar to the name of the Pigmy Militant found with the YuMuYa scroll. According to the folklore he populated the jungles of Africa with a race of pigmies that exists to this day. He also taught the Dogon tribe about the heavens and even more strangely, he taught that the Master of the universe lives in an invisible star in the constellation of Sirius. It is now known that the Dogon tribe in Mali region of West Africa has accurate data about this invisible companion star to the Dog-star, a star which is too small to be seen by the unaided eye! This invisible star was discovered fifty years ago by astronomers in the West using powerful modern day telescopes. I decided then that a good starting point would be to investigate the local Poro Society of the jungles of Africa, where the strange tales of the scroll seemed to be headed.

"The Scroll, Seth, what was it about?" Mendeleev demands.

"You kept repeating in the space station that the scroll is true and that the story of the scroll is happening. What frightened you so much?" He again asks emphatically. I could hear his voice thundering in my head as it has always been there. Mendeleev has a big heavy voice that reminds me of my dad. He makes me feel like I am being queried by dad. Perhaps it is all part of the psychological game they are playing with my head.

"I could only tell you the story the "Scroll of the Eye of M'Dulu" as I remembered it from the pieces of information I was able to put together from jungle folklore. The secret society of the Eye of M'Dulu, a Yoruba religion, which I had the privilege of joining in my research, has avowed me not to reveal its location, or its secrets." I state not wanting to continue.

My veins are now feeling empty as the drug wears off. I am entering into a dimension of Despair. I cannot control my rate of descent. It is one of those negative dimensions that you cannot really control. It starts by clenching my teeth because of the pain I feel. I do not do this willingly. It is the pain that makes me do it. Then somehow I start sucking on my molars as the effects of the drug wears off, this sort of brings me back to reality and balances the despair. It is an involuntary response that makes me suck on my molars as if that would release some of the trapped drug between my teeth. If I do not catch myself quickly, I can open my mouth in pain and suddenly enter into the dimension of hate or despair. It is a dark world full of voices and evil things. The only light that shines in the sky is a faint blue sun of Hope in the distance. There, I always encounter the shadowy images of demons I do not know. They stand far away waiting for me to get far enough into the dimension of Despair. I could hear the sobbing and torment of many people in that dimension. I once read the desperate effort of the Incas, Aztecs and Egyptians trying to escape from the despair of Quetzalcoatl, Ra, Zoroaster, Thoth, and Osiris, or is it my mind that is bungling up historical facts to make me even more desperate? For some reason, I have faith and Hope that Voyager will save me. I do not know why, but it is a strong feeling that keeps coming back whenever I enter into the dimensions of Despair or Hate.

I must have been gotten lost in my mind and forgotten that Mendeleev and the doctor are still in the room. I do not even become aware that the doctor has pulled my bed cover from my arm and is administering more Memorin. The drug shoots through my veins again making me feel calm and secure. I quickly float in Despair and find myself flying out of that land toward the blue sun of Hope. I enter into its brilliant beautiful light and then feel the heat of the sheets covering my body. I am hot, hot, and very hot. I am burning as if I am bathing in the heat of the sun of Hope. I motion to Mendeleev to open the windows and let in some air. I want to feel the cool breeze of winter on my face again. He opens the windows and turns on a ceiling fan. It is pure relief as the cool breeze comes pouring down on my face and arms. This is how I always feel when I enter into the dimensions of Love and Joy and I have not been there for a very long time. I sort of hope that Diana will come back and help me, I feel like she can take me to Love and Hope really fast.

The first time I see the pendant it is hanging on the doctor's neck like a dog-tag. It is attached to a card with the doctor's name printed on it as if it is some form of identification card. His name is Harman Cooper. I can see how they can get away with the electronic ID explanation. I could not get a closer look but I was able to get a passing glimpse as it sort of dangled when the doctor bent over to examine my eyes. The pendant looks like a funny scarab beetle with big eyes glaring back at you. It looks like an Egyptian artifact and I sort of already know what it looks like because when I saw it I remembered a dream or place I have been at before where I saw that same scarab. I must have veered off into an unknown dimension in my mind because Mendeleev and the doctor are no longer in the room when my eyes open again. Diana is standing by my bed holding a glass of water and stretching an arm to hand it to me. She realizes that I must be too tired to take it. I am thirsty. She puts the cup to my lips and I sip clumsily until I got enough to quench the thirst.

"There is gossip in the building about a war," she says.

"What war?" I ask.

"I am frightened, I do not quite know what is going on but you certainly have something to with it," she says looking at me enquiringly.

"Where is Senator Owens?" I ask.

"That is why I am here again," she says.

"Not because you already miss me?" I ask jokingly.

"That too" she says smiling.

"The Senator spoke to you?"

"Yes. He wants me to tell you that he is off to the Vatican. You must be a very important person!"

"Yes, I am and that is why you must help me now. There are things that are about to happen in the world and I need to be free from these people. They want to control the way things will turn out and I should not let them." I try to explain to Diana some of the basics. It is a risk I must take now.

"What is this talk about a war? There is a lot of preparations going on in Building 4 and they have started new emergency procedures for building 3, this building!"

"It is not a war," I try to calm her.

"Then what is going on?"

"First get me out of here. I will explain it all to you. You saw what I could do. I have special abilities that are important to mankind. The problem is I can't walk or do anything about it."

"Then why did the Senator not help you? Is he not aware of what is happening? Did you not tell him?"

"I told him some basics. He knows it is important that I be free, but I honestly thought that the Doctor and Mendeleev are working with him."

"The Senator wanted me to take particular care of your well-being. He gave me his scarab and told me not to show it to anyone but you."

"Where is it?"

She takes out a small pendant on a gold chain and points the scarab at me. It is almost alive, looks like there is a real scarab inside a small glass case. I look at it and realize that that was the scarab I saw before. I have seen in in the dimensions of Hope and Love. This is indeed all getting too real and convoluted and I am beginning to suspect that there is more to Voyager than meets the eye.

"Get me out of here," I state in a quiet voice. She looks at me and something clicks in her head.

"It will be risky and I will definitely lose my job or go to jail if I help you. This is a secure government facility."

"Where are we?" I ask again.

"In Nevada! We are in a secret facility beneath the Hoover dam."

Her words hit me like a ton of bricks. It is all making sense now. Nevada, Bilderberg Group! They are here and in control.

"You must get me out of here immediately or the entire world will be in danger. I must speak to the Senator now," I state projecting a truth aura into her field as best as I can. She removes the pendant from her pocket and removes my sheets exposing my lower body. Then, turning away as if not to look at my nakedness, she places the scarab pendant inside my underwear probing to make sure no one will easily find it.

"The Senator wants you to have it."

"Thank you, but you must get me out of here as soon as possible," I reiterate.

"I will talk to the Senator and see what we can do."

She walks over to the curtain and closes them. It is a signal that it is the end of her shift and I will not see her again until the morning. She leaves the room without turning back. I feel empty and vulnerable. For the first time I feel a dimension that is almost alien to me. It is a dimension that I can only find my brother, mother and sister in. They are the ones I love most in the world.

I think about the pendant and how best to use its shielding power. If I can only avoid their fields from controlling my answers I will be a step ahead of them. Now I know that I am dealing with the Bilderberg group and that they have captured me as a surrogate gate into the field of dreams.

CHAPTER 3
Prisoner in the Vatican

After Diana leaves, I probe the sheets to find the scarab. I locate it after some difficulty. It is a glowing pendant made with gold trimmings and very delicate threads of silver. The threading looks like electronic circuit covering a scarab's open wings. The eyes of the scarab are made of ruby. They emit a special red glow like an LED that can fill a dark room. It is definitely not an ordinary thing, it is either an electronic device or some powerful talisman that is meant for special purpose. How it got to the Bilderberg group, I do not know! The night is getting late and it must be around midnight. I cannot sleep and my eyes are focused on the cubic pattern of dots that the glowing pendant projects on the blank ceiling tiles. I hear a noise coming from the door to my room. The door opens slowly and a figure walks in behind an electric powered wheel chair. It is Diana! She comes close to the bed and whispers my name.

"Diana!" I whisper back to let her know I know it is her.

She removes the sheets from the bed and exposes my half naked body. She places a hospital robe around my neck and pulls it over my body. Then she raises the motor powered bed into a sitting position while supporting my torso from falling over. She places my legs on the wheel chair and then with a practiced move, she pulls me from the bed and swings me unto the wheel chair.

"I spoke to the Senator about what you said. He was already making plans to steal you from here. He insisted I take you to the airport before morning," she whispers.

"Thank you so very much Diana, I.."

"Just shut your mouth and pretend as if you are asleep, okay?"

She covers my head with a blanket and places a series of tubes around my mouth to pretend that I am a patient being moved within the hospital or whatever this place is. In moments she is wheeling me down a long corridor. Nervously Diana pushes through a long white featureless corridor toward one of three large steel exit doors. This is not a hospital it is a massive control bunker of some sort. She uses her thumb print to open the large door and we enter into a huge open room with many more doors. On one of the doors is written "European Central Command Chambers". What the hell is a European Command Center doing in Nevada? We enter a glass elevator that leads us to a second level with another huge open chamber. A military guard is standing by the door armed with a submachine gun. Military means secure, secure means secret, secret means important! He approaches us and demands to see identification. Diana exposes an electronic pass and after scanning me with his military eyes he opens the door into a large open room filled with conference tables and chairs, it looks like a huge meeting room. Diana knows her way around here. She pushes me into another elevator and looks back to see if we are being followed as the door shuts. The elevator arrives at the top floor and the door opens. We get out of the elevator into a wide open room with several large glass windows. I can see the open expanse of the mountains that surround the facility. Now, it is all making sense, this is the new extension to "Area 51". After all the public noise about Area 51, I know that the government decided to open up a new facility in Nevada, but I did not know that it would be under the Hoover dam. Why under the Hoover Dam? We must be on the top of one of the mountains surrounding the dam. The only exit door on the top floor is one hundred meters ahead of us. Diana pushes fast and we quickly traverse the room to what looks like an air lock chamber. Diana dials the lock and a steel door slowly opens as if an air pressure equalizer controls its motion. We enter into a small steel chamber that looks like a lift but it is different. A rush of air fills the chamber as the lock shuts behind sealing us in.

"We are fine now, we are out of the restricted zone," she says smiling.

"Ok. That is comforting," I state with a trembling voice as I feel a cold wind bites into my thin covering. As soon as the airlock stops Diana opens the steel door again and this time it opens to a parking lot full of small electric golf carts. She races us to a parked golf cart and tries to pull me into a seat, but before she could do that a large black van arrives and opens its doors. Diana tries to pull me in quickly but it is too late. The two men push her off the cart and pull me down again into my wheel chair. One of the men points a gun at Diana while the other pushes my chair to the van.

"What the hell are you two doing? We have to get to the airport now." Diana complains as her arm is pulled into the van. My chair is quickly pulled to the side of the van and before I could even protest a word I find myself being hoisted into the air and into the van. It had been specially prepared with a lift for my wheel chair and the manner in which these two executed their plan tells me that these two fellers have practiced these maneuvers very well.

"We are in danger," one of the men says "we must get you to the safe house as soon as possible."

I turn my attention to Diana, she is sitting on a seat beside me frightened but calm. She gazes back at me apologetically. Her arm is restrained to a side ring built into the seat. It is all well planned. They knew I could not walk, they knew Diana would clear me from the facility and they knew that they had to build those restrains into the van. They also practiced how to do it all. This is not just a hunch that they had. They were instructed by someone who knew I was coming out with Diana. Could be Owens?

The two men are dressed in dark suits typical of the FBI or CIA. They are the classic set of villains you find in James Bond movies. The driver is short and bulky like a consigliore in a mafia movie. The other guy is sitting behind us, so I cannot see his face, but I know he is tall and lanky. I try to penetrate their auras to find out what they are up to but my physical and psychological state of being does not allow me to even start generating an auric field. I again smile at Diana and find her smiling back to comfort me. With her, I could sense that she feels responsible for what is happening. Her aura was easy to look into but just at the surface level where it tells me that she is frightened but resolved.

"Safe house?" Diana blurts out suddenly, "what do you mean "safe house"? We are heading to the airport as per the Senator's instructions!"

"Lady, you two are in grave danger and it is my orders to take you and your husband a safe house," the driver answers. The guy behind us keeps grunting like a pig in a periodic manner. Apparently he is gifted with this empty grunt that sounds like a pig clearing snort.

"Senator Owens said we should go to South Dakota and he has a plane ready for us at the airport," Diana complains. The grunter stabs Diana's neck with a needle and grunts again looking at me with anticipation of next move. The last thing I heard before my head was bashed in by the second military man was;

"Sorry mam, these are instructions beyond and above Owens."

I awake to Diana's snoring. She must be very tired from all the adventures we went through earlier. We are both lying on the same bed with pillows neatly tucked between us. These fellows must believe that we are married or something. My state of elation makes me particularly able to dimensionally travel today. I am surrounded by a strong auric field from Diana and she is totally at peace with herself. Where are we? I try to look out a slit in the curtains but I could not see much. There are trees outside and very bright sunlight. From the shadows on the curtains I am able to see at least two armed guards exchanging paces back and forth as if protecting Fort Knox. I look around the room and find it neat, clean and rich. The walls are masonic stones that abut a masonic ceiling. It looks like we are in a Chateaux or a Castle of some sort. The lone electric bulb hanging from the center of the ceiling was an after though, with a thick cable running visibly across the ceiling down the wall and into the middle of the wall. There is a single painting of a rosary hanging on one wall. The appearance of the room reminds me of a monastery perhaps? A large wooden door leads seals us in and I can see a large old rusted bolt across the doorway. There is what looks like a bathroom adjoining the far end of the large room. A sofa seats at the far end wall facing the bed and my wheel chair is neatly packed a few feet from my side of the bed. The room is aristocratic with a large pot of rose flowers adorning the corner of a wall. On a table by the window is a large pot of still steaming coffee and condiments with two large covered silver trays that seem to indicate breakfast is ready. I could not hear any sounds other than the chirping birds that must fill the courtyard I am imagining outside the window.

I turn to look at Diana, she is sleeping peacefully and beautiful. Her mouth is partially open as if she is in the middle of an unfinished sentence. I could feel her warmth seeping through the sheets into my body. I slowly pull away the pillow between us and throw it on the floor below. She turns subconsciously and places a warm leg over my torso. She must think I am her pillow. I could not describe the sensations I feel from her being. For the first time, I feel like I am in a place that I was meant to be and that I should never leave such a place. I wonder if she will feel the same way when she realizes that we are in such an intimate semi-co-joined position. It is not often that you find yourself in a place you have never been before, where you know you are a prisoner, and where you want to be without giving it another thought. I know this moment would be just that; a moment before the real world comes back. This is a dimensional gateway to a momentary lapse of reasoning, but also a momentary gain of unimaginable joy.

There are things I need to talk to Diana about concerning the future. I have no doubt that she would seek my best interest and that her innocence would not annihilate the reality of what is to come. If she knows the truth about the Scroll, she could help me avert the greatest disaster that will ever befall mankind. I consider myself a sleeping prophet for humanity. Somehow, she must help me stabilize things again and secure the dimensional gateways to the most powerful force ever created by God. I cannot allow the Bilderberg group to find out the secret of the Scroll. The story of the Scroll is for the entire world. It is for this generation of the children of the Earth. Like Pakraban Sinhad, I will want to declare a warning that I believe is relevant for mankind. Those with ears and eyes for the possibilities and seeming impossibilities that lie ahead of humanity know and believe this prophecy. That is, those of you who believe that nothing of this world was left to chance by its Creator, God. If you do not believe in God, there will be no need for you to know about this story, it will not have much meaning for you, it will not comfort you, but will only frustrate you.

The entity YuMuYa, who the locals of the African jungles call Yumanya is said to have defined itself as follows:

"I am Yu the servant of the living God, the Force of the Living World;

I am Mu the servant of the Living God, the Energy of the Living World,

I am Ya the servant of the Living God, the Material manifestation of the Living World."

YuMuYa is a trinity and I grew up as an altar boy serving the trinity. I feel a religious obligation to the world. My reverie is broken when Diana wakes up. I have already pulled myself up in a sitting position. She sits up on the bed looking at me in utter surprise. She did not expect to find me in the same bed. I pretend as if nothing has happened and that our compromising situation is natural. She smiles and wipes her watery eyes with a loud yawn.

"What is going on?" she asks as if she expects me to know.

"I do not know. You stole me from my bed in Nevada and now I am here. What do you think happened?"

"We are in a castle or monastery of some sort. Look at the walls and ceiling. These are not of American construction. The walls are old and the ceiling is uneven masonry. It looks like we are in middle age Europe somewhere."

"Where?" She asks.

"I do not know. There are guards outside our window and there is a garden or a plaza of some sort" I explain.

"Can you dimensionally travel and find out?"

"Not like that. I can if I am channeled by an experience or a recipient such as you."

"No not me, not any more, it is like a violation."

"I agree and I am sorry. From the shape of the plug on the wall, it has large round holes and it looks like we are in Italy!"

"Senator Owens is in Italy! Does that mean he knows we are here?" I can see hope lighting up in Diana's eyes. She gets up from the bed and walks to the bathroom and closes the door. I am glad she goes first because the way my tommy has been acting, I wonder why I haven't yet disgusted Diana with my silent ones. A few minutes later, Diana emerges with a clean face and a red bathrobe hugging her body. She looks around for a moment and then sits lazily on the bed. She looks a lot older than I remember her to be. I am a bit puzzled by that. What is happening to me?

"I think we have breakfast on the table but I need to use the bathroom," I said resignedly looking at Diana.

A moment later a guard walks in and approaches us with a wheel chair. He is wearing strange clothes like some ancient warrior.

"Swiss me custodes, grata Vaticana," is what I hear as if we should understand what I presume to be Latin.

"Welcome to the Vatican," he says again in plain English.

"Ah, a Swiss guard!" I exclaim.

Diana does not seem impressed. She seems to be totally comfortable with the situation and I am at a loss as to what is going on.

"Why the Swiss guard?" She asks hoping I will shed some light into her fuggy mind.

"They guard the Pope," I explain calmly as she walks over and puts a firm comforting grip on my hand. I feel safe with her.

The guard walks over to the bed and lifts me into the wheel chair without saying any more. He wheels me to the bathroom pulls my robes and places me on the toilet bowl and walks out. At this moment, I feel completely at odds with what is happening. It is a very uncomfortable feeling. The toilet seat is cold and I could feel a sensation of some sort crawling through my being. Since I know my legs do not have any feeling in them, I dismissed the sensation as just a placebo effect or something like that. Anyway, I did what I had to do and waited a while. The toilet had a bidet attached to it so it was not difficult to clean up after the act. Again, as the water runs down my rear, I feel the same strange sensation of something on my legs. For the first time after the accident I am beginning to get some sensation back on my legs. Perhaps, I could walk again. I find myself pushing and trying hard to get my legs to react. I place my hands on the bowl and push hard to let the weight hang on my legs. For a brief moment my muscles respond and I think I was able to hold my weight! A moment later, I fall on the toilet floor with a loud thump. My face is turned toward the door and before I can react or shout, the door opens and slams hard on my cheek. The cold floor, the pain and the loud noises from Diana and the guard fill my head. It is the strangest pain and embarrassment I have ever felt, and here I am lying half-naked with my cheek bashing on the floor of a Vatican toilet and a woman I am trying to impress trying to lift my naked body and thoroughly immersed in the embarrassing sensations you find in a toilet. All I could see is her beautiful legs and the boots of a guard standing close to my face.

Anyhow, they lift my body and place me on the wheel chair and wheel me back to the room. The guard leaves to give us some space. Diana had found some very elegant clothes in a small closet that had been prepped especially for us. She looks elegant but totally different. She looks older than I last remembered. I look at her profile as she walks toward me. She is as beautiful as ever, but something is different about her now. She is no longer a young girl but an older woman full of maturity and an understanding auric field. I am sitting looking at the most beautiful woman on the planet laughing at me as she dresses me in an elegant black suit. She seems to have done this same thing many times over because she is doing it with ease that could only have come with practice. She puts a nice pair of well-polished leather shoes on my feet. It feels like I am preparing for a date. I look at her walk toward the mirror to adjust her looks. She has gained a little weight that comes with age. I do not remember her this way. What has happened to me? I adjust my torso on the chair with ease. My legs are mobile and I can feel some energy in them even though they are still unable to take my full weight. I notice a ring on my finger. It is a wedding ring sitting squarely on the correct finger! I look at Diana's hand. She is also wearing a similar ring. What is going on? Are we pretending to be married because we are in the Holy Vatican? For some reason I completely trust Diana even though she looks far different from what I remember. I must have bashed my head really bad when I fell on the toilet floor. A guard knocks on the door prompting Diana to walk over and opens the door. She does not hesitate to open the door. The guard also enters the room without hesitation. I look enquiringly at the guard. Does she know him? Diana is dressed in a beautiful blue dress marked with shiny gem stones reminiscent of wealth and prosperity. She greets him pleasantly as if they have known each other for a long time. He asks us to leave the room and follow him. Diana must have the same vibes as I do. We decided not to ask questions and to just follow his orders. We need to immediately learn more about our situation. We exited the room and enter a beautifully decorated hallway that leads to the outside courtyard garden. The large courtyard looks just as I imagined it with a variety of flowers neatly planted on large vases around tall pine trees surrounding a small fountain. The building completely surrounds the courtyard with four gates leading to the outside world. Several benches are positioned around the courtyard for maximum advantage of seating outdoors. The guard leads us through the garden toward a larger adjacent building that is conjoined to the one we were in. He knocks on a door and it opens. Another guard allows us to enter into the building after staring Diana and I with a curious non apologetic look that could pierce a hole in steel armor. The room we enter into is a curious mix of the modern and the ancient. Computer terminals with well-dressed operators line every foot of one wall of the room. The house is definitely of mediaeval origin and its décor is reminiscent of the stature and grandeur of a place like the Vatican. Large paintings adorn the empty walls with large vases full of intricate designs of a roman era. A deep sense of caution overcomes my being as my auric field feels an ancient field that I have encountered before. Diana is behind me and I cannot see or sense her reactions, but I know she will be feeling it too. I try to glance at the monitors but could not get any information from them. Moments later we emerge into a large black room lined with large black bookcases filled with volumes of black leather bound books. The guard leaves us in the black room and shuts the door behind him. There is total silence. My mind starts to race in anticipation. I have been here before. I try to enter into my memory to see what triggers that feeling but I reach a blank. I look at the books again; each book is embroidered in gold with gold laden number. They are serially marked in roman numerals. All the walls are filled with books and the only décor is a set of four wooden chairs around a large black metal coffee table. The design of the black metal table must be Roman or Greek. Each of the four legs of the table is a black cast iron horse with a carriage that becomes part of a corner of the table. The center of the table top is adorned with a thick circle of polished mahogany tangent to the table edges. Diana walks over to the library of books and tries to pull one out. The books are somehow locked unto the cases and could not be removed normally. I roll my wheel chair to take a closer look. In every man lies dormant a potential hero who needs to take advantage of the occasion and prove himself. I examine the books but cannot find a reason for them to be stuck to the case. I try to yank hard but the book would not give in and I find myself trying to solve a riddle I should not.

"They are locked in with pins," Diana graciously states pointing to a padlock hanging on a long stainless steel pin passing through all the books. Each row of books has been locked in with a single long stainless steel pin that protrudes through the walls of the bookcases. Before I could make a case for my stupidity, a black man enters the room and closes the door behind him. He is wearing priestly clothes and seems of important stature. He approaches Diana and shakes her hand as if they know each other. This adds to my confusion but I decide not to make them notice my concern. He proceeds to shake my hand. He holds my hand firmly and looks me in the eye. I look at him back and immediately feel a hint of recognition. Diana moves over to my side and places a hand on mine as if for comfort and protection. The man slowly lets go of my hand and smiles.

"The Books of Malaki!" he exclaims.

He walks over to the first bookcase and releases a lock and removes the first book.

"Who are you and why are we here?" I ask of him. I am hoping that he is a messenger from the Senator or someone that can help us.

"You are safe and I am here in person to assure myself that you are the real man and that you are safe," he states tapping his head as if he is bouncing his memories around.

"Who are you?" I repeat.

"My name is Effion, Effion Eba. I am a priest and cleric in the Vatican, a scholar of ancient texts."

"What do we have to do with ancient texts, and why are we here?" I ask. I look at Diana hoping she will ask some more questions that will shed light on my fuzzy mind. Effion places the book on the coffee table and sits down motioning Diana to also sit. His movements are somewhat eccentric and odd. He slaps his head again and wipes his hands as if they are dirty.

"Mathematics, Physics, Quantum theory, Theology and the Principle of Causal Conspiracy, all wrapped in a neat bundle by Malaki," he states as if he is proud of what he says. He motions me to move closer disregarding Diana. I place my wheel chair beside Diana and motion her to follow me to the table. Effion opens the book and starts reading out loud.

"It begins by describing the future of the world in turmoil. The date is the 14th day of June 2038 A.D., with a frantic bunch of Egyptians in a beautiful new palace. They groom a man by the name of Ayatollah Mabus Ali Resa." Effion states. He awaits a response from me but gets none.

"This man is a religious leader arising from the Middle Eastern region, in particular from Egypt. It is written that he will lead a radical and new form of Islam called "Pharaohnism." Masses of the new Pharaohnism movement will consider him the holiest man in the world. Most of the world will know him as "Ayatollah Resa of Egypt." He will inherit this title as if by the force of his very being. He will not be elected to the title. At the age of twenty-two, he will become the youngest religious leader ever to walk the face of the Earth and unquestionably the most powerful man in the Kingdom of Arabia (the future United States of Middle East)."

After reading, Effion looks at me squarely in the eye.

"You dictated all of these books to the Papacy a few years ago," he says pointing with a sweeping motion to all the books around the room.

"You must be mistaken I never did anything of the sort. There must be an error!" I exclaim in utter astonishment that this man would make such a fantastic claim on my behalf. I look at Diana hoping that she would assist me refute the claim made by this eccentric priest.

"You are a Guardian and you travel in dimensions, emotional dimensions," Effion speaks with a calm calculated tone.

"I do not understand. I know I am a dimensional traveler but I do not recall writing or dictating any books. I do not recall ever doing that. Besides I have never written any of my experiences before," I am protesting as hard as I can but something tells me that the man is serious about his claims. Diana is silent. Effion continuous to read from the first book.

"The world will undergo a drastic change that will forever change all the geography and politics of Earth. He will be the cause. After his "birth" and I put that word in quotes, because no one will ever know his parents, or where he came from, Pharaohnism will slowly evolve into a new religion separate from traditional Islam. It is written he will become a subject of great significance to the children of the West. People will be attracted to the teachings of Pharaohnism, because of the lure of the great Pyramids and Ancient Egypt. There will be many followers of Mabus Alus arising as if in need of a spiritual leader. The core principles of Pharaohnism will be rooted in the ancient Egyptian teachings that will be uncovered in the Second Book of the Dead. This book will be unearthed from its grave during excavations of the central chamber of the first Temple of Ramses II, at the city of Abu Simbel in Egypt. The book will describe a complex religion of the ancient world, when Egyptian Pharaohs were glorified as members of an elite group of angels who lived in the heavenly realm. It will be exalted as a great book, a book of similar stature to the Bible and the Koran. Like the biblical prophecies of the Book of Revelations, the Second Book of the Dead will also proclaim a future rebirth of Pharaohnic angels after a massive cataclysm that befalls the Earth in the beginning of the new millennium. It will also describe a complete system of ancient Egyptian worship of Pharaohs as angels of the one living God. This will be a shocking discovery for those studying the Pharaohs (Egyptologists), and the world. For over a century it was believed that the ancient Egyptians practiced a religion based solely on the worship of multiple earthly and heavenly gods. The general consensus of Egyptologists in this day is that the ancient Egyptians practiced a religion in which the Pharaohs and earthly gods were all folded into complex relationships with the stars, earth, underworld, sun and moon. No one suspected that ancient Egyptians practiced monotheism, and the discovery that ancient Egyptians considered their Pharaohs as servants of an all-powerful God will indeed be surprising! It will then become clear that the great pyramids are not just burial grounds for the Pharaohs, but that they also served as temples to an unnamable and all-powerful God."

"The Scroll!" I shout, "You are reciting the YuMuYa Scroll?"

I am shocked. Effion is actually reciting the exact prophecy of the YuMuYa scroll. Where did he get it from? No one knows of the Scroll's content but me. Militant is dead and I have the original Scroll neatly tucked in a cave in Hohenwald Tennessee. I do not show my emotions. I again look enquiringly at Diana and wait for Effion to reveal more information. After reading Effion pulls out a walkie-talkie and speaks a command in brisk Italian. I am at a total loss as to why he makes the claim that I am an author of so many volumes. Of course, it may be a new technique to get me to reveal the Scroll. I decide not to say much and to follow his logic until I can make some headway out of all this mess.

"Malaki, you are the first of a line of modern day prophets. You are a Guardian. You came to Pope Paul the Seventh and dictated the Malaki Books you see filling this library."

Pope Paul the Seventh. It is 2012 and there is no Pope Paul the Seventh. What is he talking about? A guard opens the door and walks in with tray of coffee. He doles out coffee and serves it in delicately decorated cups. I look at Diana again she is looking at me as if I have done something wrong. She squeezes my hand in assurance. When the Guard leaves Effion continues reading.

"For those of you who fear 2012, take heed, for it is merely the beginning of a new era for humankind. The past wars and pestilence have been mere inconveniences compared to what shall happen in the next thirty years of human existence. Worry about your faith and your beliefs, for this vision is a result of all the cumulative effects of human actions since the birth of Adam. On this 6th day of June in the year 2038 AD, the celebrations begin with "The Day of Egyptian Awareness." The man named Mabus Ali Resa is being prepared for the momentous events spelled out in the rituals of the Second Book of the Dead. His turban, blue, pristine and beautiful, gives him that extra touch that sets him apart from the men of his faith. It is made of the finest muslin fabric Damascus has to offer and its splendor is enhanced by a fifty-five-karat blue sapphire stone sitting gracefully at its center. Like the visions of Nostradamus, his deep-gray eyes will be uncustomary for this part of the world. But then, Resa will be a unique man. He will be born in Egypt, on the morning of the "Great Day of Mourning" in the year 2016 AD. On that day, not a single record of a live-human birth will be found anywhere in the world! All children born that day will die as a result of a strange confluence of events."

Effion stops to gather his breath and I feel a rush of adrenaline flow into my veins. I gasp and want to vomit as the full extent of the situation hits me. They have the Scroll! They have the Scroll!

"Are you OK?" Effion asks looking at me. His black eyes are amplified and I distinctly feel a sensation of fear run over me. I am becoming a little paranoid as the story takes hold all over again.

"Yes, I am fine." I answer.

"Why did all children die that day? What happened?" He asks waiting for me to affirm his information. I think about it for a moment. It is 2012, there is no Pope Paul the Seventh, the present Pope is Benedict XV1. The last Pope Paul is the Sixth. Who is Pope Paul the Seventh? Who is Effion? I feel cold sweat pouring down my neck into my chest. The coffee does not help. Diana is just staring at me as if I am some weird object.

"A huge comet whose trajectory is already well established will strike the planet Mars. Only a handful of people today, including those in high- authority, already know about this comet."

I feel a rush of surprise come over me, they already know. Perhaps Senator Owens already knows. They know that the comets are coming and somehow, I must have revealed the Scroll to them. After all, I have never really understood what Senator Owens does outside his duties. He was a stranger that reemerged after we separated for college.

"Did you already know about the comet?" I dare ask.

"No, we do not. We need to hear it from you," Effion states.

Effion looks at me comfortingly nodding.

"We are just as baffled about what you have said, Seth. If it is true and that important, we need to get this information out of you so that the world will know and prepare," he states emphatically pounding his fist on the table.

I know this is not just about the comets. It is about the power behind the Josephus Nambu force. It is about the power to control the dimensional-gateways to humanity, to control human consciousness and to control the human mind. If I reveal this to the wrong people they will control humanity. They will control the future. They will be able to literally control good and evil.

The powers that be have not revealed the arrival of the comets to the world. They are already preparing for its arrival. The comet was discovered in 2001 by a Russian astronomer in data obtained from the space station and the Hubble telescope. Immediately after he reported his discovery to the Russian authorities, the young astronomer suddenly died in a car crash under suspicious circumstances. Then, in 2002 Militant also died under mysterious circumstances. That set me off to really get involved in what has been going on around the scroll. I really became deeply involved when Militant called me from a public telephone in Rio de Janeiro. I have not heard from him for many years after we left the mines. He was frantic and spoke in spurts that hardly made sense. He was talking about the scroll. He kept saying that I should take the original in my possession because I was the only one who understood its real value since I studied the mysterious Poro society. He said something that really made it sound crazy back then. He said that he had been dimensionally travelling back to the camp site and was locked in some sort of time loop in which he kept going back to that night of drinking in the camp. He claims that we keep meeting there as if compelled by some strange wheeling motions in time. He could not help it, he was going out of his mind and he knew that he had to get rid of the scroll. I never realized that Memorin had propelled me into a continuous reenactment of that day. I realized then that by my dimensionally going back to 1982 from 2011, I kept locking Militant and the rest of the Brazilian crew into an inescapable cycle of time. But my travels were not a spacetime travel, but an emotional time travel that had severe consequences for those involved. You see if you travel in space and time, you simply go back and repeat the world the way it was, and it does not quite make a difference to what will be or what has happened. The past is locked and so are the future consequences of that past. However, when you travel through emotional fields the past becomes different. You can influence the emotional fields in the past just as you can make someone that hated you in the past love you in the future. When that happens, he or she has always loved you and the past and the hatred is no longer there. Bayamo's prediction about our sanity has become real. We are all going mad and the scroll is to blame. It is cursed. Militant explained that Bayano, Nelson and the rest of the Brazilian mining crew that spent that night together when he revealed the scroll have all died. They all died of mysterious circumstances and are said to have gone psychotic and schizophrenic. It was August 5th 2002. Militant had flown from Brazil to meet me in the USA the night before. He appeared at my door step with his briefcase neatly tucked into a suitcase. He was standing there by my doorstep literally crying and trembling with fear.

His entire family had been killed the week before by an assailant. It was like an evil possession gone wild. He was on a business trip to Copa Cabana and had left his family back home at Rio de Janeiro. He had left the Scroll wrapped and hidden in a safe in his bedroom. When he came home, his two children were lying naked on the floor in a pool of blood and his wife was in the bedroom brutally stabbed to death. The safe was open and the Scroll was in the hands of the assailant lying on the floor beside the bed with burn marks all over his body. It was as if he had been cooked from the inside. Militant handed me the Scroll as if it was naturally mine. Militant never made it back to Brazil. He was found dead at a secluded area in Washington-Dulles airport. I did not open the scroll immediately. I knew its contents and I knew that I could handle it because I took the time to become Poro. The Poro society is a special society that allows its members to have access to dimensional fields that can be very terrifying. They exist today as a reclusive society in Africa. I spent many years studying the scroll and learning all about its powers and its dimensional doorways. Now, here I am sitting in the Vatican with a priest who claims that I authored over one hundred books for a future Pope that still does not exist. I begin to wonder if this is just another dimensional gateway to the future that I have entered into. Could I actually be in the future?

As if we were discussing that future, Effion breaks my thoughts and starts talking to me and Diana.

"Seth, you became an astronaut in 2005 and had an accident in space in 2011. After that you settled in Hohenwald Tennessee in a secluded neighborhood that was bought for you by the US Government through the hospices of Senator Owens."

"But how did you and the Vatican and all those people in Area 51 become aware of the Scroll?"

"After the accident, you semi paralyzed and were taken to a hospital in Nevada. You were talking about the Scroll and about Voyager coming back. You were exposing details to the Bilderberg group in Nevada. Effion explains. Senator Owens decided to bring you here in 2012. Soon after your accident all large telescopes (worldwide) were immediately put under the control of a secret agency that I will tell you about later. Today, no one, outside of a few persons in the world, knows about the mysterious comets that you predicted. It is the grand conspiracy to keep this secret by our leaders especially a particular group of people."

"Is it the Bilderberg Group?" I ask shocked. How did he know about the group? Could he be one of them? This group is a mysterious group of secret and powerful people that secretly make decisions that shape our world.

"Most of us have heard of this group from the rambling of conspirator theorists such as Meltzer. What you do not know is that the Bilderberg Group is a continuation of the First Ecumenical Council of the Church which met in Constantinople in 381 A.D."

"What?" I exclaim as if a knife has been stuck a nerve.

"I have read quite a bit of history and I know that the Council of Nicaea was specifically called to end what was then called the Arian controversy. The Arian controversy was started when Arius a clergyman from Alexandria, Egypt, declared and preached that Jesus Christ was created by God at birth and so was inferior to God the Father," he states opening up his IPhone to Google search the Nicene era.

"Yes. Emperor Constantine called the Church clergy to a meeting to clarify this issue. Later, Emperor Constantine would renege the decisions that were made at the Nicene Council and by so doing he granted more freedom and conditional amnesty to the followers of Arius and exiled Athanasius of Alexandria in Egypt who was the main defender of the Divinity of Jesus Christ."

Diana looks at Effion and nods. He approaches Diana and whispers something in her ears. I feel suspicious that perhaps they know more than I know about all this. I feel particularly betrayed that Diana knows more I thought she knew.

"What did you whisper to Diana?" I ask suspecting something odd.

"Nothing of significance, I told her that these facts are available on Google and that perhaps you can read them on Wikipedia or someplace." Effion answers. He looks true and honest. I will know if he is lying, after all I am at a very heightened sense of awareness. He decided to continue.

"Pope Athanasius I of Alexandria continued to vigorously defend the Nicene Christianity against Arianism. The church has since held secret annual meetings to determine what should be creed and what should not be removed from the scriptures. Athanasius was influenced by the Ancient Egyptian teachings of Osiris the Egyptian God. The Arian Council was also established by Arius to meet annually and continue the war with the church and the Secret Nicaea Council. Ever since, the two secret councils have held secret meeting all around the world to continue their agenda. The Arian Agenda found its face lurking in the politics of Germany during the time of Hitler. This gave birth to an off shoot of the Arian council, called the Arian movement, the Nazis."

"Are you telling me that the Nazis were a product of the Nicaea Council?" I ask.

"Yes. In the year 1954, after World War 11, the Secret Council of Nicaea held a meeting in the Hotel de Bilderberg in the Netherlands, where they openly stated their goal to unite the North Atlantic nations against a rising tide of Anti-European and Anti-American sentiment. It was a front to divert public attention from their main aim of establishing a New World Order under the Nicaea creed, an International movement to re-establish the church's power in the world, while pretending that the group's intent was political."

"How does this tie in with the comets and Egypt?" I shouted.

"When the information about the comet does become public information in the future, that is in January of 2014 AD, the entire Earth was preparing anxiously for a spectacular event on the surface of the planet Mars. NASA publicly declared that a comet will strike the planet Mars. The collision of the comet with Mars occurs as expected. NASA prepares a new telescope to replace the ailing Hubble, the "Panoramic Survey Telescope and Rapid Response System" or the Pan-STARRS, to observe the trajectory of the comet on a daily basis. Several earth based observation centers were set up to observe the momentous events in 2016. The events were on TV and radio like no other events before.

"Are we in the future?" What date is this? Why am I here? Who are you people? Why…"

"Seth it will all come to you, please be patient. You are in good hands," Diana states holding my hand tenderly.

"Seth you wrote that those who read and believe in you should take heed and know that the comets bring fire and brimstone. The comets were officially "rediscovered" in 2014 AD by an innocent young French student of astronomy in Paris, France. His name was Pascal Resa. He was allowed to make public knowledge of the comet's existence. For another three years after this information became public, the comet, which were known as the "Comet Resa 1 & 2," sped on its erratic orbit about the Sun, being pulled each time into a new unpredictable orbit by the gravitational fields of the planets and the Sun."

My nerves are at a boiling point and I can feel the heat rising within my body like a furnace that has been lit insider my suite. I take off my jacket and throw it on the marble floor. This is too much for me. I push my wheel chair toward the door. I have to get out of here now. Diana follows me and holds the wheel chair. I turn to look at her. She looks like a stranger, someone that I do not know. She is not the person I remember. She is not the woman that was kidnapped with me from Nevada. Who is she? She smiles as if nothing has happened.

Effion smiles at me as if I should calm down and accept what has become of me. He walks over to the library of books and takes out a book marked "volume five". He lays the book on the table and looks at me. Diana smiles as if I will understand. He opens the book to a blank page and points,

"Look, Seth look!"

I look at the blank page. Then he opens another page and then many more. They are all blank.

"Sulum dies ut obduco, vestis pages ero repletus per vestri scientia. Is est inevitable!" Effion states in Latin. I look at him blankly as the words hit their mark.

"Each day that passes, the blank pages will be filled with your knowledge. It is inevitable".

I realize then that I am the author of the books. The blank pages are yet to be filled. I become overwhelmed with emotions.

"Take me out if here now!" I order. She smiles and nods. She bids Effion good buy and says,

"We will try again tomorrow; it is too much for him now."

Shocked as I am, I do not reveal any more emotions or concern about what she said or what she is really up to. I know now that she knows much more than I do and that there is more to all this than meets the eye. There is a conspiracy of some sort, a Causal Conspiracy that eludes space and time. It is all about the force, about the power of the Josephus Nambu field. They are not concerned about the comets the Scroll predicted, they are interested in the cubic fields that I can travel through. My heart is pounding with fear and anticipation. How did I get to the Vatican in the future? Who is Pope Paul the seventh? Who is Effion? Who is Diana?

CHAPTER 4
Field of Dreams

The YuMuYa Scroll is a codex. It predicts the future and ties in the past history of the world. It also talks about a very powerful force that scientists are now starting to investigate. This is the Josephus Nambu force. In 2012 it is still not known what this force is because the scientists have not yet read the great work of an unknown scientist who wrote two books on the Principle of Causal Conspiracy that is yet to be published by Tate publishing in 2012.

The YuMuYa codex is concise and precise. Like the Bible codes, it is a code that cannot be broken by just a spacetime understanding. In order to understand the code, one must learn about esoteric fields called the emotional fields. These are dimensional fields that allow us to travel through them just like space and time dimensions allow us to travel through them. The difference is they are regulated by human thoughts and quantum processes that we have yet to learn about as a science. The human mind thinks by lifting answers from the quantum probability firmament. This is the amorphous quantum space where the dice is all at once all six possible numbers before it is rolled. In this quantum space, the coin is in a state of flux and is both heads and tails all at once. The cards are kings, queens and jacks all at once. It is a probability fields from which the mind lifts answers. When we ask a question that has a yes or no answers, we can imagine these two answers as tied together in a dipolar state that has a net truth charge of zero. Just as an electron and a positron are tied together as a neutral charge, the yes and no answers are tied together as a neutral information charge that has yet to take on one state or the other. So when we do get a Yes for an answer, we have selected a monopole charge and lifted it from the vacuum of the quantum probability firmament. Every time our minds probes for an answer it lifts a single logical possibility from the quantum fields and gets a single answer. It is like saying you have chosen the electron negative charge and so you have thrown the positron's positive away from your space of reality. Causal conspiracy says that there are two parallel worlds of reality, one of matter and one of antimatter. These two worlds are timed together and separated by a single moment of time. So, each decision you make has a contrary outcome that is sent to the anti-world of reality. The fields that govern reality then become the fields that control how those outcomes become realities. I my crazy world, these fields are real. These are as real as space and time.

When I return to my room with Diana, I realized that I have been dimensionally and emotionally transposed to a world that has yet to come and yet has passed by in time. Each day, I would go out into the garden outside what has become my safe house in the Vatican, only to discover that there is reality waiting in the beautiful motion of a butterfly and the reflections of sunlight on its wings.

Scientists will expect Comet Resa to collide with Mars at exactly six o'clock, on the morning of the first day of June, in 2016 AD. But be warned! What no one will expect is the strange chain of events that will take place on that day. Contrary to the laws of physics and gravity, two parts of the comet will break off from the main comet and suddenly glide from the face of the Red Planet heading for a direct collision with Earth!

That day, for a brief moment, gravity will become an outlaw and for some yet unknown reason the two comet fragments will be expelled back from the face of the "red planet" along earth-bound trajectories. There will be no time to prepare for the comets' arrival on Earth. Those who believe this story will take refuge in the places I will prescribe but many will not believe. The impact will be expected in six months, six days and six hours after the collision with Mars. But what no scientist will be able to pinpoint by simple calculation will be the exact moment when the fragments of the comet will impact the earth. They (the scientists) will neither bother to listen to this story nor will they consider this in their deliberations. They will assume that this narration of mine is a mere coincidence, an aberration without merit. This is the stubborn tale of human character, and so those without regard for this story will not know where the comet will descend.

I have decided to do what I must do, that is to translate the scroll and write its story for the sake of humanity and the Earth's creators. As days go by in the Vatican, I become more and more attuned to the agenda that the Bilderberg group is preparing and the counter agenda that is being developed by the Vatican. Effion will not appear in a real spacetime environment for another 30 years. Diana and I got closer as a couple and a true love is developing between us. It is the type of love that lasts and that has a purpose.

As a dimensional traveler, I am able to go to the future and do my writing in the small black room in the Vatican library. There, each book is slowly filled with the future predictions of the Scroll. In time all these books will be filled with my writings and I would have divulged the entire prediction of the future of man without violating any laws of Physics. My encounter with Effion in the small library is not a fluke. I was dosed with a lot of Memorin that was put in my coffee that morning. Although Diana was actually there with me in the future, she was older and would not actually known that she still is also in the past, in the room where we had slept together for the first time. I have inadvertently subjected her to the same illusion that Militant, Bayano and Nelson must have suffered. I fear for her and I feel that she might suffer a similar fate as they did in

Brazil. Sometime in the near future, a scientist will study and reveal the real reason why dimensional resonance causes such strange illness and perhaps I will encounter such a theory as I travel and write.

As long as when I come back into the present there is no written knowledge of what is to come on the blank pages, no laws of physics will have been violated. Effion has not been helpful through my process of writing. He is an eccentric and sometimes becomes epileptic and schizophrenic. He happens to play an important role in all of what is to come and I have learnt to respect him despite his shortcomings and his autistic episodes. For me, there is no longer a barrier to what we call the future. I am there and here at the same time as long as I dissociate my emotions and leave some of them behind in time, the person that is in the future does not affect the person that is the past.

Each day, I go back and forth into the future to fulfill the prophecy and write the Books of Malaki. I have already written the first three volumes and hope that I did the right thing. Diana has been of fantastic help even though she feels like a prisoner in the Vatican. As time passes, we fall deeply in love with each other and learn that the important things in life are right at your fingertips. She has resigned herself to my well-being and to the purposes of the Scroll. She now understands that there is a genuine purpose to her existence.

On the morning of April 12th 2012, I write in volume five that Diana and I got married in a small chapel in the Vatican. The ceremony is performed by Father Effion Iba in a private chapel in the inner Vatican City. We are prisoners of the Scroll and she and I must work together to make sure the world survives what is to come.

Each morning Diana and I would go out to the courtyard garden and enjoy our day with the flowers, birds and butterflies that fill the garden. We would do this for two hours before the Guards come and fetch us for my prophetic episodes. I take advantage of my garden visits to exercise and strengthen my legs to get my mobility back. As time passes my strength slowly returns and I am able to slowly walk again with the help of a cane. The first prediction I made about comets is about to come true and I am worried about what will happen to Diana even though I know that I am writing the story of the future and she will be there with me in the future. I worry even though I know she will be okay.

As time passes, I rely less and less on Memorin to initiate my dimensional travels. I become more adept at entering and leaving new dimensions. Diana is learning to travel with me and with the coming of our children I hope that I can teach them to the same. My hope is that by the time the comets hit the Earth we will be able to dimensionally travel to another place and time and preserve our lives as a family. For me life is just everywhere at all times, I am in the future, past and now. When I write this story, I feel like I have lived it and that its happening as I write. The only guilt I feel is that perhaps by writing the future, I am causing it to happen!

It is now 2016 and the day is near when the comets will hit the earth. There is a consensus among scientists that the comets will hit earth somewhere in the Mediterranean, but no one is able to predict the precise location, not even I. Six months to go, six months that means so much for that will be all the time we will have to make preparations for the great calamities that will befall the earth. The estimated size of the first large comet is eight kilometers long and two kilometers in average diameter. The second smaller comet is a two kilometers with an average spherical shape. They are irregular balls of metal, minerals, ice and mud spinning at a tremendous rate about their centers of gravity. Dynamic studies show that the smaller comet is much heavier than the larger one since it is made mostly of metallic alloys. However, there is not enough data on either the mass or dynamics of the comets to completely describe their trajectories.

High orbiting space-based satellite spectral analysis suggests that the smaller comet consists mostly of iron and other heavy elements and that the larger one is made mainly of ice and light mineral dusts. This is causing great alarm as the possibility of damaging collisions with Earth become evident. I could not make out every detail of the words of the Scroll concerning this, but to this day, I can only tell you that its messages points to a great tribulation and a way to dispel it. The Scroll says that when the comets collide with the red planet Earth will be positioned a mere seventy five-million kilometers away and coincidentally, it will be at its closest approach to the red planet. It will be considered a coincidence beyond all reasonable odds that the trajectories of both comet fragments should exactly intercept the Earth's orbit in space!

Already the occasion is being monopolized by Evangelists and other radical groups all over the world, particularly in the Americas and Europe. The religious radicals claim that as predicted by the Bible the date of collision with Earth is six-six-six, for sixth month of June 2016 AD, and coincided with the mark of the anti-Christ. They proclaim that the number 666 was the same mysterious symbol of the Devil that was given in the book of Revelations.

Two months after the collision with Mars the message gets through to the eight plus billion human inhabitants of Earth that these comets are a sign of the end of the world as they know it. Great numbers of people are now being told about my prophecy. Somehow, leaks always occur. The Vatican is busy investigating where the leaks came from but I do not particularly care if the outside world should learn about their future. Paganism is being abandoned by a large number of unbelievers who now want embrace Deism (God). By the third month after the comets' bounce from Mars, Islam, Buddhism, Judaism, and Christianity, all become saturated with converts. All over the world radios and televisions blast emergency procedures detailing emergency plans. I wrote that when people start seeing three moons at night they will know that the end is indeed near for the unbelievers. The new millennium starts with a bang!

For the first time ever, the alliance of the entire world is tested in the absence of mortal enemies! Do not believe that Kosovo, Chernobyl, Bosnia, Iraq, Afghanistan, Iraq, Sudan and the Congo have seen trouble as compared to this event. For a while, love and faith triumph over hate and war. There is no visible enemy! There is no hiding place. A new threat comes from a far-off world. The enemy will be nature and the unjust deal it will strike with the heavens. The Scroll says that man has replaced God with nature. Why blame God? Everything good that exists is now attributed to the works of "Mother nature". Because of this realization and almost out of guilty feelings for the past, there is a brief period of cooperation among nations. But take heed, for shortly hereafter each man will wish that his neighbor be the one to be hit by the sin from the skies.

The general feeling is terrible. The Earth wants unity against impending doom, but to no avail. In the Middle East, African States, and the European Union, water supplies are drying out by the day as people rush to store drinkable waters before the calamity. The great riots are starting in India. They are now spreading to Africa, the rest of Asia, Europe and the Americas. The United Nations is useless in pretending to police the world, while each nation prepares itself for doom. But do not despair, if you are a believer, God will save you! A parallel world arises for each quantum decision made by the conscious mind. Those who believe and have faith will create a new reality as the world around them fragments into oblivion. I will not explain this. It will come to you through faith and study of the word of God.

The USA and Europe (G13) are holding open debates to decide whether to intercept the comets and divert them from Earth's orbit by sending nuclear interceptor rockets into deep space. There are "emergency meetings" of what is called the "World Alliance Nations Defense", (WAND) a secret organization that is now being revealed to the world for the first time. This secret alliance is made up of one hundred specially chosen men from the NATO Alliance. They rule not only the financial world but the Military world also. They make decisions about everything on Earth, including the rules of war, religions, and so on. WAND already exists today, but the world does not know it. They are the persons who hold the secrets to the alien ship that crashed at Roswell. They control the G8, NATO and caused the collapse of the USSR and even the Bosnian wars. They are the New World Order George Bush was talking about. They are the Bilderberg Group.

In February, 1992, President Ronald Regan met with Pope Jean-Paul and they conspired to help the Solidarity Movement in Poland to defeat Communist rule. It did not take a long time for that to happen. The Communist wall separating East and West Germany came down, the USSR split into many nations and the core of the disbelievers fell. The World Government had begun! On September 11, 1999, the beginnings of a group called Al Qaeda, an opposing religious and political group, started a campaign against the North Atlantic alliance or the Bilderberg Group, (the Western world), particular the USA and Britain. Al-Qaeda, Arabic for "The Base," is an International terrorist network founded by Osama bin Laden a Saudi Arabian Prince, in the late 1980s. Considered to be the most powerful terrorist threat to the world the group attacked the World Trade Center and the Pentagon and many other US and International sites on September 11, 2001 causing lots of casualties. United States led a World Coalition and attacked the group in Afghanistan. The leaders of Al-Qaeda's fled eastward into the dangerous mountain terrain of Pakistan to secure a safe haven. The "911" attack on the World Trade center in New York heralded a new era in global politics. This attack on the United States by Al-Qaeda started a global campaign to wipe out the Islamic radicals some of whom are secret clans that will eventually become Pharaohnists! It was the turning point for the divergence of Conventional Islam and radical Islamism. Many people suffered world-wide in wars and terrorist acts. The World economy collapsed shortly after this event and a world recession depressed the real estate and financial institutions of the world leaving many people destitute and wondering if it was all planned to further the plans of some secret world restructuring plan, the "New World Order". In 2009 the collapse of the banking system of the world is the first sign of global changes for a new beginning. This collapse will continue forward and eventually cause many nations to be secluded from the new world order envisioned by the Bilderberg group.

In 2015, Al-Qaeda started supporting a new movement called "Pharaohnism". This radical new movement is a continuation of an alliance of some Islamic extremist cults and the Arian group of Arius of Alexandria. The group vowed to bring a New World Order under the leadership of radical Islamists. The government of Egypt embraced it as a new drive to patronize their idealistic views of nationalism. After the Egyptian revolution in 2011, a lot of Arab nations including Syria started a new radical form of government. It will be the root of a new religion called Pharaonism. In what is now known as the "Arab spring" revolutions across Arabia will continue until 2014. It will be reported that Bashar al-Assad the son of Hafez al-Assad was brutally murdered by his people. However, he escapes to Bahrain where he will live to this day in exile.

It is the morning of June 3, 2016 AD. Television stations are too engaged with the comet shows and only radios give the world information. The one hundred members of WAND from all the New World Order nations meet at the WAND Central Command Center in a secret facility in or around Area 51, in the USA. A black military figure of great world authority emerges and starts talking about official positions of the World Government on the comets. The World Government is thrust upon us as if it had always existed for generations with no explanations and no questions allowed. This black leader starts to talk about a declaration of Maximum Earth Alert Defense Condition 4, or MEA DEFCON 4. This level of alertness allows the joint deployment of nuclear and other space military paraphernalia from the point of view of a one World Government.

Calculations by NASA and the European Space agency, ESA, show that the positions of the comets relative to Earth are too far advanced for any maneuvers. The comets are under solar gravitational acceleration and their speed will be more than three hundred thousand kilometers per hour. That is why they were not easily discovered in the first place. The tracking system needed for such fast "relativistic" comets was only available to hi-tech telescopes like the defunct Hubble. Only the Hubble Telescope and a few large telescopes around the world could have seen it coming and kept it in focus for long periods of time. When the comets were far out of the solar system, they could not be easily tracked by ordinary telescopes. As they come closer to the planetary orbits they are too fast for easy tracking using available technologies. An analysis by WAND shows that thermonuclear explosions are no longer an effective means of deviating of the comets from an Earth-bound trajectory! Computer simulations show that the comets might propel radioactive debris from the nuclear bombs back to Earth due to their tremendous momentum. It is too late! Never before did scientists observe such massive objects traveling at such fantastic speeds so close to Earth. After many hours of urgent debates, it has been decided that WAND should abandon its plans and prepare for the unavoidable impacts. Each nation is now left to their own demise. In 2014 the hasty construction of a large secret Military Command post began in Nevada under the Hoover dam, in preparation for this event. The boundaries of the restricted lands surrounding Area 51 were doubled virtually overnight. This was a sign that the YuMuYa Scrolls are on the mark. The gigantic top-secret underground facility was designed to hold in complete isolation and in full operational mode, over five thousand people. These would be key persons of the Bilderberg Group who will be deemed necessary to keep the world under control in a global catastrophe. The planned facility has been prepped to sustain five thousand people for as long as six months without contact with the outside world. From this hideout the one hundred leaders of the Bilderberg Group including the US President will assume their command post and will be able to communicate with the rest of the world via radio or satellite television. A meeting has been announced by NATO and the UN jointly. It is to be the first joint summit ever between NATO, WAND and the UN. The hundred members of the Bilderberg Group have arrived with their loved ones and they are being transported to the hideout without explanation. It is two more weeks before the collisions with Mars.

Nearly all civilian and military planes have been put into action transporting medical and food supplies to target areas where the need will be greatest. Luckily, these emergency procedures have already been put into place when the threat of nuclear war became real as a result of a rise in tensions between Iran and the USA in 2013 AD. Many countries have begun to distribute large amounts of medical supplies to suspect areas of high risk in the hope of averting human suffering. That is a good humane gesture and will help mankind in diverting total destruction of the human race! Pray!

By the night of the third month of the comets' approach three moons are alight in the night sky. The night sky appears as a dull day. All grocery stores in the world have run out of food, batteries, blankets, water, tools, weapons and other supplies. So, those who believe must stock up at least six months before that day. The religious book stores are empty, so buy your Bibles, Torahs and Korans now and read them! Churches, mosques and synagogues are filled to capacity all day long with new and old worshipers. They are being used as makeshift shelters for those who despair and the unfortunate ones who could find no other suitable shelters.

The exodus has begun. People are running away from the identified impact zones to so called "safe zones." The real trouble has started. I want to travel to find a safe zone but I am at a total loss now. Everywhere I go there are emotional dimensions filled with tormented souls. The dimensions of love, hate, sorrow, anger, greed, envy, pride, lust and hope are all so full of souls it is impossible for me to find a suitable hiding place for my future family. Everywhere I go, I see the holes in the grounds with souls of evil hiding in them. They are everywhere!

Airplanes, trains and ships are heading out of the Mediterranean areas overloaded by the migration. Despite the warnings that were issued by WAND people who are unable to access mass transportation have begun to walk or drive away from identified impact areas. The world is becoming more and more chaotic by the minute.

In the USA an anti-Christ movement has arisen out of nowhere it seems. They start showing their power and tremendous influence in the world. The airwaves are filled with discussions of the Book of Revelations and Nostradamus' predictions of the destruction of the world by balls of fire. The general feeling in the world is one of fearful expectation.

Although no one is able to explain the bounce phenomenon, many theories have been created by scientists and laymen alike, even when doom is around the corner. Even in the time of doom mankind still displays a great ability to think and distinguish itself from animals. They postulate that the bounce was caused by the comet impacting a massive field of sulfurous compounds on Mars. Such a field they proclaim exploded and propelled portions of the still solid comet as it sped toward the surface of Mars.

Another theory is based on intelligent life on Mars. This theory states that the "Martian face" discovered on Mars by NASA is evidence of intelligent life with advanced technology that may have diverted the comets to earth. There are some interesting evangelical theories that hinge themselves on Biblical and Koranic predictions. For the Islamic predictions, they take note of chapter 46 verses 24 and 25 in the Koran which states:

"So, when they saw it was a cloud appearing in the sky advancing toward their valleys, they said: This is a cloud which will give us rain. Nay! It is what you sought to hasten on, a blast of wind in which is a painful punishment, destroying everything by the command of its Lord; so, they became such that naught could be seen except their dwellings. Thus, do We reward the guilty people.

There will be those that believe and those who do not. Do not be a disbeliever, you will be saved. The wrath of God will punish the guilty. As a consequence of your beliefs, let many pagans convert to Islam if they see this fitting".

Christians rely on a quote from the Book of Revelations, chapter 16 verses 17 to 21.

"Finally, the seventh angel poured out his bowl upon the empty air. From the throne in the sanctuary came a loud voice which said, "It is Finished!"

There followed lightning flashes and pearls of thunder, then a violent earthquake. Such was its violence that there had never been one like it in all the time men have lived on earth.

The great city was split into three parts and the other Gentile cities also fell. God remembered Babylon the great, giving her the cup with the blazing wine of his wrath.

Every island fled and mountains disappeared.

Giant hailstones like huge weights came crashing down on mankind from the sky, and men blasphemed God for the plague of hailstones, because this plague was so severe."

For four months after the collisions on Mars Mankind prepares for the "thief in the night" and for the end of the World. Mankind prayed, fasted, and cried in recognition of its solitude in the Universe. Everything that could happen literally happens. Lovers marry in a hurry, wills are written, murders, plunders and suicides occur on a vast scale. Scientists have become preachers and preachers have become scientists. Television shows are being replaced by vigils of prayers and protests. Schools, libraries and other unused buildings are being converted to makeshift shelters and hospitals. A great feeling of inevitability has overcome humanity. Prisoners are being released all over the Earth to mingle with the free and save their own souls. Families are trying to come together and unite as they wait for the bitter end.

By the end of the fourth month, the commotion that is created by food and medical supply shortages starts to take its toll on humanity. Riots are breaking out everywhere on the globe as security forces abandon their tasks to save their own souls. The initial feelings of expectation and religious awareness that overcame the world in the first few months of preparation soon change to selfish self-centeredness as survival becomes the main game. Airports and train stations are overwhelmed with the desperate waiting like hopeless zombies for a chance to go back home to their loved ones. The seas are already be filled with boats carrying escapees to their endless expanse where they believe they will find a safe haven.

Since no radios or TV stations will be on during this period, I decided that I will astral travel to safe areas and oversee what is happening to the world. I will be going to different areas and reporting back to Effion and Diana about what I see as it happens. The only problem is that what I see may not quite represent the physical world in its entirety since I can only travel in emotional fields during this period. The emotional fields will be so full of souls that it will be difficult for me to get a complete perspective of the physical realm. So, I decided that I will travel only in the dimensions of Hope and Love to avoid the fantastic crowd of tormented souls that will fill the other emotional dimensions.

Effion and I had worked out an elaborate plan to make sure the Pope and the Vatican staff will be safe during this calamity. I have made several projections into the future of the Vatican, and found out everything that could happen to us when the comet strikes. The whole thing has been orchestrated to sync in with events that will occur. The first wave of winds and debris will not affect any part of the Vatican City. On the second day, some water from floods will enter into the catacombs and then proceed to enter into the selected Papal chambers. We have planned that on the morning of the second day, the Pope will be moved to a second prepared chamber where I already know he will be safe. I have given a list of safe areas to the world media, Senator Owens and other friends that we have made in the past and will make in the future. I gave a particularly safe are to Saint Jude children's Hospital and other children's organization around the world. I made sure that religious heads, including the Jewish heads, the Islamic heads and the other major religious heads got a copy of the part of the scroll that deals with the calamity, so that if they believe, they will act to safeguard their people. It was a real effort and a battle of wills to divulge this information, but it was a condition that I have placed on the Scroll translations to ensure that everyone has a chance at survival.

Diana and I are holding our ground in the Vatican and awaiting the inevitable calamity. The Vatican staff has been evacuated into the deep secret catacombs. I have already determined during my prophetic journeys that the floods and the winds will not affect this area. I am still not absolutely sure about our safety but so far if my predictions are correct, then the only safe areas left in the world are the mountain tops, caves and bunkers that have been built world-wide. The Hoover Dam was well prepared by the Bilderberg group. They knew that the waters will protect them from the winds and that the air systems needed to scrub the atmosphere would only work if there is ample water supply to do the job. A program had been instituted to use the large turbines to act as filters and scrubbers. The electric grid was converted to electrolyze water at a massive scale and generate oxygen and hydrogen for the bunkers. The oxygen was for breathing and the hydrogen was to be for fuel. There were smart and ready!

It is the night of June 6th, 2016 AD. All the world's electric grids have been turned off. We wait in darkness for the end to come. Diana and I have been situated in a bunker that I chose for us. We will be safe for now and if I need to, I could use my visions to find safe havens after the comets pass. The collision of the first comet fragment with Earth occurs at the Bekka Valley in Lebanon. The second fragment hits the Mediterranean Sea close to Italy at an angle of twenty degrees from the horizontal on an eastbound trajectory. It happens on the sixth hour of the sixth day of the sixth month of the sixteenth year, a numerology that will always be remembered as 666!

Do you wear a watch? Then, the number of the beast will be on your hand on that day. That number will be etched into the frontal lobe of your brain. It will be in your forehead, not on your forehead. A frightening wind precedes the comets, creating a huge hurricane the size of the African continent. The flash of light in the entire Northern hemisphere of the sky is the first visible sign of the impact. The glow from diffused light and heat in the atmosphere is seen as far as North America, completely circling the earth. As daylight breaks on the sixth minute of the sixth hour over the Bekka Valley, the first comet fragment breaks through the atmosphere. The Earth is surrounded by an oval shell of red dust from the surface of the Red Planet. Like a huge ball of fire, it lights up the dawn sky like a second crimson sun as it burns from the friction of the Earth's atmosphere. In less than two seconds after entering the atmosphere the collision occurs a thousand feet above the ground at the "Valley of Kings" in Lebanon. In the most spectacular explosion, one hundred million tons of debris is blown from the face of the valley spiraling into the upper reaches of space like a hundred thousand atomic bombs all going off at once!

The next thing that happens is the tremendous supersonic flow of heated expanding plasma and gas from the pressure wave that is generated by the explosion. Compressed air, propelled at supersonic speeds spreads in a spinning spherical wave from the epicenter of the explosion traveling round the curved surface of the planet. The sound wave and sonic pressure it generates flattens everything within a six hundred kilometers radius from the epicenter. The shape of the valley acts as a focusing surface for the impact and causes everything in the valley area to be blown one hundred or more kilometers to the upper reaches of the atmosphere. The tremors of the explosions reflect more than fifty times through the Earth's mantle in a decaying dance of power, giving seismologists the most updated naked picture of a traumatized Earth.

The second fragment is larger than the first. It hits the east end of the Mediterranean Sea. Luckily, it is a low tide period, and the sea level is one meter below normal. The Mediterranean Sea which is almost landlocked, covering an area of about two million square kilometers has a latitudinal expanse of four thousand kilometers and a maximum width of about sixteen hundred kilometers. The sea is shallow with an average depth of one and one-half kilometers. The display of the second comet tearing a tangential path of fire into the sky lasts but a mere second.

As hot matter falls into the sea two billion cubic meters of water is evaporated from the sea's surface into a steamy explosion of boiling water, vapor and comet debris. The resulting mess expands into a huge spherical wave which rises high up into the upper atmosphere mingling with the expanding wave created by the first impact. The tsunami created by this impact is propelled with fantastic force across the sea into the land masses that trap its fury. It carries with it hot sizzling water as it expands landward at a speed in excess of four hundred kilometers per hour. The huge wave of superheated steam and comet vapor comes rushing across the sea toward land creating massive hurricanes and tornadoes farther inland. The sky is a rainbow of mixed colors as dirt, steam, water, man-made and natural debris all mingle together to form a deadly stew. The dust wave from the collision in the Bekka Valley is also propelled by the pressure of the Mediterranean collision far into the mainland that surrounds the sea, depositing red slime several centimeters thick from the fertile soil of the valley unto the arid North African plains and the sandy deserts of the Middle East, Asia and as far as mainland China. More than eighteen million people die instantly from the pressure of the blasts alone. From our bunkers, we can hear the howling winds and the falling projectiles pummeling the cities above our heads. Very little water has entered the catacombs. When you travel in emotional dimensions, you can sometimes get real information about space and time around you. However, it is possible that you can be affected by the physical energies of the spacetime continuum and so you have to be careful not to subject yourself to hazardous events that are happening during your travels.

Two days later, tens of millions more people die as thick blankets of dust-filled air pollute the cities. No one city is spared. Meanwhile, the massive sea well cavity more than eight hundred meters deep at the epicenter of the explosion is spreading over a six-kilometer radius in a parabolic pattern is created by the evaporated portion of the Mediterranean Sea. I am standing at Mount Vesuvius and I can see a mountain of water crashing into the large sea cavity creating a huge vacuum that causes the sea levels around the world to oscillate several times. With each sweep, these oscillations result in massive floods from Tsunamis destroying entire cities and killing millions of people along the Mediterranean and European coastlines. In the dimension of Hope, you cannot hear the sonic booms generated by the two collisions. However, even Hope is marred by the high density of red steam that fills the atmosphere. I can see this steamy mess compress and expand as the sonic booms pass through them. Several more fierce explosions result from collisions of the two expanding sonic booms as they cross each other. The horrible images of the sonic echoes of the impacts thunder repetitiously across the globe before fading.

Outside the confines of the Mediterranean Sea several tsunamis sweep into the coastal areas at speeds exceeding four hundred kilometers per hour crossing the straits of Gibraltar to the open expanse of the Atlantic Ocean. These tsunamis result in a systematic array of global flash floods. The Suez Canal and surrounding land masses are completely drowned beneath twenty meters of black mud and water, killing all living creatures within one hundred kilometers of the shores of the Mediterranean.

A chain of floods and earthquakes follow the impacts rippling across the entire surface of the earth in a bid to unleash the fury of the comets. Were it not for the steep valley walls that buffer the explosion the resulting wind and debris will have resulted in total global destruction. However, the dust and debris that is trapped by the upper atmosphere contribute to more deaths worldwide than even the explosive force of the impacts. First, the terrible winds and heat waves spreading over the planet are generating huge hurricanes and tornadoes across the planet. Then, the greatest tribulations ever to be witnessed by humanity come as the polar ice caps start melting down. The sea levels are rising and massive floods are beginning to change the coastal outlines of the world.

In Egypt the Nile River floods reaching more than twenty meters above its banks and causing more than five million more fatalities in the city of Cairo alone. Standing on top of the Great Pyramid, I am surrounded by water. The pyramids stand like island sanctuaries in the middle of a massive, shallow, but evil sea covering the desert sands. People scramble up the slopes of the great structures like fleas, fleeing from the oscillating tides of the angry Nile as it sweeps their towns and cities away.

The ancient temples of Abu Simbel, once before saved by an International effort from the floods of the Aswan Dam, are drenched in four meters of flood waters. The statues of Ramses II and his family survive the forces of the floods and stand aloof, like silent witnesses from a past epoch observing the horrible tragedy that is overtaking the world.

In the central United States, in South America, and in South Europe, hundreds of millions of acres of farmland are covered in a meter thick layer of ash, mud-slime and debris. In the farmlands of the United States of America a massive exodus of people and animals toward the coastal states of Florida, Virginia, and the South Carolina has begun. But these areas will not be freed from calamity either. The massive changes in weather patterns across the globe are triggering several super tornadoes, hurricanes and dust storms from the Atlantic Ocean and Central States respectively. For four months the winds change directions erratically, making it impossible to predict patterns for hurricanes or the dust storms.

In the U.S.A the State of California suffers most. In the few minutes that followed the impacts, unearthly quakes and floods triggered eight thousand kilometers away by the energy transfer to the earth's crustal plates. Los Angeles, San Francisco and several coastal cities in the West Coast are drenched by horrible floods caused by echoes of the oscillating tidal remnants of the Mediterranean tsunami. Coupled with the powerful earthquakes, these cities have been reduced to remnants that are unrecognizable and uninhabitable. The Central and Midwest states are being deluged by gigantic unforgiving floods, as the seas spills into the great basin of the Midwest plains.

It is now over and we are again free to go outside the bunker for no more than an hour a day. For over a year after the explosions a blanket of fine red dirt and steam spread from Europe across the Atlantic to North America dimming the sun's light over the equatorial areas. Part of the world is struck by huge temperature drops, crop failures and the extinction of many animal species. Europe has been dramatically changed by the rising seas. Holland and most of Denmark are under the sea. French Riviera is in a constant state of flux as the unstable coastlines change shape day by day. Africa, the long-suffering continent has been spared most of the calamity. The continental shelf prevents the sea from rising into the mainland, except in Northern coastline along the country of Egypt. A few days after the collisions dirty red and black snow started falling on parts of the world contaminating the remaining fresh water and food supplies. Four billion, five hundred and fifty million people worldwide die as a direct result of the impacts, starvation, thirst, and diseases.

Most of the world has been without electrical power for years now, and those whose grids were unaffected by the impacts do not survive the unpredicted electromagnetic storms that follow. Fine red dust particles and other aerosols suspended in the atmosphere have changed the climate of many regions temporarily and have caused more crop failures and more deaths. Mankind has been dealt a horrible blow from the far away seemingly harmless cosmos! Nature is god, an evil unforgiving god! For man has replaced God with Nature. Nature is a punishing god. That which is deemed created without design is pointless, meaningless, without cause and without a heart. How can Nature save us when it will also destroy us like this?

Most communication satellites orbiting the Earth are not able to penetrate the energy barrier that has been created by the dusty atmosphere. The invisible electromagnetic barrier that was created by static electricity in the dust layers over the skies has prevented commercial satellites from operating properly. As a result, such communication are limited to global reconstruction organizations like WAND and WHO.

It is now two years after the disaster and power is slowly being restored to some cities that survived. If you were destined to survive, you will. Surprisingly, a lot of children and families survive this tragedy. They are spared by their faith in God, themselves and in their children and especially because they did not surrender their wills to "Mother Nature." Most of the important Earth orbiting satellites are back on-line as the dust layers slowly settle from the atmosphere. The centripetal forces of earth's rotation has begun to pull the dust and debris in the upper atmosphere and into space to form a thick equatorial belt orbiting the earth. The days have become longer as the earth's rotation has slowed down to 26 hours per rotation. The skies appear to belong to a bizarre new alien world decorated with a belt of orange-dread dust.

From space, the few satellites that are in operation reveal a bleak picture of the planet below. Railroad tracks and roads are covered by the carnage of the comets. Most airports are unusable and airplanes are not able to operate as a result of the high concentration of aerosols that still float in the upper atmosphere. The map of the Earth has changed significantly. The West Coast of the United States has lost major cities to the depths of the Pacific as the weakened Earth's crustal plates fall off into the Pacific Ocean. The Ocean waters have covered the Great Plains and most of the Central States of the USA. The ocean also yields its dead. Soon after the comet impacts, millions of tons of fish and other marine creatures were propelled by the blast into the atmosphere to rain down like plagues from the skies.

One year six months after the disaster a plague of insects and small creatures that were displaced from their natural jungle habitats start appearing in many parts of the world where they do not belong. As the temperature of the Earth settles back to normal humanity is struggling to survive. The new mix of living matter and debris is settling into a new unnatural order. With God's evolutionary ladder temporarily shaken by visitors from outer space the dance of the prey and the preyed begins to unfold. Animals roam every corner of the Earth looking for prey. Every surviving household is a target. It is a trying time for humanity.

For eight years, the recovery process continues. Diana and I have kept vigil to make sure that the world's remaining caretakers are informed about what to do and where to go. We sent out daily radio messages to the world and kept in touch with NASA, ESA and the US government. The Vatican now has very little to say about what we can or cannot do. The secret about the comets is over and the recovery process must now be allowed to continue without hindrance. So, each morning, we prepare a comprehensive bulletin from data I get the night before by travelling. This bulletin is announced over the air waves so that those radio operators and Ham operators and other communicators that survived will transmit the information to the rest of the world.

We are now deeply involved in being the caretakers of mankind. The Earth has not sufficiently recovered and we already on the eighth year after the impacts. By 2019 most major cities have come back to some semblance of manageable life.

During this mysterious genesis of Mabus Ali Resa the two comets lit up the Earth with a ghostly unearthly light reminding us of a greater birth two millennia ago in Bethlehem. During these dark years, Mabus Ali Resa was born and no records of his birth will be uncovered. Neither his mother nor his father will be ever identified and no child born on his birthday will live to see the next day. The baby Mabus was found in a pile of rubble near the palace of the future Egyptian President, Mahmoud Aziz, ruler of Egypt. After the Arab Spring, the former President Mubarak of Egypt was condemned to death and the sixty nine year old Mohamed Selim al-Awa became a temporary President of Egypt. A democratic coup d'état ensued and Mahmoud Aziz took over as President. Masked rescue workers organized before the impacts by the city of Cairo were combing the city for survivors when they found the child sleeping peacefully in a basket wrapped in warm clothing. He was covered in thick, red, mud-slime. In spite of being exposed to the raw elements of the disaster the child was without a scar or problem! In his raffia basket was a message from his parents written on the raffia prayer mat that wrapped his fragile body. The mat was decorated with the symbolic crescent moon of Islam, supported at the apex of a golden pyramid. It was to become the new symbol of "Pharaohnism," a religion unlike Islam, Christianity, or Judaism. The writing on the mat simply read in Arabic, "May God forgive us."

The child was taken to the emergency shelter of the palace which was set up for the select few in Egypt. Soon, the wife of President Aziz found the child unattended in the hurried pace of the palace. She took the infant of the cosmos into her arms and looked into its steel-gray eyes. The child looked back at her forging an eternal maternal bond between itself and its new found mother. Since God did not provide them with a child for he was also sterile, President Aziz developed a deep liking for the child and adopted him as his own. He exalted the child as a national gift and a special person in Egypt. For some Egyptians the child was to be an Egyptian "Moses" reborn, a man of great importance, for he was found in a floating basket made of straw. He was named Mabus Ali Resa, after the founder of the comet, Pascal Resa and the excavated remains of a great Egyptian god-prince, Mabus Alus from the first dynasty of early Egypt. The site of the explosion is now considered a Holy site by some Egyptians, Arabs, Jews, Hindus, and Christians. It is called the "Resa Site of Mourning".

The ring of dust and debris that straddles the Earth in space shines with the brilliance of the colors of the rainbow, changing the Earth forever into the picturesque landscape of a new alien world. The Earth is now a mini-Saturn, a planet with a new and beautiful ring of dust and debris reminding all of the day of the calamity. The ring of orbiting debris around the Earth is called the "Ring of Resa".

As Mabus Ali Resa grows up, he is referred to as the "Lucky One" and the "Saved One of the Earth". President Aziz uses every occasion to make his existence special. The child is considered as the protector of Egypt. The Resa Site of Mourning attracts lots of pilgrims who having experienced the massive cataclysm need some consolation from the psychological effects of the calamity. People from all over the world still come by the millions visiting the site and paying homage to the special child.

"For he was spared to live among us," they chant.

Some religious leaders even preach that Resa is brought to Earth by the comets to save humanity from extinction. The child they claim was an insurance policy for man, donated by God. As years go by, Mabus Ali Resa has become very popular and powerful. His appearance on special Islamic, Jewish and Christian events world-wide always draws huge crowds into the Mosques, Synagogues and Churches. He has made Pharaonism a very popular religion for both the believers and the unbelievers. Mass conversions are happening in his presence. People everywhere see Mabus Ali Resa as a "Messiah" of some sort. "Resa brought life in the midst of death" they preach.

Mabus Alus Resa's rise to power in Egypt happens in 2036 AD after massive uprising happens in South Lebanon between Israel and Lebanon. The dispute is over fresh water rights. The young Ali Resa goes to Lebanon and stands between the two armies. He confidently negotiates a cease fire and a complete retreat to peace. The military leaders draw back their forces as if by some magic spell. He then negotiates a simple settlement between the parties with the help of President Robert Hage of Lebanon. That peace has lasted for a long while.

When the Ayatollah Ali Maserati dies in Iran an uprising takes place to replace him with a moderate leader. Ali Resa intervenes. He declares that he is the new Ayatollah and that no one is better suited for the great task of leading the new Islamic movement than himself. The mainstream Islamic clerics have opposed his claim and do not support his sudden entry into the world religious arena. With the rapid rise of Pharaohnism, he is soon perceived by some religious leaders as a new and dangerous force to contend with. In spite of the objections of the religious leaders and some faithful he is crowned in the front lawn at the palace of President Aziz, his father and made the youngest Ayatollah in history.

A new Pope arrives in the scene. I have met him before. He is Pope Paul the seventh! Soon Ayatollah Resa makes his bones in the International arena by negotiating better economic pacts between the new Europe, Middle East and Egypt. His influence will spread to the United States, Europe and the Far East. Even Pope Paul the Seventh pays him a special visit when Mabus Ali Resa opens the Institute of Joint Christian and Islamic studies in Jerusalem. So, it has come to pass under these circumstances that Mabus Ali Resa's life evolves to make him one of the most important and powerful men in the world.

In the vision of the Scrolls as he gets ready for the celebration of the "Day of Egyptian Awareness," dressed in the style of the ancient Egyptian Pharaohs, Ayatollah Resa looks even more pristine and beautiful than usual. His consorts carefully and lovingly prepare each garment that must be worn by the new Pharaoh of Egypt. Laboriously, they clean his feet and manicure his nails. They wash his body with perfumed oils, following every step of the ceremony they had been taught in the ways of their ancestors. They prepare his headdress and chariot, just as the last found "Second Book of the Dead" prescribed for the ancient Pharaohs of Egypt.

As the ceremony progresses into the night, far away, in the United States of America, a translator reads the last line of a letter written in Latin by Pope Paul the Seventh to President Owens.

"It is the work of Michel de Nostradamus, Sir. His name shall be Mabus Alus."

CHAPTER 5
(August 15, 2038 AD)

I wake up to a cacophony of voices. There are a lot of people around my bed. I find myself in a large white room with large white windows. A man who looks like a doctor is probing my hand with some device. His eyes reflect a snowy grey light that makes me feel violated. There are tubes connecting to every part of my body. I slowly look around me and find a continuous trend of black suited men lining the room. I am feeling old and vulnerable.

"He is back!" the doctor shouts.

"Give him another shot of Memorin."

I feel the stabbing needle entering my veins. The voices in the room slowly vanish to become musical notes as the chemical rushes through my veins. This is not just Memorin. It is different. I hear my brother protesting in the background.

"You did not need to give him a truth serum. He is dying anyhow."

Am I dying? Surely it does not feel like I am dying. I am only feeling calm and peaceful. The colors of light fill the room as someone opens a curtain to let what looks like sunlight into the room. The problem is the light comes directly into my eyes as if aimed at me for some purpose. As the light waves away, I see the doctor's eyes again piercing into my being to see if I am inside.

"Give him a few minutes," Elaine states commandingly in a soft voice.

I do not know if I am in some dimension or in the spacetime continuum. I do not know if I am old or young and writing the story of the Scroll. I am now lost between dimensions. I can see Elaine standing by my bed looking at with genuine concern on her face. I know this is phase two of my prophecy. They now know all about the comets and either the prophecies have come and gone or they have yet to come. All I know is my reality is right here, right now, in this room. After what seems like a long silence, Effion orders me to start telling the story again. The truth serum is powerful. I could not resist its hold over me. As if by some magic, I sort of recall where to continue. I feel no resistance. The next phase of my work is to reveal the most important prophecy; the Josephus Nambu force. This will open up the science of dimensional travel to the Vatican. This will give them the ultimate power that rules the universe of the mind.

My brain is reeling and I find myself propelled into dimensional gateways that I do not recognize. It is almost as if a movie is playing in my head. It is 2038 AD and the world mostly has recovered from the effects of the calamity brought on by the comets. There is a rush of feet as people enter into the room as if to listen to my narration. The story of the Scroll continues in the small Egyptian city of Abu Simbel, within the path of the Nile River just south of the Aswan Dam. A learned man who is called Professor Abdul Aziz has become the most renowned "searchers of Earth". This I understand to be an archaeologist. The Scroll says that he spends more than twenty years studying the Ancient Temples at Abu Simbel. For the past one hundred years Egyptologists and

Archeologists believed that the temples were built by Ramses II and that Ramses dedicated them to the three gods Heliopolis, Memphis, and Thebesto. But soon after the disaster wrought by comet Resa the temple yields a new mystery. New discoveries made at the temple site point to a different and puzzling history. Massive floods have eroded one of the walls of the temple and exposed a second wall beneath the first. The wall has inscriptions that tell a strange tale that conflicted with the assumptions that Ramses II built the temples.

One of the most interesting finds at the Abu Simbel temple is that the ancient town is completely devoid of all animal or human remains. No fossil or skeletal remains from the dynasty period of Ramses II were ever found during fifty years of excavation. No pots, tools, or other man made artifacts date back to the period of the temples' supposed builders have ever been found at the site.

The front of the first temple is decorated by four great statues of Ramses II in a sitting posture each more than five meters in height. Smaller statues of the Pharaoh, his Queen and their children decorate the face of the temple. These statues are now known to have been superimposed on the more ancient structure of the temple itself by Ramses II. Previously hidden walls of the temple have been exposed to the elements by the great floods that accompanied the Resa comets. On one of these walls several unusual inscriptions and carvings are found to be alien to Egyptian culture. Evidence of past civilizations usually includes tools, weapons, manuscripts, burial sites, and mummies. But at Abu Simbel, nothing has been found that indicates a work site. It is the general conclusion of present day archeologists that Ramses II misled the world. He was not the architect of the Abu Simbel temples even though statues of himself and his Queen, Nefertiti are plastered all over the sites. Ramses II used a thin coating of plaster to cover the original inscriptions on the temple walls. Recent discoveries made by Dr. Abdul Aziz and his contemporaries indicate that the temples were built by a civilization dating as far back as 8,000 B.C., centuries before the First Dynasty of Egypt.

Many theories were proposed by eminent archaeologists and Egyptologists to explain this strange find, but none compelling enough to satisfy Aziz. Archaeologists agree that Abu Simbel is an anomaly. For some reason, two magnificent temples were built on the cliffs of the Nile, but evidently no man or animal had lived or died there during that period. With the advent of a more accurate dating technique that will be called Neutron-Emission-Spin-Bias-Analysis (NESBA), the temples are found to be more than ten thousand years old! The discovery that the Sphinx is also far older than previously thought is another startling find. This discovery is based on the erosion pattern found on the Sphinx. The erosion analyses reveal that the structure was exposed to a great amount of rain far in the past. The only period found to have such weather was nearly 10,000 years ago!

Dr. Aziz studies the site with the aim of solving some of the riddles. But it seems as if the more he probes the more perplexing questions arise about the roots of Ancient Egypt. How could two temples be carved into a sandstone cliff ten thousand years ago by some unknown civilization who left neither tools nor evidence of work and neither human nor animal remains? What was their significance? Who carved the temples and why are there no remains or evidence of their work? Why did Ramses II claim to be the builder of the ancient temples? What was their significance? Could he not have built even bigger temples for himself? Could the builders of Abu Simbel be the builders of the Great Pyramids and the Sphinx? He wonders.

The first temple of Abu Simbel is found to have astrological significance. Its design is a complex geometric pattern allowing the rays of the morning sun to illuminate a small football sized crystalline sphere, positioned on the head of a small statue of Ramses II. The statue was discovered a decade ago in 2028 A.D. by a team of International scientists who studied the site using sonar imaging. Their aim was not to discover something new but to probe and access the damage that was done by the flooding of the Nile during the disaster of Comet Resa.

It is believed that the statue had been hidden in a large oval cavity constructed by Ramses II. Over a period of five thousand years dirt and debris slowly buried the site completely hiding the chamber from the eyes of the world. Although the relative orientation of the Sun's position in the skies of Egypt changed over the past five thousand years the illuminating effect remains the same - a marvel of optical and mathematical trickery. The strange crystalline sphere illuminates the cavity with a spectrum of colors producing an eerie effect in the chamber when the sun's rays shine through.

The crystal reminds Aziz of the famous crystalline skull that was falsely said to have been discovered in a cave in South America and once featured in Author C. Clark's television series "Mysteries of the Universe." But this crystal of Ramses II is even more magnificent. On the sphere is inscribed a strange story of how a super-crystal had been lost to the great gods of Egypt and how Ramses II was consumed with the idea of restoring this super-crystal to the temples of Egypt and ultimately to God. Apparently the Ramses crystal as it came to be known is a mere imitation of a more powerful object of an even older civilization than the era of the pharaohs! For Aziz the story parallels the Bible's lost Arc of the Covenant. It generates a great interest among Egyptologists and Archaeologists who perceive the super-crystal as the ancient lost treasure of past Egyptian civilizations. Every warm-blooded Archeologist in the world dreams of finding the super-crystal, the mythical object of great beauty and power!

Meanwhile, spectroscopic studies of the imitation crystal have been performed by various first class laboratories under the watchful eyes of Aziz. These studies have neither fully determined the crystal's material composition nor confirm that the crystal is a modern fake. About ninety-percent of the Ramses crystal is made from fuzzed quartz. The remaining ten percent is a mixture of unknown compounds. Further studies show that the composition of the crystal is not anisotropic. Its composition changes depending on the line of site taken through the crystal's center! The uniformity of the chemical changes in the structure is not a result of natural phenomena. In other words, the crystal was not carved from natural stuff but was actually made by someone or something with great skill and intelligence.

The strangest discovery of all is the strange writing on the upper half of the crystal. Apart from the well-established Egyptian hieroglyphics that describe the strange super-crystal, the other writings on the Ramses crystal are not contemporary with either the ancient Egyptian history or any other civilization that existed during that period. Its symbols and designs resemble sinusoidal wave patterns, mathematical step functions, triangular wave patterns and other electronic-like signals.

Other strange writings of the same genre have surfaced in many parts of the world exposed by the great floods that followed the comets. Drawings of similar crystals have been found in Druid caves in Scotland, the Inca pyramids and in the submerged ruins of a great city washed ashore along the Italian coast and believed to be the lost city of Atlantis. All these recent discoveries have charged the archeological community. However, they are not always publicized for fear of attracting too many tourists to the sites. Archaeologists secretly postulate that an ancient and powerful race forged the history of man and spanned continents and oceans to leave their trace everywhere. Some superstitious excavators also suspect that the impacts of the comets were made purposefully to reveal the ancient sites to modern man as a message. Soon you will learn the truth. These crystals of the future will also reveal a great deal of what humanity has been through since the time of creation. The recent excavations of the Abu Simbel site indicate the existence of a series of chambers and hallways leading to a large oval cavity in the rock. Aziz labeled this still hidden cavity "The Holy Sanctuary". The Ramses crystal was discovered on the first floor of the main temple hall just above this large hidden chamber. Sonar studies on the site show that the Ramses crystal was originally positioned at the exact central vertical axis of the hidden chamber. Why? What is hidden beneath the foundation of the ancient temple?

In 2014 AD, while still a young Archeologist, Aziz becomes the head of "The Egyptian Awareness Center." His half-brother President Mahmoud Aziz, ruler of Egypt, selects him to head the secret project of restoring the temples to what they were before the comets came. Mahmoud Aziz is surreptitiously considered as a gentle and popular leader of his people. He is the spokesperson for the Fundamentalist Islamic Movement. He amasses great power in Egypt during the 2017 AD uprising. During that period, Aziz was a student of archeology at the University of London. He graduates a year later, and finds work in the newly created London-Egyptian Archeological Society in Greenwich village, England.

After becoming head of the Al-Fatiha party in Egypt, Mahmoud Aziz invites his brother to succeed him as "Head of the Egyptian Awareness Center" in Cairo in 2021 AD. Because a great sense of self-awareness for Egyptian culture and history has taken root all over Egypt, Mahmoud Aziz nationalizes all Egyptian archeological and historic sites. He replaces all expatriates with Egyptians in all important sectors of Egyptian studies.

This great future revival of Egyptian self-awareness, will be central to the power hold of Mahmoud Aziz. It will be a great political platform for that period. Egypt is struggling to compete economically with an emerging centralized African economy. Africa has united into a super-nation of small states, leaving Egypt behind as a lone nation, unable to fit itself into the African political arena. At the same time, Egypt has not found it easy to be accepted as part of the new alliance of the United States of Middle East.

The Middle East nations, including Israel, have united into an economic power. Recognizing the future loss of revenues from oil sales resulting from the newly emerging energy technologies, the Arab oil states have consolidated their oil wealth and used it to build a technological society whose base is economically independent of oil revenues. In fact, the new pioneers of today's technologies on the fuel cells and electric cars, arose from Arabia. Most of present day money funding the research of these new technologies actually comes from these states. Israel has become part of this integration process, working side by side with the Arabs. Having little to offer to the new emerging Mideast alliance, Egypt was left behind to fend for itself. It neither belongs to the brotherhood of African nations nor to the new Middle East alliance.

Mahmoud Aziz seizes the opportunity to arise a sense of pride in Egyptians. By the age of twenty-seven, he had carved an image for himself as a leader in Egyptian issues, and has become a symbol of Egyptian self-awareness. Just a year after he assumes power in Egypt, the Museum of Demur in Paris loans the Museum of Pharaohs in Egypt treasured Egyptian artifacts. These artifacts had been removed from pyramids and tombs in Egypt over the past one hundred years by French excavators. In a bid to restore them to Egypt, Mahmoud Aziz seizes all these precious artifacts, and places them in the Museum of Pharaohs under strict military guard. He refuses to return them to France. Aziz claims that the artifacts belong to his forefathers, and therefore, should be rightfully returned to Egyptian soil.

Egyptian self-awareness has heightened when the remains of the oldest fossils of Homo Erectus were found in excavations on the upper Nile basin in 2018 AD. The country is now recognized by the world as the place where ancient man first stood up and walked. When the uprising of the people against the government started in 2014 AD, Mahmoud Aziz seized the opportunity to start a coup. With the help of nearly half the citizens of Cairo, he marched into the Presidential Palace of his predecessor, President Maron al Mubarak and arrested him and his family. He publicly burns the Egyptian flag and replaced it with a flag carrying the symbol of Egyptian Awareness, - a pyramid, crowned with the symbolic crescent moon of Islam.

Then, slowly, Mahmoud Aziz built a fatherly image for himself among his people. He has learned how to deal with the people. He learnt from the mistakes of Saddam Hussein and other dictators of the past. Consequently, he is now seen as a strong unifying force in Egypt after the Resa Comet disasters. By focusing on Egyptian self-awareness and pride, he slowly rebuilt Egypt without international aid. He nurtured Egyptian self-awareness to a religious frenzy, placing an unforgiving grip on the people of Egypt, following the example of the notorious Ayatollah Khomeini of Iran, decades ago. He brought back the greatness of their Pharaohs to the people of Egypt.

Now, twenty two years after the horrendous impact of the comets, there is no better place on Earth for an Egyptian archeologist to work. The temple of Abu Simbel has transformed the nation into a dreamland for every young archeologist of the twenty first century. Coupled with the modern technology of lasers, robots and satellites that promise an instantaneous unleashing of amazing discoveries, Egypt has indeed become an archeologist's wonderland. The floods of 2016 A.D. open up a Pandora's-box of information with the erosion of soils that exposed great ancient artifacts. Every few years thereafter, a new find placed Egypt at the seat of human civilization, envied by all. Abdul Aziz's life-long ambition is to make Egypt a focal point for the world and revive the glory of the ancient Pharaohs.

Abdul Aziz is living his dream. He is celebrated in Egypt. His name appears regularly in the newspapers. He knows that the temples of Abu Simbel had been saved from definite calamity by some unknown set of circumstances, during the Resa comet disasters. Once before, the same temples had been rescued from the destructive flooding of Lake Nasser through international cooperation. It is no coincidence that the temples have survived the test of time. In his most remarkable role as "Preserver of the Treasures of Egypt", Abdul Aziz designs an elaborate engineering project. He cuts the temples into small "Legos", and reassembled them on a prepared site, ten meters above the river level. When the secret cavity beneath the temple floor was discovered by sonar, the entire temple site is classified as a "Top Secret Facility" by the Egyptian government. Abdul Aziz's involvement in the project will remain unknown and hidden by his brother. Despite several inquiries by suspicious people, curious as to the top secret nature of the project, the Egyptian government keeps silent. Descent will rocket through Egypt and its people start to demand to know why the flow of information has been squelched. They have grown accustomed to the constant resurrection of their Pharaohs and ancient gods. But the Egyptian government continues to say nothing about developments at Abu Simbel. Guards and military personnel are posted around the temples to prevent any unauthorized persons from entering the temple grounds. Aziz declines invitations to speak at world archeological conferences regarding recent findings at Abu Simbel.

The new project to enter the cavity beneath the first temple kicks off on the morning of July, 12, 2038 AD. Two months into the project, Aziz still finds no clues as to the contents of the secret chamber. Each morning, Aziz arrives at the site to take stock of progress, spending an hour at the site to collect data from the robots. Then, he proceeds to the University of Cairo to give his daily lectures on Egyptology. There, he holds the "Chair of Oman," named after the original founder of the school of Egyptian Awareness in Cairo. At Abu Simbel, the site is a stark contrast to the daily grind in the ancient city of Cairo. A line of specialized robots slowly excavate the temple sites, recording the exact place each piece was found amidst the massive temples. The pieces are cut to exact specifications, and prepared for reassembly at a safer site on the upper banks of the Nile river.

A control center operated by a computer ran the entire operation. Aziz is the only person who has access to the program codes of the robots. Each robot is designed to do a specific task. In all, there are twenty robots working at Abu Simbel. A majority of the robots are designated for excavation work. They are equipped with laser cutters of pinpoint accuracy. Some of the more specialized robots that utilize electro-fluid muscles delicately dissect large portions of the temple walls for transport to the safer site. With the robots in place, Abdul Aziz can find time to spend preparing "Hololectures" and meeting with important state officials in Cairo. He is a devout Moslem, faithful to the traditions of Islam.

Every Friday Aziz and his wife visit the magnificent Temple of Resa and pray to Allah, the one true God. The temple stands at the top of the Hill of Angels, just outside the city limits of Cairo. This Friday, he is to give a speech on behalf of both Imam Mabus Ali Resa and his half-brother, President Mahmoud Aziz. As the Head of the Egyptian Archeological Awareness Center, it is his duty to construct a speech that will intertwine the themes of politics, Egyptian pride and religion. Ever since the September 11 terrorist attack on the World trade towers in New York, he thinks the Muslims have been victimized and demonized by the West. Islamic fervor has gained momentum after the weakening of the Western world by the comets. It is fascinating how the Reza disaster has become an advantage to Muslim nations across the world. He believes that this was a direct consequence of the act of God.

Today is the national holiday of Egyptian Awareness. It is a great day of celebration and of Egyptian pride. As Aziz's wife Leila dresses, he watches her jealously. This is not the first time he has observed her wearing that sexy red dress. He has told her repeatedly not to wear that dress to the Mosque. He hates hearing comments about his wife from the older Imams in the mosque. Leila slips on the tall dress over her head and slowly covers her naked body, adjusting the delicate frills as if they were about to break open and expose flesh. She knows that Aziz is watching her.

"Leave me alone!" she shouts dashing out of the bedroom into the living room. Aziz follows her and points to the long slit that exposes her beautiful long legs.

"That," he says is what I am talking about," pointing to her naked exposed leg.

"That can expose your leg when you bow down to pray and I cannot have that."

"You are not man enough to take it. That is why you are so afraid of what others see in me," Leila shouts back gazing at her image in the illuminated mirror.

Aziz angrily storms out of the living room and heads to the magnetically propelled vehicle (MPV) parked in the garage. He realizes that time is short and fighting serves little purpose other than delay their arrival at the Mosque. As he waits in the MPV for Leila, his thoughts go to the rumors he has been hearing from some elders about his wife and Reza. Perhaps there is more to the rumors than he wants to believe. Many a time, Leila has disappeared into the Reza palace after attending prayers while he was held up in political discussions with elders. She would make excuses about her disappearance claiming to be with the wife of his brother, the President. Lately, she has not been too nice to him and their relationship has been slowly waning. He knows it must be Reza. The rumors have spread throughout Egypt and he feels ashamed and used by his wife and by the Ayatollah. He bites his lips in anger. The Ayatollah has taken control of his own family. He controls Egypt and has already taken his brother's mind. Now he is taking over his wife. He is glad that they have no children and that he is sterile. At least he has no children who will have borne the shame of such a mother. He could do nothing about it. He could only bring the issue up to her father and perhaps he can put some sense into her.

As his wife Leila drives her magnetically propelled vehicle through the crowded streets of Cairo his thoughts continue to roam around Reza. He feels a hate for the young man who is now the most influential man in their world. His thoughts break when the speech he had written falls from his hand. He picks it up almost in disgust. He starts reading it over to make sure he phrases the words correctly. He will do this one last time for his brother's sake, for his country. After this speech, he will start doing something about this political power the Ayatollah wields. He knows that some of the Imams secretly do not approve of Reza's political power. Perhaps there is room for some actions. Imam Mahdi, Imam Shamon and Imam Saladi are still politically strong and have a lot of influence in the Muslim world. Perhaps they too feel like he does. He has to take the time and find out who is on his side. He has to find a way to build a secret consensus for a political change to help Egypt.

It is difficult driving through the streets of Cairo on a Friday. Although it is the holy day of rest for Moslems and Muslims should not work, most of the common people disregard the law continue to trade and do business on the streets. Despite several outcries from the Ayatollah Resa, a lot of locals seem to be too involved trying to make a living and so they care little about keeping the holy day holy. This has considerably angered the Ayatollah Mabus Ali Resa and he constantly preaches against the ways of the Western World and paganism. Aziz has constructed his speech well. It does not remind the people of the ways of the West nor does it anger the people by reminding them of their sinful ways. He is going to be the diplomat who compromises between the religious cravings of the Ayatollah and the real political needs of his brother. "Compromises," he thinks to himself. He must find a way to get the attention of the elders in his speech. After all it is time to act.

They arrive at the crowded Mosque early. Outside, separate groups of men and women stand talking and awaiting the arrival of the Ayatollah. More than ten thousand people have arrived to pray with Ayatollah Resa. Leila parks the MPV in a VIP section of the parking lot and waits for Aziz to get out of the truck. She does not want to deal with him right now. She places a small veil over her head and steps one foot off the vehicle ramp exposing her beautiful thighs. Aziz is waiting and watching her. He again points to her legs,

"You see, just getting out of the MPV exposes your private parts and you think you have come to the Mosque to pray?"

"I don't care, I want you to leave me alone right now," she blurts avoiding his eyes. He has been standing by the vehicle door to prevent anyone from looking at her as she gets off the vehicle. Their emotional aura is very intense. Aziz in the dimension of Anger and Hate, and I can see his auric trajectory toward the evil land of Despair. Leila disregards him and walks to a group of women gathered at the designated entrance for women and children. He watches her long flowing hair peeping through the small veil. She is a beautiful woman and he is very jealous at this moment.

Aziz walks over toward the main entrance of the temple. He is recognized by some people who crowd around him trying to pay him some respect. He avoids bias, for here in the temple of prayer, no man is greater than another and all stand equal in the presence of God. The temple is the symbol of Egyptian rebirth and a new found glory. It is an important factor in the integration of Islam and Pharaohnism. Covering more than ten thousand acres on the hill top the mosque and home of Ayatollah Ali Resa flank one another like monuments of the past visiting the future.

He enters the temple and finds his seat at the designated VIP seat beside his wife. The temple dome is carved out of fine Italian marble and other stones of the finest grade. Its thirty-meter diameter dome is lined with a two meter wide band of pure gold leaf, decorated with drawings of ancient Egyptian Pharaohs. At the top of the dome is a crescent moon held by the apex of a golden pyramid. The symbol of Islam is held by the symbolic pyramid of the ancient Pharaohs of Egypt. The shape of the temple is a perfect cube, a strange design even for this modern period. I know what that design is and I am sure that is meant to simulate the cubic Josephus Nambu field. It is an imposingly large structure surrounded by a huge ten meter-high concrete walls spanning four hundred meters in length on the sides and rising forty stories high into the skies of Cairo. The great hall beneath the dome covers one hundred and sixty thousand square meters of marble floors, with ten thousand automatic electrically powered prayer mats that unfold and fold at the command of the faithful. The North and South walls of the temple are lined with a frame of marble beams forming a symmetrical structure supporting the temple shell.

Two diagonal beams cross the opposite ends of the cubic structure joining opposing corners and intersecting at the center of the structure. At each of the eight corners of the cubic frame, is a perfectly spherical crystal decoration similar to the Ramses II sphere, only, they are ten times larger. At the center of each of the eight edges of the cubic structure is a sphere of the same kind. A similar frame is set in the center of the temple between the North and the South walls. In all there are twenty-seven spherical crystals in the temple each supporting square frame that holds the temple carries nine. The Eastern and Western walls mirror the North and South walls so that the temple is perfectly symmetrical with respect to the central sphere. The entire temple is like a giant Lego held together by huge slabs of marble, concrete and steel.

The huge crystalline spheres are made from locally smelted quartz. When the rays of the sun hit them a spectrum of colored lights transforms the temple into a beautiful lantern. At night, the temple can be seen throughout Cairo, glowing like a jewel in the sky. It is said that the sacred structure is a replica of the original ancient Temple of Abu Simbel. About four billion US dollars were spent by the Egyptian government to build the temple in the name of Mabus Ali Resa. The Egyptian builders claim that the temple will outlast the ancient pyramids due to superior architectural design and strong cubic structure.

The holy seat of the Ayatollah is lined with gold and silver trimmings and placed in front of the altar where prayer and sacrifice are offered. Each aisle is decorated with written praises to Allah the one true God, interwoven with drawings of ancient Egyptian glory. The ceiling is decorated with a huge painting of Amman-Ra in prayer, kneeling to a crescent moon and a pyramid. Four gates open the temple to the outside world. The Resa Temple is one of the great modern wonders of the world at this time. It rivals the Vatican in beauty and grandeur. As a result, security is tight to protect her and each gate is armed with automatic infrared scanners, counting, and identifying each person as they enter and leave the temple. After Comet Resa, the whole world has initiated a programming of electronic identification using an electronic tag beneath the skin of the right hand. This helped control the crime and chaos that followed the comets.

When the Ayatollah arrives, he is dressed in the manner of an ancient and powerful Pharaoh. His face is covered by a blue veil made from silk of the finest quality. His body is covered with a blue silk robe made from the finest muslin Damascus had to offer. His feet are decorated with blue sandals made from the polished hide of now extinct elephants. His head is crowned with a blue turban graced with a big, bright, blue sapphire stone. He is the elegant representation of the glory of Egypt.

A contingency of bodyguards surround him as he moves toward the temple. They are dressed like ancient Egyptian warriors, almost like reincarnations of ancient Egyptian warriors with an imposing presence commanding the respect of the people. They part the crowds as they approach the temple doors. They carry an assortment of strange looking weapons some of which look like spears shaped in the form of snakes into which are embedded lazar-powered rifles of the modern world called Lazukars. On their heads are helmets carved perfect replicas of the great Sphinx. Their loin barely covered with a flat skirt-like silk garment and lined with painted raffia cloth. Their sweaty skins glisten in the sunlight as they escort the great Mabus Ali Resa into the temple. They look savage. The scene is a mix of ancient and modern Egypt. The Ayatollah looks every bit like a reincarnation of Amman-Ra himself.

As they move past the crowd, a sudden cry echoes from a man wearing a long black robe. The man falls to the ground and prostrates in front of Resa and his guards blocking their way to the temple. A sharp spear pokes his head and causes him to bleed profusely as the guards kick him away from their master. The man rolls over and quickly runs away from the scene holding his bleeding forehead. Not a soul could glimpse the Ayatollah's face as he enters the temple through the special entrance beneath the holy dome. Moments later, the gates open to a flood of worshipers each anxious to pay homage to God and party alongside their new Pharaoh.

Unlike modern day Islam, Egyptian Pharaohnism is taking on a new direction. The glory of ancient Egypt and the teachings of the Holy Koran have been mixed to form a new religion that has a powerful hold over the people of Egypt. Followers of Resa worship Allah through the greatness of the ancient Egyptian pharaohs. The new religion offers a new found esteem to the Egyptians. Complete submission to God also means submission to the authority of the leader, the new Pharaoh-Ayatollah Resa.

The ceremony starts with the appearance of the Ayatollah at the Pulpit.

"Allah ou Akbar. God is Great," he proclaims. The people respond in like manner.

"Allah is merciful to his Pharaohs," he proclaims again in sharp treble. The people respond in a cacophony of voices. He goes on to preach about the word of God, reading from the verse 70, chapter 5, of the Koran.

"The Food," Resa announces turning his gaze slowly in every direction at the crowd in front of him.

"Certainly, We made a covenant with the children of Israel and We sent to them apostles; whenever there came to them an apostle with what that their souls did not desire, some (of them) did they call liars and some they slay.

Certainly, they disbelieve who say: Surely Allah, He is the Messiah, son of Mary; and the Messiah said: O children of Israel! Serve Allah, my Lord and your Lord. Surely whoever associates (others) with Allah, then Allah has forbidden to him the garden, and his abode is the fire; there will be no helpers for the unjust."

He pauses and bows,to the Holy book raising his head and throwing his steel grey eyes around the crowd of worshipers. A woman cries out loud with ecstasy as her gaze falls upon the Ayatollah. Two men walk over to the woman and remove her out of the temple. There is absolute silence as she is removed.

"This is the rift between the Nation of Egypt and the Christians, my dear followers. God has no son!" The Ayatollah proclaims pounding the pulpit firmly and jolting the microphones from their stands. Some of the worshipers are visibly shaken.

"Even as we speak, I see the destruction of the world with the arrival of He who calls himself "Son of God." Let us not be deceived by the wanting of the Western World." Resa preaches smiling.

"We have a grand aim, for we are the founders of mankind. Remember Amman-Ra. He is a son of the true god of Egypt, the same god of Islam. For god has made a covenant with our Egyptian ancestors, with Amman-Ra and all the gods of ancient Egypt. We are the chosen ones whose birth right was stolen from Esau by Jacob."

Most devout Muslims do not understand the preaching of this revolutionary Ayatollah. Many prefer to think of Islam as a religion separate from Pharaohnism. But after ten long years of continuous indoctrination by the ever powerful movement of "Egyptian Awareness," Egypt is becoming a cultural center for the new religion.

The new religion of "Pharaohnism" is used to describe the mix of Egyptian and Islamic ideologies. But it separates Egyptians from the rest of the Arab world and makes the follows of the movement a despised people of Islam. Egyptians loyal to Pharaohnism consider themselves the new "chosen people." Pharaohnism is likened to the Ethiopian Rastafarian movement. Weren't the ancient Egyptians the greatest race that lived on earth? There were findings of ancient fossils of man in Egypt. The Nile was the cradle of man, the original "Garden of Eden." Weren't the pyramids still standing in opposition to all the forces of man and nature? Egyptian awareness movements sponsored by the

Ayatollah and the government of Egypt have slowly infiltrated the annals of Islamic faith in Egypt, just as the Western cults infiltrated the Christian faith.

The Imam goes on to explain to his people the connection between Ancient Egypt and Islam. Sometimes, the holy man preaches the existence of the ancient gods and their connections to the pharaohs. The more concerned Muslims do not consider this proper to the Islamic faith. But on every occasion the Ayatollah has he preaches about Pharaohnism, especially on feast of the Pharaohs.

Abdul Aziz is sitting beside his brother, President Aziz, in the front row of the temple. They are surrounded by a contingency of military police. It is his turn to speak. He walks to the pulpit to deliver his sermon where Reza awaits him with outstretched arms in greeting. Aziz feels a surge of hatred boiling within him. He steps away from the Ayatollah and fixes a stern gaze on him. Then he composes himself and hugs Reza. Reza leaves the podium and seats himself in a secluded throne facing the audience. Aziz assumes his position on the pulpit and prepares to give his speech. To the Egyptian government and to the Ayatollah he is the symbol of Egyptian awareness. He is the man who revived Egypt from its two thousand year old slumber. Abdul Aziz knows all about Egyptology and about the ancient glory of Egypt. He preaches about the great pyramids of Egypt, about the Egyptian Self-awareness Society, about the great findings of Abu Simbel. He makes all the faithful aware of the greatness of Egyptian culture.

"Egypt," he preaches, "has become a center of learning for the world. No prouder Nation stands on the face of the Earth today. Our symbol, the pyramid of Cheops is a true mark of our greatness. Never before in the history of mankind has so much glory been bestowed to so few a people.

"We should not neglect the fact that today we are neither Africans nor Arabs in the eyes of the world. We are Egyptians! Great is our glory and even greater is the glory of our Ayatollah Resa. Allah has blessed us in every way possible and we should not forget our Muslim ways. He has made us aware of our roots and has given us a very young leader when we needed him most. A leader that tells even our elders what is right and what is wrong. A leader that proclaims truths that very few understand. A new kind of a leader. Man, that needs no help to rule and govern. A man that even my brother, the President of a Nation, Mahmoud Aziz must listen to," he preaches pointing to his brother.

He looks over at Leila sitting on the far right in a front pew of the temple. Her eyes are fixed away from him. She is smiling and Reza is looking at her. Neither notice Aziz. His blood boils in anger. He puts his written speech aside and starts speaking from the heart.

"Oh, you elders and Imams of our land, this is the moment that we must act together. This is a moment that Egypt must become aware of its roots and humble beginnings in Islam, a clean and honest religion. This is the time that Islam asks us not so much pay attention to great pyramids and massive temples and power. For only Allah gives power and takes it away. This is a time when Egypt must examine itself and help our leaders make the right decisions in government.

"Oh, elders of our lands, Allah has appointed you in special places. In special positions even before our Ayatollah was born. Do not forget that. Do not forget the suffering of our people under Hosni Mubarak and the religious zealots. Do not forget it is our generation that made the great revolution possible and corrected our destiny to greatness. Now, I call upon you all. It is our generation that must keep this truce we have made with circumstances in line. Wake up! Wake up and walk with me! Walk with me and my brother to a promised land that we must build. A promised land that we live in. May Allah be merciful and may he bless us all".

He is a very proud man that walks down from the pulpit. Amid cheers of the people, an occasional "Ra, Ra, Ra" is heard from some. As Abdul Aziz approaches the sitting area, an infrared switch trips, activating a magnetically propelled motor buried beneath the temple floor. The carpet comes forth and covers the bare cold floor of the temple. Abdul

Aziz sits down on the carpet and bends his head forward to touch the floor in reverence to the one true God, Allah. He is a devout Muslim at heart. He no longer has faith in the power of the ancient gods of Egypt. Deep down, he feels that there is something amiss about this mix of Islam and Egyptology in the name of Allah. Allah has no rivals or sons. He thinks of himself as a complete servant of Allah. He wants his two children to follow these steps. He does not bring his children to the Resa temple. He does not want them to put Amman-Ra either before Allah or even close to Allah. This Amman-Ra, as a "sub-god" of the Almighty Allah is not quite right. But that is the new politics of Egyptian Awareness. No doubt, Ayatollah Ali Resa is a man of God, a special Ayatollah sent to rejuvenate Egypt. There is no doubt of his faith in the Koran. So, when Abdul Aziz lifts his head to look at the pulpit, his doubts fade into anger as the over powering and penetrating steel-grey eyes of Ayatollah Resa once again cast a shadow of doubt upon Aziz. He is now sure of his unconditional faith in God and God alone. There is none equal to Allah and there is none between him and Allah. Allah is great, Allah-ou-Akbar.

After the prayers at the Resa Temple, Abdul Aziz is approached by Imam Bangura, also a Senator.

"For a man with good family values that are being violated, I admire your courage!" he says. Aziz nods without saying a word.

"I know many that respect you. I know many that understand. Let us meet tomorrow by the Waterfall Restaurant near the river. My friends and I would like to spill water over the dry land".

He is interrupted by a Pharaohnic guard who asks Aziz to join the Ayatollah in his private chambers at his elegant palace which is located just beside the temple in the same grounds.

"Tomorrow in the day, I will be too busy with the excavation program, but we can do it at night after work." Aziz says smiling. He walks over to the exit of the temple where Leila is standing with two other women.

"I have to stay for the cooking," she says.

"What cooking?" He asks in doubt.

"We have to prepare charity food for the poor and distribute it throughout the city. It is Ashura," she says.

Ashura! He thinks to himself, some Ashura that she must celebrate. He is sure she has a rendezvous with Reza again. This time, it will be investigated. He continues walking toward the Palatial grounds were Reza lives. He uses his cell phone to call his father.

"Dad, I did not see you at the ceremony," he says.

"I am sick, Abdul. My spleen is acting up again".

"Have you seen the doctor?"

"No."

He thinks in silence for a few seconds, then asks,

"Is mom in the temple?"

"Yes, she is. She should be there doing some cooking for Ashura."

After making his father promise to see the Presidential palace's doctor he says goodbye. He walks to the rear gate of the Reza palace and searches for his mother. He had already confided in her about his doubts about Leila and Reza. His mother had discarded them insisting that he is a jealous man. He sees her and approaches with a huge smile and a hug. After kisses he explains his situation and what he saw in the temple between his wife and Reza. He asks her to keep an eye on the movements of Leila at all time and try to find out the truth. She promises to do that. In the back of her mind, she thinks her promise will just appease him and put his mind to rest so she can get on with her project of cooking.

Five guards stand at the entrance to Ali Resa's home as Aziz approaches. They are all dressed in the manner of the ancient warriors of Egypt. They salute Abdul Aziz in Arabic pointing their snake head decorated riffles up in the air. Abdul Aziz walks into the most elegant receiving room of the palace. Everywhere in the room Egyptian paintings have been placed to remind the visitor of the glory of Ancient Egypt. A replica of the Tutankhamen's tomb hangs in the middle of the visitor room reminding everyone that sees it of the eternal life of the ancient kings of Egypt.

The floor is decorated with complicated verses from the Koran, written in hieroglyphics incomprehensible to the lay. Massive gold-lined ceramic vases stand majestically in every corner of the room. There is a strange air of mysticism and power in the palace of Resa. A strong sense of history fills the space with a past presence, a power beyond time and space. Deep within his being Aziz feels the vibrations of a powerful presence within the great temple palace of Resa. This is a different place from out there. A place for dead Kings and living Pharaohs. It is not the wealth of the palace that affects him for this palace is no better than his brother's palace. But somewhere within these temple walls he speculates, must be hidden a secret known only to God and Resa.

Abdul Aziz seats himself in a comfortable leather chair in the huge waiting room and waits in silence searching his thoughts for a reason for this invite. Why would Resa call upon him when he has hardly ever dealt with him before? The quietness of the palace is very comforting, yet mystifying, reminiscent of the pleasures of his brother's Presidential palace of Egypt, but different. There is a sense of the existence of extraordinary mystical powers.

His reverie is interrupted by the appearance of Mabus Ali Resa. The stately Ayatollah is dressed in the garments of an ancient Pharaoh. He is young and well-built for his age. His sparse garment is an exact replica of the stately robes of the ancient Pharaohs. The details were uncannily accurate. Resa's tense muscles protruded through the garments, exposing a white skin very different from the typical Egyptian. His steely grey European eyes shine soullessly as if they are empty. Aziz hates to look into them. They make his bowels move.

"My brother," Ali Resa speaks in a soft but strong voice as he approaches Abdul Aziz with outstretched arms. Abdul Aziz coldly embraces the Ayatollah and kisses his outstretched hand with a fake love.

"Why did the Holy One seek to see me?" He asks.

"But let us do away with the formalities, my brother," replies Resa.

"I would like to thank you for your help in today's sermon".

"It is my duty, your Holiness," Abdul Aziz replies.

"Tell me about your fantastic work in Abu Simbel. How is it coming along? I heard you have discovered a hidden chamber beneath the temples?" The holy one asks.

For a moment, Abdul Aziz stands transfixed at the sudden intrusion into the most secret affairs of the temples. How did the Ayatollah know about this?

"Yes, yes indeed, your holiness," he forces an answer.

"We have not entered the chamber yet. We are still in the process of preparing for entry." Abdul Aziz is perplexed by the Ayatollah's knowledge of events that only two persons in Egypt know. Perhaps, he thinks to himself, Mahmoud told him about the inner chamber.

"Abdul, my brother, this is a great time for our nation, a great time for mankind. Ever since the disaster of 2016, people have looked to God for guidance. I praise Him for giving me the opportunity to help. So far, we have emerged with pride as the one nation under Pharaohnism and Islam and we should not forget our glorious past. You and I, we have a lot in common. You, standing as the keeper of Egyptian glory, and I, standing as the symbol of Egyptian glory and faith in God and the Pharaohs. Perhaps it is time for us to work together. I intend to visit the temple site as soon as possible."

The Pharaoh moves closer toward Aziz who instinctively backs away. There is a lot of tension in the air. The Ayatollah demonstrates his intentions by patting Aziz on the back. Aziz has been taken by surprise. No one else but his brother knows of the entry date.

"Surely, your holiness has no interest in the petty work of his servant," Aziz quickly urges, steering his eyes away from the Ayatollah's penetrating stare.

"Let me show you something Abdul. This is why I called you here my brother."

The Ayatollah walks into a room adjoining the waiting room. He emerges a few moments later with an envelope in his hands.

"We received information that the Secret Service of America is in our midst my brother. Our Ambassador in America has been informed of secret meetings between the Vatican and this American Agency over the last few months. You see our North Korean brothers have informed us of a secret plan by the US government to infiltrate our government in order to destabilize your brother's regime. They believe that Egypt has become too destabilized by the growth of Pharaohnism and they fear the great revolution will spread to the entire United States of the Middle East."

Resa pauses and waits for Aziz to say something. But Aziz remains silent. He continues.

"The Vatican is also concerned about the health of the Glory Pact. You know, the Pact that was signed to remove tensions between Islam, Christianity and Judaism after the great disaster of 2016."

"My brother has informed me of this fact, your holiness," Aziz remarks quietly, "but I cannot understand what Pharaohnism has to do with the Glory Pact or the United States of Middle East. It is purely Egyptian business."

"There are now four billion Moslems of all sects in the world, my brother. Pharaohnism is the fastest growing religion in the world. We now enjoy the patronage of more than five hundred and fifty million followers worldwide. Since the tragedy that gave me life twenty two years ago, the world has recognized my birth as a symbol of hope for not only our people, but for the world as well. I have converted millions to the faith of Islam while recognizing the glory of our ancient people. You see, it is just the same with the Jewish people. The Jews are the chosen people of Yahweh. No other people were chosen by Yahweh. To the rest of the world, that is unfair. But here we have a chosen people of Allah. What better people should be chosen over all others than the great Egyptians?"

The Ayatollah speaks as Aziz glances through the letter.

"I cannot agree more, your holiness," Aziz answers almost half-heartedly without looking at the Ayatollah.

"The Glory Pact gave credibility to all faiths. It affirmed the rights of each Nation under Islamic Law, Christian Law and Jewish Law. The Jews own Yahweh and His covenants, the Muslims own Allah and His covenants and the rest of the world own Jesus and His covenant. It is a fair deal. We all preach these facts in our temples and schools, so why are they afraid of us?" Aziz asks awkwardly still avoiding the Ayatollah's gaze.

"Come, come. Sit with me my brother," the Ayatollah motions pulling Aziz by the hand toward the sofa beside him. He sits beside the Professor.

Abdul Aziz hands back the letter to the Ayatollah and shifts away from him as if the Ayatollah is too close for comfort. Aziz feels a charismatic radiance flowing from the Ayatollah that is overpowering.

"How young, powerful and beautiful this man is," Aziz thinks to himself. With merely a few inches between their eyes, the moment is tense when their gazes meet. He wants to yank his hand free from the Ayatollah's, who is still awkwardly holding his. The Ayatollah's naked leg touches the covered leg of Aziz. The Professor's stomach churns with anxiety. He hates it. A loud fart echoes in the room. He did not do it intentionally. The fart comes so quickly and Aziz had no control over it.

"Damn Leila and her beans," he thinks to himself embarrassed. He stands up quickly to excuse himself from the Ayatollah, walking a few feet away, making sure he does not offend the holy one. He is visibly embarrassed and finds it impossible to look at the Ayatollah. But the Ayatollah consoles him in a casual manner.

"I must tell you that I would like to visit the site of Abu Simbel as soon as possible," the Ayatollah states disregarding the embarrassing incident.

"I believe the holy temples of Ramses II must be kept holy at all times. We cannot allow infidels to enter the holy site before you open it Abdul. I must pray in it first," the Ayatollah states moving toward Aziz, but then smelling the fart quickly retreats a few feet away. Aziz smiles shamefully pretending to understand.

"You see, we also have to start looking after our covenant, just as the Jews have hidden their Ark and Jesus's Shroud has kept the faith of the Christians. We must protect our secret pact with the past. The Arabs have the Kabba to keep them warm with money of the faithful pilgrims. The people of Egypt need this very much. It is a good way to make our religion visible, to fuel our efforts in the glorification of Egypt," the Ayatollah speaks while Abdul Aziz listens intently.

"I have asked Mahmoud to restrict all activity on the site, and please forgive me for the intrusion, but it is necessary. During the next week, I will visit with you to bless the site with the words of Amman-Ra, just as our ancient kings did."

"But your holiness, we are not doing anything religious or special at Abu Simbel. For me, it is work that keeps my family's belly full. Politics should be far from Abu Simbel," Abdul Aziz pleads softly, searching for the Ayatollah's reaction.

Ayatollah walks to one side of the room and focuses on the great painting of Ramses II on the wall. Surrounding Ramses II are three brown pyramids standing tall, proclaiming the glory of Ancient Egypt in the setting sunlight. The holy man places his hands tenderly on the wall and starts to caress the painting.

"Oh, holy Egypt!" he softly says, shaking his head.

Abdul Aziz follows his movements. A sudden fear overtakes him while he considers the Ayatollah's state of mind.

"What is it that you do not understand about me Aziz?" He asks.

"I said your brother has approved that I take over the site as soon as possible. Do you think this is about politics?"

"No," Aziz replies, his forehead soaking with sweat as he realizes the possible difficulty he is in.

The Ayatollah stares at Aziz,

"Mine is the task of the Pharaohs, the great divide between man and God. Mine is the will of the Pharaohs, the great inclusion of the gods of Egypt. You are working for the good of Egypt. You must listen to my will and allow me full access to the Chamber of Ramses, my own brother!"

"But," Aziz starts, but the Pharaoh interrupts him.

"I will only ask you once to let go of the keys to the Chamber of Ramses' burial place, and allow me access to bless it before you defile it with your blood!"

"Burial chamber?" Aziz asks surprised.

How did he know that Ramses is buried there? He thinks to himself.

"Yes, burial chamber of my own brother, Ramses. You must let me bless it before you defile it or I will have to do it without your help."

It is a threat. Aziz looks at Reza, he is red in anger. His face betrays no deception, it is on fire. He decides to quickly resign himself to the situation.

"But I know that we will be blessed by your interest in Abu Simbel. And surely, I cannot deny your wisdom. So, I will do as you have bided me to do."

The Ayatollah turns to Aziz and says,

"Thank you, my brother, you have chosen wisely, for you are a valued member of our brotherhood. Surely you must not forget that."

He abruptly bids Abdul Aziz farewell and returns to his chambers, leaving the Professor alone to find his way out of the holy palace.

It is getting dark for me and the lights have been turned on in the room. I am now feeling a little pain in my ribs and the images of my visions are becoming strained and I am slowly returning to the reality of spacetime. There is no more sunlight. I am weak and tired. I must have been talking continuously. I do not even remember what I said. Elaine is rubbing my forehead as the sweat pours out of my glands. As the pain returns, I grunt to let her know that I need some medication. I do not notice the crowded room as I narrated. I forgot that there are a bunch of others in the room.

"Who are they?" I ask Effion.

"Don't worry about them. These are representatives of the US government and President Owens." He says.

President Owens. How did I miss that? Owens has become president after the comets. Perhaps it is written that he will, but I have not yet learnt of this in my astral travels.

"They are here to help you," he adds.

To help me, I think to myself. It is more like they are here to help themselves. I no longer care about who or what they are. I only need to rest and sleep…and to get all this out of my aching head. I slowly sink into the depths of sleep as the drug wanes off. I need to sleep to keep the pain off. I could hear them debating about what to do with my body. I am certain thy can do nothing with my sleeping mind, after all I am a traveler and if I sleep, my mind could still watch over my body!

CHAPTER 6
(August 8, 2038 AD)

I wake up to a burst of Beethoven's fifth symphony. It must have been playing all night and I kept dreaming my old self jumping from rock to rock protruding through the clear blue waters of some big unending lake. It was almost magical as I jumped high from one rock to another never missing a beat. I awake to a pleasant surprise when I see my brother Edward, standing in front of the bed holding on to the headboard and Diana pointing a bottle of water at me to quench my thirst. Edward is a Professor of Geomorphology and is the President of the University of Dunkirk in France. For him to be here, things must be really bad. The old doctor, whoever he is, is there again looking at me as if he is about to extract my brains or something. The music softened as someone turns it off, but it continues playing in my head as it slowly becomes a rhythm that matches my heartbeat.

"Edward," I try to shout out the words as hard as I could, "what are you doing here?" Edward puts a warm hand on my head and smiles.

"Just visiting Seth," he says.

"I wanted to say hello before leaving for Belize on a project."

I can see that he is not telling me the entire truth. Edward has always known that I could read his thoughts and ever since we were kids, I had always had the advantage of knowing things that he would rather I not know. I do not pursue things further because I know that he will not be here if I am okay. A man dressed in the uniform of US Army General approaches my bed. His face is familiar but I cannot place it just yet. I feel pain running through me once more only this time, the pain is coming from my chest. My heart is beating fast and I do not enjoy breathing. I try to move my legs but they do not respond easily. Then I recall that they must be paralyzed. If they are then I am back I time before the calamity. I am in a predicament. I am under the control of these people and my wife or future wife is allowing it. I feel a rush of anger and I yell out in pain throwing the bottle of water Diana had given me at the doctor.

"He is delirious," I hear the doctor say.

"What can we do?" Effion asks. He looks far-older now. He looks like he is dressed in the trappings of a Roman Catholic Cardinal.

"We cannot wait we must channel him now. The President is waiting for the report," I heard someone speaking.

"Give him another dose," the General commands.

"He is in some sort of shock we must first regulate his pulse before he goes," the doctor advocates.

"The dose does that just give it to him now!" The General commands.

My fate is in the hands of these people and there is nothing I can do about that. I need some drug to slow the pain and my body is sort of gyrating to a rhythm that has no beat. I am probably a dead man lying in wait and I wonder when the light will show up. I wonder if the afterlife will be there waiting for me in the dimensions of Hope, Faith and Love. Will some Angel come to get my soul or will a devil show up in Despair, Anger and Hate? I have been good all of my life and cannot recall ever doing anything that will make me deserve Hell. My reverie is interrupted with a stabbing pain on my arm. I feel the drug entering my veins once more and slowly relaxing me as my irritation starts to fade away. I look around the room for Diana but she is nowhere. I feel left out, forsaken and I start entering a world I am beginning to like. My mind opens up to a very empty space and slowly an image fills the screen as the doctor's face fades away. I can hear my mouth rambling again but I cannot control it as I enter into a strange world of dreams and confusion. I am confused and my identity is not clear to me and I feel like I am a different person. I am an Astronaut and it looks like I have channeled myself into someone else's identity. I can walk freely and it is as if the future has been altered. This has happened to me before. My mother was cooking dinner and I was seating on a kitchen stool looking at her and talking to her about school work. I did not go to school that day and she was worried that my grades will falter. As we discussed, I heard what sounded like hundreds of cats meowing through the front lawn. My mother heard it too. We ran out to the patio to see what was going on. Over one hundred cats were matching though our yard all bellowing the same strange meow. It was shocking. The sounds penetrate my skull and I felt myself transported into a different place and time. When I came back I was surrounded by my father, brother, sisters and neighbors. They said I was talking about my rights to a farm. I had taken the identity of an old dentist who bought a farm in Hohenwald, Tennessee. My name was John Pickering. Someone has stolen my mineral and water rights during the sale and I had inadvertently signed purchase agreements for the Dibble House and land without mineral and water rights. My father, who was a Lawyer, was finding it difficult to defeat my arguments about my rights. He was representing the other side against me. I was vehemently defending myself against a band of neighboring farmers who have diverted the only creek that fed my farm. For two hours I had assumed this strange identity and defended my rights in front of the Dibble House which I had bought with hard earned money. I was posing for an artist named Grant Wood, who painted my image and entitled it "American Gothic". My sister was standing beside me in the painting. The painting became real and also survived the comets in the Art institute of Chicago.

Now, I no longer remember a past different from who I am and I am not sure that there is anyone else inside my head but me. I consider myself a good Catholic and a decent family man. My first priority is to raise and protect my three children from the claws of a world that has seen just too many catastrophes. I believe this world is an angry world without mercy. My second priority in life is my job as Professor of Holo-archeology and Physics at the University of Maryland. Holo-archeology is a new discipline I helped create to facilitate the study of archeology by using holographic tools to visually simulate the past.

As I eat breakfast two hundred kilometers above the Earth's surface on the Geo-Space Orbiter, I watch the Earth roll by at two thousand kilometers per hour. I enjoy watching the Resa Ring of debris that has now become a huge belt straddling the Earth's equator. Over a period of twenty-six years, the ring has settled into a huge one kilometer-wide by ten meters thick belt of rocks, dust, man-made objects and other interesting natural objects that were thrown into space during the terrible impacts of comet Resa with Earth. The ring reflects sunlight in a brilliant spectrum of colors that can be seen clearly from the planet's surface.

Whenever the United States of America passes by, I take out my Apple-5 data link and reorient the huge space telescope in the orbiter to search for Frederick, my small hometown in Maryland. I radio my wife Elaine and ask her and the children to stand outside the house and wave to the skies hoping that on a clear day I would catch a glimpse of them before they pass by. Sometimes it works. Far below, I would see them waving back at the satellite for a brief precious seconds. I have amassed a collection of photos that keeps me connected to them. In space, I am working on a new experimental station dedicated to space-based archeological and geological studies of Earth. I have already spent twenty-two days on the Geo-Space Orbiter which houses the new experimental stations. Although it is very lonely up here, it is serene and beautiful. The Earth appears as an ever-changing canvas of colors and patterns. It is a work of art that evolves spontaneously and is always different from times before. But I am no stranger to the Geo-Space-Orbiter. As a young scientist, I was involved in the construction of a Geo-Space-Orbiter arm of the huge Alpha Space Station which now orbits five kilometers away. I have visited space before and in 2011 I was an astronaut in a Soyuz spacecraft that was supposed to connect to the then space station.

Just two days ago, I had detached the Geo-Space-Orbiter from the Alpha Space Station to begin conducting delicate observations of Earth's most active crustal thermal currents. The electronics of the Alpha Space Station is too noisy for the delicate sensors of the experimental station. Recent studies of the Earth's crustal thermal currents have shown a strong and noticeably increasing crustal cooling effect in twenty-seven different sites around the Earth where I was told the US government, in collaboration with certain other powers, are building a new weapon based on something called the J-N field! Sometimes, when my alter ego sees a possibility of great importance to control the powers predicted by the Scroll, it forces me to assume the identity of the most influential person in my travel episodes so that I can make sense and control what is going on as well to control the future holds. It is a physiological mechanism that we all have built into our being.

The impact of the comets has severely changed the substructure of the Earth's crust migrations, resulting in some funny business with these cooling and heating effects. Recent studies show that these cooling effects were no natural results of the comets' impacts. I was told that for more than ten years, a US government and Anglo-Russian joint venture have been developing a new strategic weapon based on the revolutionary Josephus-Nambu theory, which redefined the relationship between matter, space and time. I suspect that the symmetry of the Global Cooling Effect (GCE) is no accident. I believe that the cooling effect is a direct result of the work the government is doing to create the new weapon. I have never really been to any of these military sites, but I am finding out strange things about these sites or sites near them.

I found strange archaeological artifacts emitting z-particles associated in all the GCE sites. I had called a meeting of the highest levels of government to discuss this strange finding and has concluded that the phenomenon is unnatural and is not due to the artificial J-N field I was told was being built.

I have read recently published articles on fundamental z-particle emissions from archeological objects in research done by one Professor John Finkle and a Professor Arvin Nath of the Indian Super-Cooled-Particle-Collider. He learned that z-particles can only exist in the presence of specific spacetime distortions which can be produced by energy generators that create Josephus-Nambu forces. I know that these newly discovered forces are more powerful than nuclear fission or even fusion, and that only researchers in the USA, Europe and Russia had the theoretical knowhow to generate such force artificially. Now, I am seeing some z-particle effects around the world in archeological sites. It is as if the government or the Bilderberger's Group has deliberated chosen ancient archeological sites for the manufacture of the global weapon.

I am overwhelmed with curiosity. Could my government be lying about the sites? If the archeological finds produce these patterns, is it possible that the J-N field is not man made but natural? But then, how can the field be natural if it is found in archeological sites only. Using the gravity wave sensors on the Geo-Space Orbiter experimental station, I am finding very high distortions of the gravitational constant around Abu Simbel and the twenty-six other areas on Earth. The most highly concentrated distortions occur deep in the jungles of West Africa. I am startled by the eerie observation that twenty-six out of twenty-seven GCE maximas seem to be moving toward a single point in the Moroccan Desert! That is unnatural. Unless the Government is moving the various parts of the weapon to Morocco, the phenomenon cannot be natural. The trajectories of all the twenty-four centers will make them all coincide at the same point in the Moroccan desert in two years! That will make the Moroccan Desert the coldest place on Earth! It does not make sense. How could this be explained by natural forces? I do not want to believe that Bilderberg Group is behind this. The planned assembly area has not yet been decided upon, and the Europeans have still not agreed to Area 51.

The Josephus-Nambu force is too powerful to be fiddled with by man. Nuclear energy is enough, and now the Earth has to deal with an even more powerful force. If there is a concerted effort to develop J-N weapons all around the world and bring them to the Moroccan desert, then, the Europeans, Russians, Africans, Arabs and Israeli are in on it. That means the whole world is planning a military maneuver of global proportions that does not include the United States. There must be a global conspiracy for the technology against the USA!

I had been involved in setting up special experiments to detect z-particle emissions using satellites. After the weapons are constructed, I was to measure the field intensity of each site accurately down to the micro-watt level to precisely determine the rate at which the weapons will be brought together without a nuclear explosion occurring. If the weapons are brought together too fast or too slow, there will be a global nuclear-like explosion. No one really knows how such an explosion will influence the world. Also, I am to develop the exact codes that will be used to activate the weapon. This can only be done from space, where I am to set the complete coding system on the Geo-Orbiter that had been built specially for that purpose.

I discover that there was a very high concentration of z-particles in Nevada in Area 51 and this surprised me very much. I know that Area 51 was a preparation site for war and new technology but never did I suspect that it was also a place where z-particle experiments are being carried out. Not that z-particles are unsafe. They affect the neural pathways of human brains and so can cause Bell's syndrome, a disease of the brain identified by a hypnotic trance. With the correct goggles and protection, Area 51 will be very safe for humans. What worries me is the fact that the emissions in Area 51 are very intense and directed as if there is a source of z-particles that is being manipulated there by humans. Z-particles have a wide random dispersion angle particularly when they are emitted naturally. I am beginning to believe that the US government is also developing the J-N field weapon in Area 51 separate from the concerted effort around the world. I trust my government, but only to a certain degree.

I have gathered enough data to verify my claim that there is a conspiracy and realize that I must make sure that the Bilderberg Group acts on what is going on. I have been given a secret direct line to the President of the United States, a major member of the group and my mission is no longer just a scientific mission, it is now a highest priority classified mission. Armed with my findings, I contact the President on an unmonitored communication channel of the Orbiter. President Cooper (Owens was before him), instructs me not to discuss the findings with anyone other than himself. Except for a select few including I and the President, a know-die classification is ordered for the mission. Anyone who learns of this mission will be assassinated by the US Military upon discovery. As I speak through the secret channel a sudden jolt jerks the space craft into a wild spin. I am forced to terminate the communication and focus on the bay window of the craft. I cannot see what is going on outside of the craft. I am becoming dizzy as the rotation slowly stabilizes into an awkward and uncomfortable rate. All I can see is the Earth and the stars taking turns past the small orbital bay window. I decide to dislodge the artificial gravitational field of the craft and push myself to the central air space of the craft. The craft lets go of my body and continues spinning at a faster rate. I am at least free from the centrifugal forces. I search to see what caused the impact and do not find any visible damage to the inside of the craft. Momentarily, I think I see a blue stripe pass across the bay window. The orbiter must have collided with some debris, I think but I am not sure. How could I not have detected the object with my extra sensitive sensors? I wonder. I swing to the control console and detach the remote-control pad that is made just for such an emergency. I coax the craft to a relatively stable spin and slowly adjust the thrusters to compensate for the spin vectors. The craft comes to a slow but regulated stop. The clouds far below steady as the Geo-Orbiter module synchronizes with the Earth's rate of spin. I let out a sigh of relief and hope to investigate later. I decide to resume communication with the White House.

"What happened John?" President Cooper asks me.

I explain the incident to the puzzlement of the President.

"Are you sure it was debris?" He asks worried.

"Well it is the best guess because there is no indication of thermal sources outside the craft other than my own boosters."

"Ok, try to pack up and leave, we are anxious here, the Special Forces are being briefed as we speak and we must find out the truth".

I decide to continue to analyze the data and wait for the correct rendezvous orbit with the Alpha Space Station. It will be in the next dark orbital trajectory. During the day orbit, the sun's heat can distort the connection hatches slightly and cause air leaks that can disrupt the pressure equilibrium chambers between the orbiters and the Space station reception hatches. In the past four years, two astronauts were lost in sunlight rendezvous and so it is only done during the dark orbits. I decide to rest and induce a two-hour sleep to relax from the transfer stresses. Two hours later I wake up to a terrible headache. The air mixture must have been contaminated by the collision. After wolfing down a breakfast of dried cereals and fruit, I race into the control room of the orbiter and activate the autopilot to reattach the orbiter to the Alpha Space Station. As the orbiter speeds toward the Alpha Space Station I activate the data transfer program to download all information in the orbiter's data banks into a cubic crystal embedded in my wrist-watch. The data-storage system that never comes off my wrist since it is permanently welded around my wrist and cannot be removed except with a special key. I wait impatiently as more than twenty gigabits are transferred through the optical link. When the transfer is complete I set the codes for the activation of the weapon and then begin the process to jettison the instrument's data crystal from the orbiter into deep space.

I stop the process and erase the codes I was given to put into the satellite by the Bilderberg group. I am not too sure about all this anymore. My alter ego seems to be controlled by someone else and I feel like everything I am doing is automated. How I came to work for the Group, I do not quite recall. The natural occurrence of z-particles around the world cannot be explained by what my superiors have told me. The emissions are natural and not man made. There is much more than I know at the moment. If the entire weapon is heading for Morocco, then something is not quite the way it should be. Morocco is not in Area 51. I cannot allow the Bilderberg group to act against the interest of the United States. I instinctively change the codes to new codes. No one but I will know the new codes. Until I get a good grip of what is going on, why make it easy for the enemy to activate the weapon.

A blue stripe swipes past the orbitals window, as it veers to rendezvous. For a brief instant I think I see a blue striped craft veering off in the direction of the jettisoned data crystal. It is too late to know what it is. The orbiter is already rotating to prepare for rendezvous. What the hell is going on? Am I dazed, crazy or what? I am feeling very confused and very disturbed. The orbiter orients its main transfer port to link up with the Alpha Space Station entry port. In less than twenty minutes, I am back in the Alpha Space Station pressure equilibration chamber. In five eternal minutes the ion beam sanitizer decontaminates me and my belongings before I am allowed to enter the huge Space Station. I hate the damn monotonous buzz of the sanitizer as it resonates through my brain to clean out my dentures. There are rumors that the damn thing causes impotence. I try to ignore the irritating sanitizer by entertaining myself with thoughts of evading questions from the scientists inside the station and of my upcoming return trip home. I cheer myself up as I imagine that in a few days I will be back with my family.

I walk to the end of the corridor where I was assigned a room in the Space Station. When I arrive at the door, I find a name tag "John Davis" on my door. I do not recall that I shared a room with John Davis. I try to remember my name but cannot. I put my bags down and open my briefcase confused. Am I becoming insane? Am I losing my sanity? Is there an illness in space that they are not telling us about? I take a briefing from the briefcase and look at the memos in it. They belong to John Davis. How did I get such a name? What is my name? I do not remember. It is as if a new life has been thrust upon you. I know I married Elaine and I have three girls but I do not know my own name! I decide to look at the manifest on the Space station bulletin. I approach the Monitor and I see my arrival listed at the correct time. The craft I was piloting is an SV class 2 robot suit and I am the only one that should arrive now. The name on the manifest is John Davis! I am John Davis. If anything, that is what they know me as, and I cannot afford a medical leave of absence from the projects I have involved myself in. I will deal with this day by day. I call Elaine on the house phone. It is secure and I know it is not tapped because of security reasons. Elaine answers excitedly,

"John! I am glad you are back in the Space Station." She says. I spoke to her for a while to make sure my mind is intact and all in all my conversation was normal and consistent with the world view. My name is obviously John Davis and my severe psychological trauma with Bell's syndrome must have made me forget my name. I decide to shelve the incident away and not make a fuss. If the Space Medical Board finds out about my memory loss I will lose my license as an Astro-archeologist and that is the last thing I need now.

As I pack my bags and instruments in the confines of my room, I couldn't help thinking about the strange geothermal changes taking place on Earth. I also think about the weird collision with whatever that thing was out there. I reason that if it was a craft and that it must have been at least a two-cabin craft with enough momentum to put the orbiter into a spin. I will ask the CIA and NASA to investigate later. I was lost in my thoughts about the mission when the holographic news reports on the TV screen break my thoughts. Some snippet of a journalist's story jolts my very being.

"Discovered by the Russians,...could be a real threat," is all my brain managed to register. What could it be? My heart skips a beat and I throw the last unpacked bag on the floor and dash out to the lounge, startling the few people with my abrupt entrance.

"What did the Russians do?" I shout desperately, but no one answers my question. They disregard me but stand up protectively.

"What did the Russians do?" I repeat shouting irreverently looking at the crew. I must look half-crazed, I am disheveled and sweating. Security personnel approach me.

"What have the Russians discovered?" I repeat again to my startled colleagues.

"Nothing . . . they found hundreds of dead seals in the Baltic sea and they believe the cause was a chemical spill," one scientist explains looking at me with a consoling smile. I sigh.

"Thank God. Thank God," I whisper to myself glad that that was all it was.

CHAPTER 7
(Egyptian Awareness, October 12, 2037 AD)

Professor Abdul Aziz is preparing for the day of entry into the great hidden chamber beneath Abu Simbel. It is a day of jubilation and worry. Worry because the Ayatollah is going to be there and perhaps hamper his delicate archaeological efforts. Jubilation because of the anticipated discoveries that are waiting to be exposed in the hidden chamber. For two months the Ayatollah Ali Resa jealously places the mysterious chamber under heavy guard so that no eyes other than his own and Aziz's could peer into ancient Egypt without his blessing.

Despite all the details he has to attend to regarding security, Aziz enjoys his work and considers himself lucky. Ever since his daughter was rescued from the jaws of the great Nile River by a kind American tourist who performed CPR following a swimming lesson gone awry, he feels that Allah the Almighty smiles upon him. The American had breathed life back into his daughter's lifeless body. Second to God, his loyalty is forever with the American tourist. For two months the foreigner had been staying as an honored guest at the Professor's suburban mansion in downtown Cairo. Johnny Carlton, the American, came from Ohio State. He is a Math Professor on study leave at the University of Cairo. He had spent time familiarizing himself with the city and its many ancient wonders. Although Aziz kept Abu Simbel a secret to the world, he explained some of his exciting finds to Johnny. He had to let out the pent-up excitement to somebody far from the Egyptian everyday way of life. Johnny has even helped him write some of the more complex computer programs for the robots at the site. He is a very good mathematician. Aziz is content with the help he receives from his new friend and he focuses diligently on his duty to protect the Abu Simbel site from unlawful entry. He had posted military personnel at the site around the clock. Each morning after meeting Aziz at the temple, Ayatollah Resa would visit Abu Simbel with an entourage of Egyptian guards dressed in ancient Egyptian attire. Despite the strong security, rumors are already spreading that some great find has been made at Abu Simbel. This worries the Ayatollah considerably. To dispel such rumors, he orders the Egyptian government to finance a public campaign to popularize discoveries from other ancient sites to divert the public's attention from Abu Simbel. This was a strategy the US government used in Area 51 and Roswell.

Finally, the day of entry into the great chamber has arrived and Abdul Aziz is very excited. He leaves his house early and drives his MPV to the Cairo University campus to drop Johnny off. Then he heads off to Abu Simbel so that the hidden chamber could finally be exposed. Ayatollah Resa, surrounded by a dozen armed guards is already waiting impatiently at the entrance. The moment is here at last. Together, Aziz and the Ayatollah drive off into the main temple compound to begin the journey into the unknown. Aziz is now more comforted by the daily presence of the Ayatollah, after being convinced that Resa will protect him from any evil powers that might lurk within. The case in point is the famous curse of the Pharaohs when Lord Carnarvon died on the 5th of April 1923, seven weeks after the official opening of Tutankhamen's tomb. This discovery created media frenzy when the lights of Cairo went out at the moment of his death. Carnarvon's dog Suzie was said to have howled and died at the same instant. The novelist Mari Corelli warned of dire consequences for anyone who enters the sealed tomb and so for Aziz, Resa presence offers some measure of protection against such evil spirits.

From a distance, the first temple of Ramses II appears as a magnificent mix of high technology and ancient mystery. Huge sections of the temple wall had already been removed by the robots and replaced with equal cuts of composite foam material to keep the integrity of the structure intact. Military vehicles patrol the perimeter of the temple site at a frenzied pace and everywhere inch of the Temple grounds is filled with armed Special forces.

The Ayatollah draws particular attention from the crowd of guards and military personnel outside the temple. They bow reverently to greet him as he passes them in the MPV. When they arrive at the temple gates, Aziz notes that new personal guards of the Pharaoh have been placed at the temple door to replace the military guards that had been under his own special command. The guards are dressed like ancient Egyptian warriors.

"What happened to my military guards?" Abdul Aziz inquires as the warriors salute the Ayatollah and open the main door to the temple.

"I thought it wise to bring my own men to the holy site to protect us," the Ayatollah quickly explains. The warriors of Resa are everywhere inside the temple grounds, making sure that no one comes within a few hundred meters of the temple walls. Not even the Egyptian military are permitted close to the temple. Huge ultrasonic planar fences line the perimeter of the structures preventing even ants from crossing the powerful and invisible ultrasonic barrier. The only open door to the first temple of Ramses II, is a small security gate installed by robots. Abdul Aziz places his hand on the digital palm reader of the door knob. The door responds by announcing his name, rank, and serial number. Then, Abdul Aziz enters his secret password on the key pad, and the door opens. The smell of dirt fills their nostrils. The Ayatollah takes a deep breath, holding in the aroma to savor its taste. He is smiling.

A near infrared beam sweeps across the room, scanning for the spectral signature of living tissue in the temple. A female voice announces from a door speaker in Arabic,

"Good Morning Doctor Aziz, there is a second presence in the temple, but I could not read its signature by spectroscopic analysis, please confirm."

Abdul Aziz looks at the Ayatollah, surprised. He authorizes a second scan. Again, the female voice announces,

"No detectable spectral signature. Un-confirmed presence within the temple."

A screeching siren fills the temple chamber and moments later the site Commander calls Abdul Aziz on his personal wrist communicator.

"Is there any problem Professor? Please confirm," he calls in Arabic.

Abdul Aziz is puzzled. How could the AZ-5000 security system make such an error? It is the most sophisticated spectral scanner on the market. If the machine could scan his spectral signature, why did it not detect that of Ayatollah Resa's?

Aziz answers the call.

"No, commander, there is no problem. There must be a glitch in the system. The only other person here is Ayatollah Resa."

He walks over to the door and enters a code to stop the alarm. The incident does not sit well with him. For two years, this system has never made an error. For a moment he idles at the door wondering. Should he try it again? Cold sweat runs down his face.

"What is the matter, my brother?" the Ayatollah asks looking at Aziz from one corner of his eye.

"Nothing," Aziz replies. "It must have been a problem with the system circuit, but I have corrected it."

Abdul Aziz asks the Ayatollah to stand by the left wall of the temple while he climbs into the control dome of the circular excavator that hangs suspended from the ceiling of the temple. He closes the glass dome of the excavator and turns on the huge magnetic impulse engine. He feels more protected inside the cockpit of the machine. The controls light up as the excavator slowly hums to maximum power state while performing an initial diagnosis. A metallic female voice announces from the diagnostic center of the control consul of the robot:

"Room width: fifty meters.
Room height: fifty meters.
Room length: fifty meters.
Floor thickness: thirty-five centimeters.
Humidity: eighty-five percent and rising.
Turning air systems ON. Then, there is a pause.
Life forms detection within work zone perimeter: none.
Clear for excavation."

"None?" Abdul Aziz repeats silently to himself looking to see if the Ayatollah is still in the temple. He turns off the speaker of the excavator to prevent the Ayatollah from hearing the results and turns the life form monitor in the control console to Life Form Priority Scanning Mode. This mode will repeat the scans every twenty seconds to make sure that the excavator does not accidentally kill someone in the vicinity of the temple. To avoid any directional disturbances, Abdul Aziz slowly rotates the excavator through a full circle, scanning every part of the temple in segmented sweeps. The results are the same, no life forms!

Confused, he turns on the infrared scanners to see the heat emission spectra within the chamber. They reveal no detectable heat sources in the room! Abdul Aziz decides to record and telefax the scan results directly to security headquarters of his brother through the communication system of the robot. He will analyze them later. He remains calm and proceeds with the excavation of the temple floor while Resa eyes him intensely, his robe flying about as the wind generated by the excavator intensified. The robot carefully clears the top soil of the temple above the hidden chamber, vacuuming the dirt through a huge hose that empties to an outside conveyor. At each sweep, the thickness of the floor beneath is read with x-ray vision to ensure that the thinning floor will not collapse and destroy important evidence below.

Slowly, the robot etches away the soil above the chamber, taking a centimeter at a time while the Ayatollah Resa excitedly paces around the outside perimeter of the cavity. The weight of the robot causes the suspension cables from the ceiling of the temple to squeak as the excavator works its way around the room. A noisy air blower removes the stale musty air from the chamber to the outside world. As the soil of the floor slowly erodes, the color beneath the excavator slowly changes to the dull orange of a strange dye. There are now only twenty centimeters of soil left before breaking through into the center of the ceiling of the mysterious chamber below. Aziz's heart races with expectation as the Ayatollah urges him to proceed. Aziz thinks about the strange scan incident. He has no explanation for it. He decides to dismiss it as the excitement of entering the chamber overwhelms him.

With nearly all the soil removed, it becomes evident that the reddish-orange hue is not in the dirt but is coming from something that is glowing beneath the chamber. As the layers peel of the floor the glow intensifies and the temperature in the room starts to drop dramatically. Aziz is becoming a bit worried by the strange events unfolding in the chamber. First, there is the Ayatollah's lack of a body spectrum, second is the strange glow and now it is getting very cold in the mysterious underbelly of the Ramses' temple. It is as if some huge freezer is being unearthed beneath the robot. The monitors on the control are catching some interference and the image starts to blur. It is getting crazier by the minute. Aziz is sweating profusely despite the chilly air in the chamber. His thick beard is shining yellow in reflected light from the glowing chamber below. As his heart races, he thinks excitedly about the great crystal and what it might reveal to him. Perhaps this is the real crystal that eluded humankind for eons. All these strange happenings must be because of the magnificent crystal that the ancient pharaohs wrote about. This is it! In a few moments, he and perhaps, Resa, if he is human, will be the first to find the lost crystal. Again, he turns his attention to Resa. Resa is grunting loudly as the yellow rays intensify. It is an amazingly exhilarating place and it is now becoming automatic. Aziz's eyes catch a beam of light reflecting from the steel grey eyes of the Ayatollah now resonating with the yellow reflections of the light coming from below. A cold chill flows through his body. Should he stop the excavation? Without waiting to analyze his question, his fear acts for him. He abruptly puts the excavator in neutral as a mist starts forming inside the chamber. The humid Egyptian air is condensing from the intense cold. He looks suspiciously at the Ayatollah. The holy man is smiling with his hands raised high in air.

"Do not fear my brother we must not fail to stand up to the glory of our past, keep going!" The holy one urges with his hands waving. After hesitating for a few seconds, Aziz re-engages the gears of the excavator and removes the final layer of the soil that seals the chamber. The ropes that support the excavator are freely swinging in a slow rhythmic dance, indicating that the weight of the machine is fully suspended on the cables and the excavator has freed itself from the floor beneath.

Aziz leans over the console and looks out into the cavity that is now consumed by a fantastic yellow pulsing glow. The Ayatollah waves his hand at Aziz to stop the operation but the controls are already failing as the mysterious rays from the chamber below bombard its circuitry. The robot comes to a halt and stands lifelessly above the chamber. It was eerily silent. The robot is swinging in a slow rhythm on the cables in the middle of the opening and it is now bathed in the intense yellow pulsing light. Aziz now visibly shaken with fear scrambles to turn the excavator on again so that he could extend the access ramp to the solid ground around the hole, but the ramp does not respond.

The robot is dead.

He hurriedly opens the plastic dome of the excavator and manually extends the ramp toward the solid perimeter around the gaping cavity below. Quickly, he raises his heavy body over the excavator door without opening the door. He wants to get out as fast as possible. His hands fumble to catch the rail of the ramp and as he scrambles onto the ramp, his incredulous eyes catch Ayatollah Resa stepping into the thin air above the abyss.

"Stop, stop, you will fall . . . die," Aziz struggles internally with the warning as the Ayatollah levitates toward him. The Ayatollah seems to simply float above the mysterious chamber!

"B-b-but you are walking on thin air!" he shouts in utter amazement, as the Ayatollah floats toward the excavator robot.

"Allah Ou Akbar," Aziz starts praying in Arabic, as anxiety seizes him. The Ayatollah is standing beside the excavator on an invisible canvas above the chamber beckoning Aziz to come down and join him. Startled, Aziz jumps back into the control chamber and attempts to turn the engines on, but his hands are shaking uselessly and he cannot concentrate. He looks at the Ayatollah in resignation. He knows what must happen will happen quickly. The holy man's eyes are now one with the mysterious yellow glow from below. Transformed by the mighty force unleashed from the abyss, the Ayatollah stands like an ancient Egyptian god ready for battle.

"Close your eyes my brother, and believe in me, for the power of the Pharaohs your eyes should not see, else you die!" The Ayatollah proclaims raising his hands into the air.

Aziz prays out loud shielding his eyes from his horrible former master.

"I am a Muslim, Oh holy one, but you have shown me a power that only Satan should know, let me die, but let my soul live, Oh holy one, I pray to Allah for forgiveness."

"Blaspheme!" the Ayatollah cries out angrily, his inhuman eyes penetrating the depths of Aziz's soul as his voice thunders across the room like a lion's roar. He appears larger than normal, more powerful and almost inhuman.

"Look below and see the ancient masters of the world, who sleep in the eternal light of youth, frozen for the age that has come."

Ayatollah Resa jumps high into the air, falling back into the strange thin canvass beneath his feet. His body bounces off the invisible membrane into a high parabola that lands him in the control area of the excavator beside Aziz. He is an agile cat, his wild eyes blazing yellow with hell's fury. He holds Aziz's head and turns his face toward his. Their eyes are in full contact. Aziz knows the moment he feared has come. He knows this is the end.

"Thy faith has faltered in the ways of your ancestors, Oh foolish one!" He roars at a limp, terrorized Aziz.

"You have blasphemed against the blood of Zoroaster and Amman-Ra, which even now is flowing around you. Do you not see the might of the Pharaohs around you? The same power that holds the pyramids aloof? Do you not see the wrath of Jesus and of Yahweh as we spill the blood of the light that makes souls for men? The light of Creation? Cursed are you and your religion, Oh follower of Mohammed. You have disregarded the faith of lord Amman-Ra, of lord Ramses and of Lord Zoroaster. Their blood has reddened the planet Mars in battles for the soul of Mankind. Even as we carved the face of Amman-Ra on Mars for all to see his glory, as verily NASA did find! We have visited you in our own machines of time, and over the centuries, we have changed your race to look like ours. You have called us aliens in your world. Your genes we have removed from Adam and his children and in their stead we have placed ours, even as we have manipulated the evolution of aggressiveness and survival. But you deserve this not!"

Ali Resa holds Aziz's head, and lifts his two-hundred-pound weight into the air as easily as a child would lift a toy. The shouts of terror from the Professor resonate through the mysterious temple echoing into a cacophony of twisted pain. Resa twists the human neck as if it were a piece of spaghetti, throwing the dead body high against the solid temple wall twenty meters away. Aziz let out a loud cry of pain as his neck snaps under the strain. The robot is swinging wildly. The archeologist falls into a heap beside the temple wall painting a portrait of red terror as his blood spits on the ancient paintings of Ramses and his family. The light of the crystal below pulsates with increasing energy as if fed by the life of its new victim. It is resonating with the heart of Mabus Alus, that who is known as the anti-Christ.

Then, pushing the robot with the force of a hurricane, Mabus Alus Resa exposes the full wonder of the spectacle below. The head of security flanked by two other guards arrive at the scene. He opens the temple door and runs toward the excavator ramp now swinging wildly back and forth. He jumps onto the ramp and runs toward Resa, hoping to assist the holy one, thinking he is in trouble. The holy man extends his unearthly hands like an elastic band, folding his grip around the unsuspecting guard in an ungodly manner. The guard is strangled in an instant.

The other two guards not suspecting the absence of a temple floor, dash directly onto the invisible canvas of the temple. They bounce off the invisible elastic membrane completely surprised by the mystery unfolding before their eyes. Then, clumsily trying to run off the invisible floor they bolt out to the safety of the temple grounds crying in Arabic, "Shaitan! Shaitan!"

Alone in the chamber of the ancient Egyptian gods, Mabus Ali Resa stretches his elastic limbs like a rubber band into the twisted spacetime of the cavity below. The wrapped bodies of Aman Ra and Zoroaster lay in their pure gold coffins suspended in thin air just below a huge spherical glowing crystal object.

Bowing to his fellow gods, Resa opens the caskets and strips of the wrapping, easily lifting the limp bodies of the gods so that they also bathe in the full power of the light of the pulsing crystal. The bodies start wiggling with a new-found life force from the crystal sphere as they slowly awaken from their five-thousand-year slumber. Resa holds the spherical crystal floating at the center of the oval abyss and pulls it toward him with his flexible arms. The entire temple vibrates and the yellow light from the crystal is amplified a thousand-fold exposing the power of the cosmos to the Egyptian drawings on the temple walls below.

Five body guards outside the temple see the fleeing security men. Following their example, they run away from the temple as the structure screeches from the strain of Resa's strength. It twists and turns in the shadow of the distorted spacetime as it warps into an elastic mass of reddened bricks and sand. Outside the temple no one waits to see what is going on. They scatter in all directions away from the cold hell unfolding, but it is too late for them! Like a huge vacuum, the crystal and its occupants pull the entire base of the temple into the now boiling and erratic cold spacetime that engulfs the crystal sphere in Resa's elastic grip. The ground around the temple swells as the crystal's force field reaps through its base pulling the structure and half a square kilometer of ground around it, up into the air. Debris flies hundreds of meters into the air falling back into the great cavity at the call of gravity. The explosion is heard for several kilometers as tons of temple rock, dust and the metallic remains of the massive robots come crashing back into the open cavity.

Like a giant invisible spaceship, the crystal and its passengers stand suspended above the temple grounds with Resa defying gravity as he manipulates the crystal field. Faster than lightning, the crystal sphere and its occupants are transformed into a beam of light that temporarily paints a panorama of colors into the skies of Cairo.

No one really knows what happened that day. According to the official reports released by the Egyptian government, scientists broke into a huge cavity underneath the temple site, weakening the temple floor and causing the base to collapse. Five scientists died including Abdul Aziz, fifty body guards, the Ayatollah and several dozen military personnel.

Several eye witnesses spoke of a strange line of lightning that spanned the skies of Cairo. The various newspaper reports were inconsistent. No victim of the explosion is expected to have survived. Many newspapers around the world report the Ayatollah's cause of death is unknown. Others accuse the Egyptian Government of backing a conspiracy to kill the Ayatollah. Other religious organizations claim the holy man was murdered by militant Muslims in a bid to reinstate Egyptian Islam to what it was before.

The faithful view the loss of Ayatollah Resa as a devastating tragedy. He is to be the "Al Mahdi" of the faithful, the Messiah to save the world. Officially, the Catholics, Protestants, Jews and other religious denominations sent their representatives to the funeral of the Ayatollah, although they secretly celebrated his demise. Ever since the great explosion at Abu Simbel, the Egyptian Government receives daily bomb threats from the now leaderless Pharaohnics. Most of the high ranking Islamic leaders are happy with the change in power. To them, the Ayatollah Resa is a tragedy for true Islam and represented an unforgivable shift in their rigid ways. Generally, Islam regards godhood as unique to Allah, the Almighty God. Nowhere in the Koran does Islam condone the use of Pharaohnic powers as a vehicle for true Islam. It is a new beginning for the continuation of truth and integrity of the ancient and good ways of the Holy Prophet Mohammed.

The government of Egypt did not tell its people that Ayatollah Resa's body could not be found. Instead, the best experts were used to reconstruct his face and body onto the dead body of a commoner thus fulfilling Resa's prophecy;

"They shall not cover my bones with the Earth of this world for I shall hide myself in the cloak of my ancient brothers."

No acceptable explanation is given by the Egyptian government for the events at Abu Simbel. It is declared a "National Disaster Site," and sealed off from all civilians. The Egyptian army is ordered by President Mahmoud Aziz to take over the activities of the site and to secretly investigate the death of his brother, the Ayatollah and the other victims of Abu Simbel. Several other investigations attempt to uncover the mystery of what occurred at the temples. One, published by the Washington Outpost, describes a fantastic soundless explosion which hurled bodies and debris into deep space. Evidence for the theory is substantiated by satellite pictures taken from the Alpha Space Station. Other reports conclude that the place was attacked by an alien UFO. But since no living eye witness has been found to tell the true story, the mystery of Abu Simbel continues!

CHAPTER 8
(Space Orbiter September 2038)

President Alvin Cooper wakes up at 5:00 am each morning. He always goes into the Jefferson room next to the presidential suite to have coffee and read the newspapers. Ever since he was ridiculed by reporters of the other conservative newspapers for the so called "Big Snafu", his psychologist has demanded that he not read any other paper except for the Washingtonian, which had defended him vigorously. The Big Snafu is not only about politics, it is also about his personal life and his affiliation with the Church of Scientology. A few months after he was elected to office he personally donated five million dollars of his family's wealth to the Church of Scientology. This news became the headlines in every paper, except the Washingtonian Post, which defended his actions diligently for well over six months.

That would have gone away, but there was more. The Scientology Church announced publicly that the scientific mission called Voyager 1 that was launched by NASA back in September 5, 1997 is returning home! This created a frenzy of attack on the President for supporting a church with such wild claims. In several interviews, the President reiterated his faith in the church and denounced those that do not believe as skeptics of science.

In his first State of the Union address to the nation, the President made the Big Snafu by announcing that he was allocating a substantial amount of research funds to study the return of Voyager 1. This caused a great deal of anger among members of his Cabinet and Congress. Half of the audience left the halls of Congress immediately after he made that statement. The Vice-president resigned and called for the impeachment of the President. His wife divorced him and he became isolated by the Republicans and most of his own party members. Although he was not able to get the funding, the issue remains a hot topic, and to this day he is bombarded with outrageous editorials from most newspapers, television stations and the Internet. He became depressed and isolated. While impeachment proceedings are still on their way, he still hangs on to the strange notion that his church is right and that there is evidence to support the claim.

The problem is that the Voyager 1 issue is supposed to be kept a Top Secret. Scientists at NASA and the SETI program, together with many other space agencies around the world have determined that a strange signal has indeed been received from Voyager 1. The signal is supposed to be long dead since Voyager 1 left the solar system for deep interstellar space many years ago, and the intensity of the signal had died out over two decades ago. NASA has requested that the information be kept a top secret, and in collaboration with the European and other Space agencies, a treaty of secrecy has been signed to keep the information silent from the general public. The CIA, NASA, ESA, and all agencies that signed the treaty, have vehemently denied that such a signal exists, but the church of Scientology saw an opportunity to preempt the consequences by predicting the arrival of Voyager 1 on Earth before the information becomes public. The Church of Scientology surmised that if the general public later finds its scientific predictions to be true then there would be many conversions to the church.

With all that has been going on around the world, the announcement that Voyager 1 is coming home would disrupt every sensibility that man may have gained after the comets. The Bilderberg Group immediately went into action to implement the impeachment proceedings and the publication of information about the issue. They have to control the way this information is presented upon final release.

The President has a brief meeting this morning with Dr. Mel Howard, chief scientist of NASA.

"Mr. President, it is a mistake to declassify the information about V1," he pleads.

"It is not a mistake it is to my advantage to do so. The public does not believe. We are hiding crucial information from citizens, and now, your organization has confirmed that it is true," Cooper argues.

"My worry is that we will break the deal we made with the Europeans and the Russians. We agreed that we will capture and study, but not reveal. I still think that this is the correct policy," Mel states emphatically.

"We did not agree that I will take the political fall for this!" Cooper shouts.

"I and many others will be exposed as liars if the information is made public. You are going against the grain of the Bilderberg Group," Mel warns.

"I do not care. I am the President and this is what I want!" Cooper emphasizes.

"We can at least limit the information to the Congressional hearings on the impeachment. We can ask for a closed room hearing," Mel pleads.

"No! I want public disclosure. This is the only way I can be vindicated. Anything less than that will not do!" Cooper states angrily.

"But, what about the people?"

"What about the people?" Cooper repeats.

"If they know that Voyager is returning, there will be total chaos. They would ask questions about why it is returning. They would want to know who or what sent it back," Mel explains.

"All they would want to know is that a malfunction sent it back on a reverse path," Cooper states.

"Sir, Voyager 1 disappeared into deep interstellar space over twenty years ago!" Mel explains.

"But it is coming back!"

"People will think that there is nothing out there and that there is a barrier out there. The Resa comets have done enough psychological damage already. We cannot allow them to think there is something more out there that NASA and scientists do not know about." Mel pleads.

"I am psychologically damaged. I am politically damaged. I am personally damaged. What about me?" Cooper asks.

"Then, at least let us construct a story. A story that it never went beyond the solar system and that it was designed to return," Mel pleads.

"Do you think we can fool the people? I don't think so. I think there are scientists out there that already know the truth. The Bilderberg meeting is not till December. That gives us eight months to prepare and explain our actions to the group," Cooper says.

"The Bilderberg Group will not let you get away with it. Your impeachment will be certain and then they will reverse it all on you!"

"Look, Mel, we lied about the data on Voyager 1. We did not make it public that it carried a huge data base about every scientific dream we have created. We did not tell them that the satellite will teach aliens or other beings about all our knowledge. We lied about its mission. We did not tell them that it was designed to reach out and send deep space signals to the Orion Belt. We did not tell them about the Resa comets, we did not tell them about the alien SETI signals we received in response to Voyager 1. We lied, we lied!" Cooper shouts.

"That is TOP SECRET. We cannot tell the world that Voyager 1 is designed to respond to alien signals we received through the crystal in Area 51. That will expose everything. The whole world-program we have been working on for twenty years will be made public!" Mel shouts.

The President stands up and looks at Mel Howard.

"They can kill me if they want, but I am going public. I will only go public on Voyager 1 and not the rest of it. I have to vindicate myself." Cooper states emphatically as he walks away from the table.

Shatangra Shang has been the presidential psychologist for almost ten years. He knows Cooper well. He is interviewing a new patient today. His name is Monseigneur Rodini Maserati. He is a Cardinal in the Vatican, on a visit to the United States. The recent phenomena of cases of Bell's syndrome have been on the rise and Shang is beginning to believing that perhaps Rodini has being exposed to the syndrome and has the telltale symptoms.

"Your eyes are dilated a little bit. Did you consume alcohol before coming here?"

"No," Rodini answers. "I had holy sacrament but not alcohol," he states.

"The dreams, tell me about them."

"Well, I have been seeing things lately. I saw the President of the United states sleeping in a Ghetto in New York!" he states without emotion.

"Mmm. Is there more?" Shang asks knowing his relationship with the President.

"Yes. He was nearly killed by a bullet. Then, I saw a vision of a man holding this handkerchief," he pulls out a well wrapped handkerchief from his robes.

"You saw this same handkerchief?"

"Yes, the handkerchief saved the President," Rodini states.

"How?"

"It has in it very strong checks worth over ten million dollars in certified Swiss Bank Checks that cannot be traced. These checks are strong enough to stop a bullet before it reaches the heart," Rodini states. Shang thinks for a while. The mention of the President is a bit confusing. Is this man a nut case? He wonders. He decides to probe more.

"And how does that save the President from death?" Shang asks.

"In my dream it saved him when he was pronounced incompetent by his psychologist and removed from office during the congressional impeachment hearings! The psychologist had this handkerchief held to the President's chest and that prevented the bullet from entering!" Rodini states without showing any emotions.

"Mmm," Shang thinks for a while looking at the stern face of Rodini. He understands where this is headed. He is a bit shaken now and he feels like vomiting. He rubs his hands and warms his forehead all the while looking at Rodini's unyielding expressionless face.

"And if the psychologist does not do that what happens then?" He asks probing.

"I dreamt that he also dies and so do a lot of members of his Church!"

Rodini is talking in the present tense, as if his dream is happening now.

Rodini hands the handkerchief to Shang. Shang hesitates but takes it. He examines it and finds three bank checks totaling ten million dollars.

"How does his psychologist know that the handkerchief has the power for real savings?"

"I dreamt that he deposited it in Milan in the Central European Bank. He also verified it by spending some of it and transferring some it to other banks in his name."

"This is before the Congressional hearings?" Shang asks clearly curious about the strange deal he is finding himself in.

"Yes, the psychologist was able to take the handkerchief and verify that it is a real savior. He was also given a different passport to keep him safe. He was also given a safe house in Rome, near the Vatican. He was also guaranteed safety by the European High Commission and Scotland Yard," Rodini explains.

"Then, I assume his dream comes true!" Shang states smiling clearly seeing the offer being made.

"Yes, ours and his come true!" Rodini states also smiling.

Shang walks over to a cabinet and removes a packet of medicine. He knows the cameras must be watching him. He knows that they were installed for the Presidential visits. They have no audio recording capability but his actions will be scrutinized closely by the watchful eye of the CIA.

"Monseigneur Rodini, I think your dream has meaning for me and I do not think you have the syndrome," he states.

"Take these pills, one a day, and I hope I will never have to see you again," Shang states pocketing the handkerchief in his Rufus suit. The Camerlengo thanks the psychologist and bids him farewell.

CHAPTER 9
(The Trip to Africa, January 2, 2040 AD)

I am packing my bags for a trip to the West African jungle and I have not told Elaine how long I will be there. What she does not know is that the data I have amassed from the Space station has created an up roar in Washington and in the higher circles of government. The Military is on high alert and everywhere there is heightened sense that something is about to happen and the world alliance is becoming fragile. The secret facility in Nevada is piling up with important Military and Defense personnel and I have been instructed that my family must be taken there immediately.

It is dinner time and my three daughters, Michelle, Nadia and Jessica are standing by the kitchen door watching their mom operate the potato chip maker. Elaine looks tall and beautiful and all my students love her. They call her "the caretaker," because they know she loves to take care of everything, from the children and the house to my personal business including managing my Digi-prints, compubanks and the maintenance on the MPV. Elaine's influence even permeates my lecture room. Every day she posts electro-mails messages to the telereporter in my lecture room. Sometimes as I lecture on "The Physics of Sunspots in Ancient Times," holographic animations from Elaine would burst into the vision field of the holographic images in my lecture room. My students have learnt to welcome the interruptions since it gives them time to rest their eyes. The merging of the telereporter from Elaine with the lecture images causes the messages to form overly magnified holograms in three dimensions. Although some of the students wear their vision field correction glasses, the holography equipment has still not been fully developed to accept individual eye codes for myopic vision defects and so this causes eye-strain. Laughter always fills the lecture room when a huge Greek Coliseum is intercepted by Elaine's notes to me. I have learnt to be impersonal.

My mind is busy with my next action as I watch Elaine cooking. I mull over the situation I discovered in the space orbiter earlier. Am I being played by either the US government or by the Bilderberg group? Hal Shea, my second in command in the United States National Science Center and my very close friend must be aware of what has been going on in my absence. Hal had been heading the programming group that developed the programs I just entered into the satellite. He is a mathematical genius. He is also the US liaison to the Bilderberg Group's programming team headed by Dr. Nairo Nambu one of the discoverers of the Josephus-Nambu effect. Hal and I have been working together on several related government projects for the last four years and I have learnt to trust him. Professor Hal Shea is also my assistant as Chief Scientific Advisor to President John C. Cooper. If anyone can understand me it will be Hal and so I decide to call him on my cell phone and find out what he is up to.

I pick up the cell phone and dial his number.

"Hal, this is John, how are you doing?"

"John! I am happy to hear your voice what a surprise. I thought you were already gone and I did not want to call you until you come back well rested from the mission. I am glad it all went well," he says.

"So, what is going on? Are you available for a quick talk before I leave for Africa?"

"Well, yes, I can come over. I thought you already left this morning," his voice sounds strained as if he is avoiding telling me something.

"Can you come down and see me at the house before I leave. I don't think I have time to catch-up with the staff at the office, so you will have to come here, OK?"

"Ok. I am just trying to clear my schedule for my vacation in Alaska," he replies.

"Alaska?" My protective being senses something about what he says.

"Who told you to go to Alaska, may I ask?

"I thought you arranged the trip for me as a Department expense since I have been saying I want to go and see the melted ice-caps. I received a memo two days ago that it was authorized and I could use a research facility in Anchorage for free."

"I certainly did not authorize that nor did I suggest it. Then, who did?" I am now suspicious that someone does not want Hal or myself around here and Alaska will be the last place I would authorize Hal to go spend a vacation even though he thinks it will be such a grand adventure.

"Was it the President?"

"No. I suspect General Martin is behind this, what do you think?" Hal asks

"Mmm…I am shocked. I thought that the papers were signed by you. I got confirmation that the Obama Research Station in Anchorage is waiting prepared for me. Then who signed the papers?"

Hal is confused. I am not supposed to be in touch with him at this point, I should have left the USA before he came back from his mission in Beijing. So, someone does not want Hal around when I am away!

"Hal, you are not going to Alaska. You know that the Resa belt has blocked any communications by either satellite or radio from polar ice caps, so you would have been out of touch with me completely!" I explain.

"You think someone or something is trying to scatter us for some sinister purpose?" I can hear the concern in Hal's voice.

"General Martin?"

"He is too weak to stand against the Bilderberg group," I retort.

"No, he is not. He has been holding several meetings in Nevada, and he is actively sending troops to Andrews Air force base in Maryland. They are preparing for something big John, very big." Hal reports.

"I need you to come over here right away. There is something I have to tell you and only you."

The urgency in my voice is evident, in a few minutes I am headed into the thick of the African jungle and Hal would have been totally out of touch. It is obvious that Hal is not aware of the Defense statement that was issued to President Cooper.

"Ok, I will be there in a few minutes," he replies.

I am not aware that Elaine has overheard the conversation and now she is thinking that it is going to be another one of those secret missions with Hal again.

"Why are you asking Hal to come over? What is it John, what is going on?"

Elaine does not yet know that she and the kids will be heading for a secure Nevada facility. I have not told her because I was instructed not to. She and the kids will be escorted and taken to go to Nevada at the prescribed time after I leave for Africa and not before. If I tell her now, she might refuse to go without me and I must go to Africa to investigate the fields. Elaine is a tough lady, so when she breaks down in tears in front of the children, I know it is a serious matter.

"John, Monday is our anniversary, I thought we will spend the next two weeks together. I wanted to plan a vacation with you and the kids," she sobs softly.

"What's so important about going to the Jungle that can't wait?"

Almost simultaneously, the children start chanting

"Virginia Beach, Virginia Beach, Virginia Beach, magno-surfing, magno-surfing," as if they had rehearsed. I silently and methodically continue to pack my garment bag. My eyes are also tearing up. Ever since our recovery from the strange illness that struck us a few years ago, Elaine and I cannot completely disregard our emotional fragility. I was told that Elaine and I caught a strange virus and suffered from a strange syndrome during which we lost our mental activities. In fact, they have put it in a mild way because from what we have gathered, we actually lost our minds and identities to Bell's syndrome. I in particular have been through many so-called psychologists and modern day "Resologists" who believe that the comet catastrophe caused our psychological problems and that the virus caused a new syndrome that has yet to be fully understood by science. The syndrome is supposed to eat your psyche and because amnesia followed by an identity crisis. As an Astro-archeologist my career had been on hold for four years and I was confined to just the lecture rooms here on Earth as a Holo-archeologist. For four years my identity as Dr. John Davis was in question and I was not allowed to resume my space station research until just a year ago. As a catholic, I was told that a Priest had even performed an exorcism on me in the hope of clearing the strange identity that had overtaken my being and I have been in a fragile emotional state ever since the incident. Elaine has been very supportive. Throughout the episode, I was protected by the US Department of Defense who desperately needed my services as the foremost Astro-archeologist in the world. Science was unable to pull me out of it and now I am who I am and I have recovered enough to resume my duties again, but only under the guardianship of the Bilderberg Group and the President. It is an unofficial position and I am never to reveal my work to the outside world except to the President and a few others. So, my work is very secretive indeed and I know Elaine does not like secret missions. It has been trying times for us.

Carefully, I pick up my field vision glasses and wipe them clean with mysol paper the hundred wipes no scratch lens tissue. My mind wanders for a moment to the TV Ad, a beautiful model wiping her jewelry dangling on her bare breasts with Mysol paper. I wipe my lenses vigorously as if the motion will erase the tension in the air. Elaine puts some raw potatoes on the chip maker creating a burst of angry noises from the hot Canola oil. She marches into the bathroom and locates a packed toiletry bag. She cautiously looks at me as I ready myself for the long trip across the Atlantic Ocean. She offers me the bag without making eye contact and almost as if she knows more than I think. I reach out and pull her close to my body. She too had gone through her episodes of identity crisis in the past but I know she is strong and able to cope with what I have to do for our own good.

"Please Daddy can we all go to the beach tomorrow instead?" Michelle my daughter breaks the silence. My heart sinks and I feel guilty that I am about to leave them for a while and that they are about to be taken to a secure facility without me. I will not be there to protect them and explain all that is going on.

"This trip is at the request of President Cooper himself. It is an urgent and top priority mission. I could not refuse to go even if I wanted to. I have to go but I promise I'll make it up to you when I get back."

"That is what you say before every Space mission. What do I have to do to make you understand that we also need a break? You need a break too," she grumbles. At least she thinks it is a space mission. My heart is pounding and I feel very guilty as her blue eyes fill with tears. The children have entered the room and I could not say anything in front of them. I try to stay strong but I know that as a good husband I cannot put government before my family. I must explain to her so that it will not come as a surprise to her. I would hate it if my children and I were suddenly picked up without Elaine by a bunch of black suited men I do not know and then taken to a strange facility in Nevada. I decide to tell her a little of what is going on and that she and the girls will be taken to Nevada.

"I have arranged for you and the children to go to a safe facility in my absence. There are things happening in the world that I cannot explain to you because I do not myself know what is going on," I state looking at my watch to avoid her eyes. That makes me feel even guiltier.

Her jaw drops in surprise.

"What do you mean, something is happening and why should they take us to a safe facility? What is going on John?" She asks very concerned.

"Elaine, I have acquired some data that the Russians and other countries are about to test a powerful new force as a weapon. This force is thousands of times more powerful than the atomic bomb. We do not know what it is or how it will be used, or whether it will affect the whole earth. All I know is that the President has opened up a clearance for the evacuation of some of the political leaders, many scientists and his own family to a site in Nevada. You and the children have been given a Class A clearance to the site. Only a few people know about this. So, when the time comes and if I am still away you must take the children to the site and be safe."

"Is there going to be a war?"

"No," I hesitate, "but if these tests are real, it could happen quickly."

"What about Papa and my sister and your sister? What about the rest of the people, what do they do?" Elaine asks.

"I do not know. You cannot tell anyone about this. You must promise me that you will take the children when they call upon you. What happens after that is only between Man and God. This is serious stuff and I cannot help anyone but you and the kids at this point. Clearances were refused for the Obamas, the Hiltons, the Gates, the Buffets and many other powerful people. Only a select few are allowed to the site. So, please do as they say when they come for you."

"What about you? We cannot leave without you," she says with tears welling in her eyes.

"I will be okay. The government has made all the plans for my evacuation in case of an emergency. Here, hold on to this crystal data bank. It is the only other copy I downloaded from the space orbiter. I want you to hold on to the crystal and not hand over the codes to anyone until the government makes sure you and the children are safe. If anything should happen to me this is your insurance to get to Area 51 in Nevada."

I hand her the data crystal, it is a clear cubic crystal about the size of a sugar cube.

"What do I do with it?"

"Guard it with your life. The code is the first and last letters of your name and the kids name and then your dates of births in sequence. Do not forget that. I have an original copy with me. If for any reason they refuse to take you and the kids to safety in Area 51, you must call the President personally and tell him you have the correct data codes and will hand them over only if they guarantee yours and the kid's safety. Be alert and stay by your cellphone at all times."

My eyes betray my concern and as I hug my beautiful Elaine, I could not help also crying with her.

"What happened in the orbiter? I read the order they sent you. Why did the doctor ask that you go for a medical evaluation?" She asks.

"You know what that means. They are going to find a way to exclude me from future missions and restrict my clearance after my mission in Africa. That is one reason why I do not care about the consequences of this mission, and what it is all about" I state.

"Did something happen in the orbiter?" She asks concerned.

"Elaine, when I was in the Orbiter, I found that there is something going on around the world that you must know in case something happens to me," I pause thinking about the highly secret information I am about to release.

"I could die for telling you this, but I will have nothing to lose if you and the girls are not safe. I was sent in space by a group known as the Bilderberg Group. They are a powerful group that controls much of what is happening around the world. They told me that they are building a new weapon around the world. I was to place codes that activate this weapon on the Orbiter's computers. But, when I measured the particles emissions, I found that they correlate to archeological sites around the world and that is not what I was told. I do not believe that a weapon is being built by the group, I believe that a weapon has been discovered by the group that is scattered all around the world and it is in the hands of our enemies."

"What do you mean scattered all around the world, by Who?" Her question is very difficult to answer.

"That is what I need to find out in Africa. Whoever or whatever built the weapon is moving parts of it to Morocco. It is neither the US government nor the Bilderberg group. They do not operate in Morocco. Morocco is a terrorist state. "

"So why don't you just ask the government. You are the chief advisor to the President."

"I did not because they have been deceptive about my mission in Space. They told me a weapon is being built, but the evidence points to natural sources of emissions. It is too lengthy a process to explain to you. Just be careful and make sure you keep the codes I gave you. They are the correct codes that will be needed to activate the weapon by satellite, if they are correct. I love you Elaine, I love my family very much,"

All I could do is steal a smile to comfort her. I reach into my pocket and remove a small jewelry box.

"I did not forget our anniversary it is the most important day of the year to me," I state smiling. I place the jewelry box into the palm of her hand and help her open it. She removes a green emerald pendant from the box. It is a scarab beetle pendant on a lovely gold chain.

"It is beautiful!" she exclaims smiling at me.

She giggles as a memory floods her mind of our first meeting in Egypt. My memory has been altered by my strange illness. Although I am fully recovered and it is behind me, I do not recall the meeting in Egypt. It is one of those things about the past that I relied on Elaine and others to remind me of. She was a nurse back then and was young and beautiful, barely twenty-one years old. I was much older and in my thirties. I was vacationing in Egypt and I was on an archeological mission. It was the birth of Astro-archeology. We did not know each other but had stayed in the same hotel just two rooms apart. I had had too much to drink and was stumbling down the hall when she exited her room to go for ice. A beetle had lodged itself on her skirt and I saw it. Being the hero, I saw Elaine terrified of the creature trying to slap it off her skirt. I hurried and took the beetle in the palm of my hands and started reciting a poem,

"Beetle, beetle in my palm,
a beautiful lady I have found."

She had laughed hysterically amused by the smiling man who seemed to be drunk and smelled of alcohol.

"The beetle," I had said, "pushes the sun along its course in the sky"

"It is just a beetle," Elaine had said.

"Khepera, the scarab-beetle god amulets are placed over the heart to outweigh the feather of the truth scales during the final judgment day. I have put an inscription on the amulet that reads, "do not stand as a witness against me when I am judged"." I am hoping that this will bring her back to reality. I can feel her thoughts as memories of the meeting flood her mind.

"It is a creature of great significance to the gods and when it lands on you, you have been chosen to be the goddess of love," I had told her.

"And who are you?" She had asked.

"I am the immortal soul of the great Khepera, who pushes the shining sun like a crystal made for you," I had laughed stumbling against the wall. She fell for my simple charm and childlike humor. The next day I brought her a red rose and asked her on a date. She had accepted the rose and had pulled it close to smell, when a scarab beetle flew out and landed on her nose. She had jumped with fright and I attempted to smack the beetle off her nose when I mistakenly slapped too hard.

"Beetle, Beetle on a rose,
An archeologist broke my nose!" She had said poetically rebutting with her own poem. Ever since that day, the scarab beetle had become our symbol of love. I take the chain and place it around her neck and smile kissing her forehead. She grabs me and hugs tight and she says,

"I never could stay mad at you for long.
Beetle, beetle on a rein,
My heart is bonded to a chain."

She walks over to a closet and takes out a jewelry box.

"I," she says pointing to herself, "also have a surprise for you!"

She is smiling and pointing to me. She hands the small jewelry box to me and I open it smiling.

"A scarab!" My heart fills with emotions that I cannot describe easily. I am suddenly missing her already.

"Yes, a scarab for you to keep.

"I am the immortal soul of the great Nefertiti, who shines like a star for you," she laughs pretending to stumble against the wall as she repeats my drunken words a long time ago.

I remove a ring from the box she hands me. It is a scarab beetle with a single green emerald sitting on its heart. I am happy. I place the ring on my finger and feel connected. My memories go to the mythical scarab and its magical place in Egyptian history. We are bonded and we feel rekindled with our love immortalized in the intimacy of the moment brought on by the magic of the holy scarab of Egypt meant only for the final days of judgment.

I can see Hal as he enters the front door. He approaches the kids affectionately as each of them hugs him with returned affection.

"Who is my Booboo?" he calls out to Nadia, repeating a ritual he has done with her for the past ten years. She smiles and hugs him with affection. He takes some French fries from the plate Elaine had prepared and eats them while waiting for me to come out of the bedroom. Elaine walks out of the room to greet him.

"What is going on Elaine?" he asks.

"Oh, hi Hal," she says smiling, "the usual, he calls you and you two take off for a week of adventure."

"Oh no, I have to go on vacation in Alaska. I am not going anywhere with John."

I come out of the room and greet Hal. There is no happy look on his face as usual, he seems irate.

"We have to talk," he says urgently motioning me into an adjacent office. I enter and close the door behind us.

"Hal what do you know about J-N that I don't know?"

"What do you mean?" He replies innocently. I am hoping that if he actually knows something he will open up and tell me about it. .

"Well, there is more to what is going on than I know, so I assume you must have gathered more information in my absence," I challenge him.

"John, what can I tell you?" He replies innocently. I suspect he know nothing.

"Do you know about Morocco?"

"No. What about Morocco?" He queries back.

"Damn it Hal. Stop playing with me, do you know about Morocco?"

Hal paces about the room and then walks to the half open door and closes it.

"Yes, I do," Hal answers turning away from me as if he wants to exit from the room.

"My hands are tied on a "need to know basis" and I only know what you know," he explains.

"I know that the J-N assembly stations around the world could not have been taken by the Pharaohnics as President Cooper and that damn General have said, so where is the damn weapon going?"

"John, as far as I know, certain actions are being taken during the next couple of weeks to avoid a possibly final war."

"A War? What has happened in the three weeks I spent in space?" I ask. It is high time I get some answers and Hal seems willing to talk.

"They have commenced the ingress into the secure facility in Nevada." Hal says.

"Is that why you are going to Alaska?"

"Yes, sort of."

"If there is a war, Alaska should be the last place you go and you know it, so what is up man?"

"General Martin did a simulation of the war scenario with the J-N field on our side. There were no survivors, no plants, no animals, no humans!" Hal explains.

"Is the Group in Nevada?" I ask.

"Yes, they are preparing for war John. After they got the information from you about those things moving into Morocco, they started DEFCON 4," Hal explains.

"Defense Condition Four? Well that is next to launching! My family, Hal! What about my family? They are going to die if we don't do something. Democracy, humanity, women, children, hope and everything under the sun will vanish if we don't do something," I emphatically state pounding a mahogany desk top in the room.

"John, the Bilderberg group is assembling in Nevada as we speak. They have already activated CODE RED after DEFCON 4,"

"That means there will definitely be a war!" I shout in anger. This is totally unexpected and I feel like I have been left out of the loop. My family is out of the loop!

"I will die for telling you this, but I feel like I must."

I look deep into his eyes and Hal returns the stare. There is a real bond and respect that we have developed for each other over the years. Hal is a young professor and I am like his older fatherly mentor.

"There is no manmade weapon, John."

"What do you mean, there is no manmade weapon?"

"The weapon is not manmade. We have been told it is manmade, but the truth is we placed military bases next to each site where those things were uncovered, but we are not building anything anywhere."

"All those military bases were just put there to follow what is going on?"

"Yes".

"Then, who or what created these J-N fields?"

"We do not know. All I learnt is that these fields spontaneously appeared after the floods of the Resa comets uncovered them and that the ones at Abu Simbel and other places have disappeared and vanished into thin air!"

"Then, what was the purpose of my mission to the program in the Space Orbiter?" I ask.

"The intensity of the J-N fields must be measured precisely for correct assembly of a J-N weapon."

"How so if we do not have the weapon in our possession?"

"They say that the weapon is like twenty-seven parts that have to come together precisely. By measuring the intensities and locations of the sources, we could determine the exact stage of the weapon arming by whoever or whatever is putting it together," Hal explains.

"So, if the exact sequence is not followed and they are brought together and assembled there will be an uncontrolled explosion similar to but far more powerful than nuclear?" I summarize.

"Yes. In a nuclear device every charge has to come together at the same moment to generate a critical mass for nuclear explosion. In this one, the fields must be precisely tuned when they come together otherwise there will be uncontrolled explosions everywhere."

"Is that why the hologram programs were installed by me on the Orbiter?

"Yes," Hal confirms turning away from me and looking out a window as if he is avoiding my eyes.

"And I was not told the truth by anyone. Etu Brute!" I shout angrily.

"Need to know basis, John, that is all, need to know basis. You cannot blame me."

"Well Hal, the cat is out of the bag. Are you still going to play the game and die or are you going to help me do something about it?"

"What can I do? I am all set in Alaska. I have a small cottage with enough supplies for six months and a floatation sub. That was my plan before you called. I wish you had not called me John." Hal says regretfully.

"I wish I did not call you either Hal, but I did, and now we know the truth and only real actions can make a difference now."

"You can start by telling me about the J-N weapon. What is the cause of the explosion in Egypt?"

"We do not know. The J-N theory states that the result is a spacetime distortion, but the effects could be unphysical, almost like a nuclear explosion in unseen dimensions of spacetime. A ghostly explosion, a sort of spiritual explosion. I think it will be the end of man and this world," Hal states.

"The codes I programmed into the satellite, what do they really do? Are they not for activating the weapon?"

"No! They are meant for the satellite to generate a comprehensive mathematical hologram to simulate and change in frequency and color of any of the J-N crystals if they are in enemy hands."

I am puzzled, and I wince.

"If it is not manmade, how can we change it when we do not fully understand it?" I ask.

"My take on the real plan is, at least according to the Bilderberg group, if there is a J-N weapon controlled by some enemy, then, we can change the color of one of the crystals so that upon assembly, it will detonate in one of twenty-seven ways depending on how it is assembled. Twenty-six of these explosions will be physically limited in four dimensions of spacetime and will destroy only the vicinity of the weapon, and the last way will be the real explosion that will be a spacetime event." Hal explains.

"So, the satellite map I recorded and gave to the group gives a precise GPS location of the crystals so that we can track them? What is moving them toward the desert in Morocco?" I again ask hoping for answers.

"We do not know. Your data gives the precise location of the known J-N centers around the world and by GPS we could precisely track any of them and activate a holographic color change on any J-N crystal, at least in theory."

"What has the color got to do with it?" I am really puzzled by all this new deep science that Hal knows so well as a mathematician.

"There are twenty-seven colors to the J-N symmetry. If the colors are not assembled in the right way, there will be different resultant forces, just like different nuclear atoms can be made by assembling electrons and protons in different ways," Hal explains.

"Nevada!" I find myself biting my lips real hard, "I have always suspected that there is a crystal in Nevada, Area 51 isn't there?"

"Yes, there is a crystal in Area 51. It is the one Father Josephus Mulcahy and Dr. Nairo Nambu, the J-N team in the Vatican labs have been studying. It has been kept a "to die for secret" by the Bilderberg Group and the US government for over twenty years," Hal explains.

"Where did the crystals come from?"

"It started with the famous Roswell incident," Hal answers.

"Roswell?"

"It was an alien craft that crashed into the desert. They found a glowing green crystal where it crashed. This is not an ordinary crystal," Hal explains.

"What do you mean by that, Hal?"

"For some strange reason, the rays of the green crystal had a strong hypnotic and emotional effect on people close to it. It completely erased memories, and it is almost as if the people working on it fall in love with it. When people have negative emotions, it responds to that. After they brought it to Area 51, it has done strange things to the area," Hal explains.

I feel a stab into my heart as his words propel confusion into my head. Memories, memories. I had lost a lot of memories and I am beginning to feel that perhaps that was the cause for my so-called identity crisis. I know I must have worked in Area 51, but I never recalled working on a crystal of any sort. Did I?

"What strange things?" My question is not specific and I hope that Hal will just shoot answers everywhere into my brain.

"It attracts the Amish. They have scouted the site since it was brought there."

"The Amish? That is weird".

"Yes, it is. The Amish would frequently travel close to the site and just sit there staring on their boogies. They would give no explanation."

"So, who guards the crystal?" I ask.

"No one knows. There is a thick brick wall around it and the few persons that have entered the walled area never come back. So, the government has restricted access to it. It is now stored in an icy underground silo in Area 51, close to and under the Hoover Dam. We studied it from the outside, using a minimal number of rays that penetrate a designated access hole on the wall."

"The Bilderberg group, are they behind President Cooper?"

"I do not know. All I know is that a mathematical and scientific theory based on the Principle of Causal Conspiracy has been built around the strange field by Josephus and Nambu, the two scientists that work for the Vatican." Hal paces around the room as if he needs an escape from the questioning. I am relentless and curious and I need to get some more information.

"What is this psychosomatic effect the rays have on people?" I ask hoping to enlighten my past episodes.

"In less than ten years after it was recovered, over one hundred scientists and engineers have been mentally damaged by the field of the crystals and no one knows why. The crystal seems to be part of a weapon or a propulsion system for a massive craft powered by the J-N field. That means it was a craft that traveled through spacetime and God only knows from where."

"Who are they?" I ask,

"Who are the Bilderberg group? Who do I really work for? Who acquired my mind after my identity crisis?"

"We had a briefing before we were all dismissed from duty. In 1947, after World War Two, it was thought that the Germans developed a weapon that missed its mark and crashed in Roswell. A commission of the Vatican Council called a meeting of leaders around the world to secure peace among nations. They met in Hotel de Bilderberg in Germany. Soon after, in the fifties, the group selected a site in Nevada they named "Area 51" to study the crystal. They developed the J-N theory by studying the crystal. They supported research to develop special containment shields for the rays. It causes a debilitating disease called Bell's Syndrome. They succeeded." Hal explains.

So, I suffered from Bell's syndrome! Why did they keep this from me for so long?

"So, whatever is taking these remaining crystals to Morocco needs the Nevada crystal to complete a weapon?"

"Yes, we think so, but we do not know why or how they manipulate the fields. According to J-N theory, the blue spectrum is the most powerful emission of the J-N field."

"The satellites show that there is a blue emission in West Africa that seems to be fixed and not moving unlike the others. Do you know about that one too?" I ask.

"Yes, the Blue Crystal. It is the most powerful crystal of all. The Vatican labs call it the Armageddon crystal. We already know about it. They are assembling a military team to take it over," Hal explains.

"Is that why I was asked by the President to go West Africa to find out about it?" I can feel my anger rising.

"Yes. If they are an alien technology then we are dealing with a powerful force that we neither have control over, nor do we have the technological knowhow to manipulate it."

"You must come with me Hal, this crystal in the Jungle has a very profound spectral signature that is not yet moving toward the desert. If we can somehow find out what it is all about, we will have a better understanding of this unnatural weapon. There may be a chance for mankind after all. If we do not try, we will fail our children and families and all we have to live for." I am desperately pleading knowing I am asking Hal to risk his life with me.

"I know that there has been some speculation about a religious connection to the crystal. It is rumored that a lot of Rabi's and Priests of several different religions have been allowed to enter the field in Nevada!"

"To what end? What could be the religious connection?"

"They have not come out, John. Those that went into the crystal are still in there. We do not know what they saw or what they do in there."

"Does the Government know?"

"No John, I do not think so. I would know if they know and so would you to some extent."

"Is Nevada under the US jurisdiction, or is it under the control of the Bilderberg Group only?"

"There is a split I think. I thought the Bilderberg group controls everything, but the US military presence is just all around the site, and now I am begging to believe that General Martin is a separatist!" Hal explains.

"A Separatist? That can be very dangerous indeed. I never suspected that he would be Separatist. I know there was a build-up of separatists after the Snafu, but beyond that I have not been able to follow up. I have been kept busy by the NASA programs."

"Well now you know. I am getting out. I must protect myself. You should get out too. If there is a split in the Government, I will take the Bilderberg group, because they will survive. They control Nevada."

"Hal, I know that the archeological sites around the world are finding these strange things, so, there must be some connection with Nevada. Why do you really think I was sent to code the satellite?"

"The codes manipulate some laser generators on the satellite," Hal repeats, "I think the lasers tune the frequency of the rays from the crystals and somehow make it oscillate. When the first people attempted entry, there was a great fluctuation in the frequencies of the crystal. It went through a spectrum of colors. We were able by using the J-N equations to exactly calculate the frequency that would make the crystal in Nevada oscillate," Hal explains.

"So, was anyone able to go in and come out?" I test Hal again.

"No! No one goes in and comes out, yet there is evidence that no gasification of bodies has taken place. There is no smell to indicate that they died, unless they are fried, but even that is not the case, we would have detected the gases," Hal explains. His hands are trembling. He has revealed too much and he knows there is no turning back now!

"Is there any scientific evidence that can point to them being dead or alive?"

"Yes, the fluctuation of the frequencies of the crystals shows that there are people alive in there and sometimes we sense movements that can only be attributed to living deliberate beings."

"The Amish, what do they really want?"

"No one knows. They have been scouting the site since the Roswell incident," Hal answers.

"John, we know so much and yet so little. I have been puzzled myself about all this strange Vatican and Bilderberg business. If what you say is right about Africa, then there may be a chance for us."

"It is our only chance. There is a site in the Kabba in Mecca that is stationary and there is the other one in West Africa. If these are stationary, we could find out what is going on and perhaps find a way to avoid war. Perhaps we can find a way to deactivate them," I explain hoping that would send a ray of hope in Hal.

"The one in Africa is the US target. The General has assembled a force to capture it," Hal says.

"But what if there is no weapon and they start a real war by capturing it? What if the fields are natural archeological or religious artifacts and are not meant for war?" I speculate.

"Then, we would have created a war for nothing!" Hal concludes.

"Yes, we would have destroyed the world for nothing. We must find out. We must go." I try again to convince my friend.

"John, you are right. I must go with you! I have nothing to lose. I have no family but yours, and if something like that is going to give us a chance then I am willing to die for it."

We discuss and try to formulate a plan for the mission. I learned that for Hal, his curiosity to actually see the blue crystal overwhelms him. He has studied the J-N field in Nevada and has met the great Nambu himself. But one thing has always bothered him about the great Nambu. Who does he work for? Is it the Vatican or the Bilderberg group? What are they up to?

Elaine knocks on the door and hands me my bag. Obviously her ears must have been glued to the back of the office door and there is no more need for explanations. I want her to hear it all, after all she must know what is going on for her to know what to do when I am not here.

"You will need this," she says and walks away to give us space to talk. Hal and I decide to exit the room. I need to have time with my family before I leave.

"Elaine, there is too much going on right now," I explain,

"I need you and the kids to understand. I have to go and find out for myself, why I have been sent to set the satellite codes and what is going on in Africa. There is much going on that I do not understand and it is only right that I go and find out."

"I heard about the explosion in Egypt, and about all that stuff going on in Nevada. Are we in a war?" Elaine asks.

"I do not know. If we are, then, it is my duty to find out how to stop it from escalating. This is a scientific war, perhaps even a spiritual war. I do not know yet and I am glad my dear friend is coming with me," I state smiling at Hal.

Elaine turns and looks at the panic in my eyes.

"Nevada! Hal, if anywhere is safe it will be Nevada right? I was told that the area is being secured and that it is going to be a safe haven for the group. Is that true, or is it just another lie?" I ask rethinking my plan for my family.

"Yes, from what you have said about the other fields moving to Africa it is safe there," Hal supports, "there is a definite build-up of the Bilderberg Group there."

"Yes, they Elaine and the kids should definitely go there for their safety. If CODE RED has been activated, then there will be lots of dignitaries that are being flown into the area, so I presume it is the place for them to be. But why did you decide to go to Alaska?"

"I was not given access to Nevada. They refused me a pass," Hal lies.

"The bastards, they were doing me, my family and you too. They were going to let us just die while they wallow in safety. Thank God I held on to the real satellite codes," I said without thinking about what I wanted to say!

"You mean you did not hand over the real codes?" Hal asks.

"No! I did not."

"Well, the Military has taken over and you can thank God all you want, but that was why I cannot go to Nevada. The entire J-N team with the exception of Nambu and myself were refused entry. Even you, John were refused entry. I was able to steal three extra passes for your kids and I decided to give up my pass and go to Alaska. That is why I was headed for Alaska, not because I was going on vacation, but so Elaine can have access. But did not want to tell you that until you leave for Africa," Hal admits.

"Oh my God, you did what?" Elaine exclaims, crying.

"They gave me access but not your family. I secured access for the kids by switching their names for three old farts that they included in their program. Believe me a lot of money has crossed hands with officials to let in some of the wealthy. When I learnt what was going on, I was supposed to give passes to Father Mulcahy, Nambu and Senator Owens. Instead, I sneaked into the pass program which I wrote, and stole their passes and instructed them to go to the Vatican as a safe haven. I replaced their names and mine with Elaine's and the kids'. So, I was not being as devious as you thought I was!" Hal explains, looking at Elaine and myself. Elaine and I are stunned. We are silent for some time realizing what a true friend Hal has been.

"You knew all along didn't you? You knew you were going to your death to save my children?" I ask.

"No, I only knew that it was the best thing to do and that there was no time to explain things to you. You were in space and you could not have fended for the family and they are my family," he adds.

"I only wish that you can understand how we feel." Elaine states kissing Hal on the cheek and hugging him tightly.

"Whenever you leave us and go on these dangerous missions, I always worry so much. I hope you two will be OK, I hope God will look after you until you return," Elaine prays.

Elaine starts to sob uncontrollably and the kids heard her and start to join in. It is time for them to leave. We share kisses and Hal and I walk out of the house to the waiting vehicle that is headed for the Reagan International Airport in Virginia.

"We will need special goggles to be able to approach the crystal. The Nevada team gave me some. We must pass by my house to pick them up."

"The Nevada crystal is red and the intense crystal in Africa according to you is blue. The crystal in the Kabba is green. Will these goggles work for all and any of these emissions?"

"I do not know. We can only try. They said that they have a broad spectral range. What range? I do not know."

We pass by Hal's house and he stuffs some luggage and supplies for the trip. He also picks up some instruments that will help us in our search for the crystal. On our way we discuss excitedly about the trip to Africa. Hal Shea is a tall, tanned and heavy-set man. He is wearing a blue stripped suit with no tie. When the President first introduced Hal to me four years ago, he had called Hal "a bull with a kid's heart."

The bull is an old hand at these expeditions. I need him. There is no turning back now, we must sort this strange stuff and be sure that my family will be safe if something happens. Should any form of either an escalation of hostilities or a war should occur, at least my family will hopefully be safe in the confines of Area 51.

The magneto-field grumbles as the MPV's permanent magnet is forced into motion by the linear propulsion field wave buried beneath the tarred roadway. The pulsing and traveling electromagnetic field propels the vehicle to maximum road speed. As the angle of the permanent magnet changes, a greater force is created between the traveling electromagnetic field beneath the road and the permanent magnet that is the sole engine of the MPV.

"No Engine, just a permanent magnet, how ingenious," I contemplate as we are effortlessly and noiselessly transported to the airport. How far science had taken the world!

I remember the good old gas and hybrid electric cars of my youth, puffing smoke and releasing poisons battery chemicals into the delicate atmosphere. Back then, I was an advocate for cleaner fuels and for funding of MPV technologies. I recall the first test of the MPV by Enertron Incorporated, in Florida. A pilot program was undertaken to test the first MPV technology in the small town of Coral Springs. The road track electromagnets were buried one foot below ground in more than five hundred miles of intersecting roads within the city.

As a young science professor studying to become an Astro-archeologist, I marveled at the new high-tech machines that could work with the infrastructure of the old highways. The propulsion fields are buried beneath the roads and all the old engines were taken out of cars and replaced with large permanent magnets! It was and still is a great idea! The field beneath the roads is pulsed from a remote power plant far from the city. By aligning the permanent magnets on the MPV with the propelled field beneath the roads, the MPV moves forward. To reverse the MPV, the magnet is simply reversed through a half rotation. No connections to the roads, no fuel charging stations, no problems, no pollution.

The MVP transports Hal and I to the VIP sector of the airport where we are escorted into a ten-sitter, free spin, Ion Flash Propulsion Jet. The seats of the IFP 200 series jet are placed in a circular pattern allowing each passenger to benefit from the slow counter rotation of the fuel dome of the jet.

I always marvel at the jet's saucer-like design. I like technology. I am up to date on everything new that is going on. The new technology of Ion flash propulsion was pioneered by the Indonesian Transport Company back in 2022, after the comets devastated the airplane industry. Jets could not travel through the smoggy thick polluted airways. The first set of commercial ion jet crafts were developed as an alternative technology because they do not need intake air like a jet. A powerful stream of ionized particles creates a propelling force on rapidly rotating circular discs surrounding the passenger dome of the craft. By performing a screw-like motion on the air, the ion-flash propulsion disc supports the craft with a thin layer of ionized particles that holds its weight against the force of gravity. The pollution free Ion Flash Propulsion Unit is quickly established as a reliable technology for commercial air transport, and the entire world united in making the new technology commercially available. The world can unite when the need arises! There is a lower, middle and upper deck. The lower deck carries the quiet ion generator and propulsion unit, the middle deck carries passengers and supplies with a rotating food and drink tray. The smaller dome-shaped upper deck is home to the pilot and his control center. The large view ports of the middle deck allow each passenger to see the flight path, as well as other crafts passing by at great speeds. The seating is arranged circularly and broken at one point in the craft where a service room is located in one of the radial arms of the disc of the jet. Although seats face outwards they can be rotated in any direction for an easy three-sixty view. The craft is capable of traveling at speeds of up to eight hundred kilometers per hour. The passenger dome is stationary except for a very slow rotation which compensates for ionic discharge variations across the circular field of emitted ionized particles that propel the craft. The slow rotation also allows the passengers a full view of the air space around them.

The craft takes off with a soft hum, rising at a steep angle into the dark skies until it levels out at about ten thousand meters above sea level. I can sense Hal feels inhibited about discussing the project in front of the Captain. We make small talk for several hours before landing on a fuel island platform in the middle of the Atlantic Ocean. It takes about twenty minutes to charge the craft thermal batteries with plasma and to check the magnetic flasks for leaks. We reload the supplies for the remaining flight and after re-boarding, we take off for the African coast.

After we are settled in the craft, I take out some papers I had been given by the President himself.

"We both know about the events at Abu Simbel. I assume that you know it was no accident?" I ask Hal.

"I know very little about Abu Simbel, John," he replies.

I study my friend's facial expressions carefully. He is a very handsome young man, very lucky to be so close to the President at such a tender age. Perhaps it is unlucky at this point, only time will tell. Sifting through my briefcase, I take out a bunch of papers and hand them to Hal.

"These are classified and I have not read them yet," I said in a whisper.

After giving Hal a little time to study the papers, I start to explain what I knew about the mission at hand.

"The US Government suspects some military activity took place at the Abu Simbel site before the supposed accident occurred that supposedly took the life of Ayatollah Resa and some of the world's most noted Egyptologists. I am only telling you my perspective as an Astro-archeologist involved in the space mission."

I know that Hal knows about the incident but would not know the details. I watch for a reaction and Hal reads on, silently listening.

"A now defunct Green Peace satellite has shown heightened thermal cooling at Abu Simbel and other locations we are now currently investigating. We think this cooling has something to do with the presence of z-particles of the J-N fields."

Hal nods acknowledgment, he already knows about that.

"We can no longer neglect the z-particle connection. Work done by Dr. Harry Finkle and Dr. Arvin Nath on fundamental particles at the Indian Super Cooled Particle Collider indicates special spacetime distortions must be present for the stability of z-particles. Measurable distortions of the gravitational constant around the Abu Simbel sight were performed by a Military Atari satellite recently. The result from Atari tells us that there is a correlation between the phenomena at Abu Simbel and some gravitational flux in that area."

I sigh and continue.

"Nath! I know that bastard," Hal states laughing youthfully as he recalls a date that had gone sour for him because Dr. Nath showed up and the girl favored Dr. Nath.

I disregard the abrupt remark and explanation.

"I am no relativist. I am an archeologist and a fundamental particle physicist. So, I cannot postulate further on these matters. We must find a link between two sites of heightened global cooling activity, namely, Abu Simbel and that of Sefadu, in Sierra Leone. If there is a link, then we can find out if they are indeed some sort of J-N weapon."

I unfold a map of Sierra Leone and points to the location of Sefadu in Sierra Leone.

"The recent report sent to the White House by Finkle's team indicates heightened activity in this area and I was able to confirm that in the satellite data," I state pointing to the small town on the West coast of Africa.

"According to Finkle, an ultraviolet temperature spectrum in the region showed a fifteen-degree Fahrenheit temperature drop before the so-called accident that took place in Egypt. That corresponds to an energy loss of ten megawatts per hour. After the accident, the drop is only five degrees."

Hall seems surprise.

"Do you know anything more that I don't?"

"If I recall correctly you found a temperature depression in twenty-seven such places, one of which is fixed in Mecca in Saudi Arabia and the other is fixed in West Africa. The rest are moving from all over the world towards Morocco. Why are the Abu Simbel and the Sefadu sites so important?" Hal asks.

"What more do you know?" I query smiling and wondering if Hal knows more about the archeological connection. Hal does not answer, instead he reiterates his question.

"What makes these two sites special?" he asks.

"They are not special. They happen to have exhibited the same behavior before the Abu Simbel incident. Yamandu, for example, is still active, but Abu Simbel went silent after the incident."

"So, we can get a clearer comparison of the before and the after?" Hal surmises.

"Exactly! If these are natural z-particle emissions then there will be no war. If they are artificial, we are in trouble," I state.

"Did we get clearance from the government of Sierra Leone?" Hal wonders.

"Yes, I got the clearances through Senator Owens. But no one other than you and I know the true reason for our visit. We have told the Sierra Leonean Government that we wish to study the impact of unorganized mining activities of the 1980's. We can help the government turn the old mining sites into agricultural centers to grow genetically engineered Monsanto patented grains. We have a two-month permit to do this. Meanwhile, the rest of our mission has to remain Top Secret."

While the small craft speeds toward the green jungles of West Africa, we talk late into the night. For me. I believe I know the truth about what happened at Abu Simbel, but a plethora of perplexing questions still plague me.

Who is the author of the strange writings on the spheres of Abu Simbel? What language did the ancients use? Why were there spherical cavities at the secret Holy Chambers and why, after one hundred and fifty years of studies that preceded the accident at Abu Simbel, there were no bodies found? Were these ancient structures ticking time bombs from our past?

For the past two years, several spy satellites have monitored activities at Abu Simbel and other sites in Egypt. One such spy satellite is the World Satellite Mapping Resources Station, sent by the International Green Peace Organization to study the Global Cooling Effects or GCE. With the help of the US government, I had assembled secret teams of leading scientists to study satellite photos of the site obtained from the Geo-Orbiter. We learned that several Egyptian artifacts were removed from the site during transfer to the safe zone on the upper Nile river. From US intelligence sources, I learned that some of these objects bore the strange encryptions of the spheres of Ramses II. Ever since I met Professor Abdul Aziz five years ago at the World Symposium on Holo-Archeology, I have become very interested in Abu Simbel. I know that Abdul Aziz was also searching for an answer and at that time, we had a joint desire to share our findings. I am still struggling to understand the force that spread the spheres around the world. Human remains were found everywhere accept at Abu Simbel. One thing I have discovered is that the global cooling pattern is related to the form of a regular octagon and triangle. No one but myself and President Cooper know of this design and I wonder if anyone else suspects this. Now, Hal also seems to know. I only hope that the word had not yet leaked, otherwise, it could jeopardize our mission. I imagine the headlines,

"Alien Writing Found at Mysterious Sites Connected with Global Cooling Effect!"

"Abu Simbel Spooked by Aliens!"

It is no coincidence that all the land sites where the global cooling effect occurs coincide just too neatly with the location of the spheres and the myths of ancient people. The correlation is uncannily accurate. But for now, the focus of our work is to quantify z-particles. If global cooling is indeed caused by the z-particles of Egypt, then I have found an ancient power plant, a weapon, or something. But my mission goes beyond that. I need to discover the mechanism that is causing a phenomenon which is contrary to the laws of Thermodynamics. Things heat up naturally. To cool you must use a lot of energy or somehow absorb the Sun's energy into something. Nothing can absorb so much energy without being sensed by instruments. Only z-particles could explain this phenomenon. If ancient texts on the spheres are correct, then, I should find the true spheres that cause the effect. If there had been a similar accident in the West African Jungles of Sierra Leone, there would be no doubt that the spheres caused the explosions. If there is a record among the people of the jungle of an explosion, then, I expect to find a record of the sphere hidden there as well. An international body of archaeologists is suddenly shut out of Egypt after the Abu Simbel Site is uncovered. But I have seen enough, I have already made the connection between the GCE centers and the spheres. I also know that the spheres had been found at all twenty-seven sites. Three imitation spheres exist in Egypt. In Rome, another three spheres were uncovered in the foothills of the powerful Mount Vesuvius. They were all the size of a football and made from fused silica exactly like the Egyptian ones. In Scotland, the ancient Druid caves contained drawings depicting the same spheres. In Lima, Peru, an ancient building is excavated whose walls are decorated with facets of the spheres and the enigmatic language. Some of the Greek encryption at Abu Simbel talks of crystal spheres that emanated a strange force. The knowledge of the spheres is all over the world but why and how? Why the sudden surge in knowledge about the sphere after the comet floods exposed them? What could the real spheres be? How could, for example, the pyramid drawings of the spheres been duplicated in caves found in Scotland? Easter Island is in the middle of nowhere. How did spherical stone balls with the same inscriptions get there?

The President had explained to me that Area 51 is only a safe haven that was prepared in case of a war or catastrophe, but now I doubt that. Surely they will not tell me all they know, I am just a little pawn the scheme of things. They hid information from me before. I am briefed that I am on a need to know basis only. Need to know! Well I need to know! "What Power!" I speculate as I gaze through the small window of the IFP jet into the blackness of the night. Perhaps the real spheres are a huge energy pump that sucks in heat energy converting it to another form of energy. Perhaps, it is a massive weapon that must be deactivated. I am heading for the dark jungles of Africa to find this out. I feel like James Bond, the famous detective character back in the late twentieth century. I feel like the entire purpose of my being is being fulfilled. I feel myself tumbling over a political and religious landscape of an unknown world.

CHAPTER 10
(The Black Pope, January 15, 2040 AD)

The first African Pope won the holiest seat in the Vatican two years ago. When Pope Gregory passed, the Vatican was in a frenzy to elect a new Pope and there had been a lot of controversy as to who that should have been. My job as soothsayer is to direct the Vatican in the right direction and make sure that the future of the Papacy is secure. It has been 28 years since Diana and I were captured as prisoners of the Vatican City and I am now fifty-five years old. I am starting to feel my life passing away and Diana has been very strained by the situation and yet we could do nothing about it. We have become tools for seeing into the future and into the past. We have attempted to escape from the Vatican on many occasions but our efforts have been useless against the expertise of the Swiss guards. We are now resigned to serving our last days as royal instruments of Hope for the world. Diana and I are treated like royalty, like queen bees. We can have anything we wish for but leave the city for the outside world. We are sometimes allowed to exit the city but only if we are followed by an entourage of Swiss guards. Diana has learnt to astral travel with me and I we are now able to effectuate an escape through the dimensional portals to spend time outside when we can find an opportunity. I no longer know what the real outside world looks like except when Diana and I astral travel and assume different identities out there.

Recent developments in mind control and other mental studies have shown that there is a special gene in the human genome that controls how we think about God and Faith. The v-neck-2 gene, also called the God gene, controls how humans enter into a state of faith. It seems to light up the parietal lobe of the brain in MRI studies. These studies were done by Dr. Andrew Newburg of the Thomas Jefferson University Center of Integrated Medicine. They found that the v-neck-2 gene actually controls the perception of the dimension of faith. Just as we perceive space and time, we can also perceive Faith and other emotions as dimensions we can travel through. In his experiments, Dr. Newberg found that people that express faith in God seem to have the ability to light up MRI scans of the parietal lobe of the brain while those that do not believe in God do not. This opened a new door to investigate the dimensional gateways of the human mind and for me it places my dimensional capabilities on a firm scientific footing. In 2035 the Vatican started synthesizing my genetic profile and what they found surprised them very much. I not only have a specialized version of the God gene, I also carry a series of strange genes that act in a similar manner to the God gene. In one study, they tried to splice one of my genes that had a lot of mono amine allyls of v-neck-2 into a rat and see how the rat would behave over time. They placed the rat in an empty multi-chamber cage and allowed its movement to be recorded over time. The video of the rat's motion was not allowed to be watched until experiment was completed. The second phase of the experiment was to actually place food in some of the chambers in the cage. It was discovered that the video showed that the rat visited the compartments with food much more often that those without. The video was taken before the food was placed in those compartments and so it appeared as if the rat knew ahead of time what chambers were going to hold the food in the future! Well, my genes seem to have influenced the rat in some form of time travel. Today, the science of Quantum Theology is taking off and it is becoming a major discipline in Physics, Mathematics and Biology. The Principle of Causal Conspiracy has become a spring board for new evolutionary theories to develop. In a two-part series of books, it describes how the human mind interacts with the quantum environment to generate information. At last, I am also beginning to understand the science behind the J-N field and the dimensional landscape that I travel through.

Today, I have been isolated from Diana and I have been placed in a small room in the Sistine Chapel. It is not the isolation that bothers me it is the statement that was made by the Camerlengo Monsignor Rodini Maserati that bothers me. He said that I am particularly vulnerable today because of circumstances he cannot explain to me at the moment. Rodini said that I had to be in isolation until a certain hour passes and then I will be taken for an audience with the Pope. It is obvious that Pope Paul the Seventh had a very special message for me today and so I was not feeling too badly about my short isolation. In fact, I am excited about what is to come. I have met Pope Paul the Seventh before in my dimensional travels and I am sure that he will remember that this date is the day I gave him a vision of what was to come and what is to come. Two years ago, as the Cardinals gathered in St. Peters Basilica for the conclave to elect the new Pope, Monsignor Rodini had locked me in a small room similar to this one to probe my brain about the future Pope-elect. I had explained again that the new Pope is to be named Paul the Seventh, and that he will be of African Origin.

"What do you mean by African origin?" Rodini had asked.

"In my vision, he was stout, black and spoke with a crisp African accent," I had replied.

"Did he write with his left or right hand?"

"With his right hand," I replied.

Rodini thought about that for a moment, and then he pounded his hand hard on his chest. He seemed angered and disappointed by my reply and I could not figure out what he was angry about.

"Did he talk with a lisp?"

"Yes, he had a strange habit of banging his head with his hand as he spoke."

I was not shocked that the new Pope is from Africa since I already knew that of the two hundred or so Cardinals that are left in the Church, Kotubu Iba and his brother Effion were the most powerful candidates. The problem is everyone knows that Effion Eba is slightly autistic and that he is probably not the candidate that should become Pope. Later I learnt that Kotubu Iba had been selected Pope. Pope Paul the Seventh was born in Nigeria in 1990. He and his identical twin brother, Effion Iba were raised in a poor Catholic "Ibo" family, on the outskirts of Lagos. His father was a farmer and owns a small banana plantation with several other farmers.

Kotubu Iba, the young man who became Pope, attended a Catholic school and then went on to college to study engineering in his hometown. In 2010 Kotubu Iba earned an honorary Doctoral Degree in Astrophysics. In exchange for a scholarship from Cambridge University in London for the Doctoral degree in Astrophysics, Iba agreed that when he completed his studies he would return to Nigeria to set up an observatory for the World Astronomical Society. Kotubu Iba also studied particle physics and received a doctorate in Physics. He was profoundly affected by the findings of the mysterious Father Mulcahy, Nambu and other Scientists, regarding the nature of mind and matter. But Iba also felt a strong calling to the faith and at the age of thirty-two he studied theology and became a Catholic priest. Upon returning to Nigeria he was given a very strong diocese to shepherd in Lagos. Finally, he was able to reconcile his love of science and religion by defending creation theory through astrophysics and quantum theology. He then advanced to the position of Bishop of Lagos at the young age of forty-five.

It was not long before his international fame as a man of science and God won him a visit to the Vatican before a private audience of the Ecumenical Council that consisted of the Bishops of Rome. Kotobu Iba was elected a Cardinal of the church and for five years, was the most outspoken Cardinal on religious and scientific matters, taking an active role in scientific debates on theology. He coined the phrase,

"I see hundreds of species of evolving monkeys and apes, but only one species of un-evolved man," alluding to the wonders and ironies of evolution.

His twin brother Effion Iba is autistic. Brilliant in mathematics, he could hardly do anything else of value. He is an idiot savant. His brother, Iba, looked after him until he became a priest. To keep his autistic brother from the public eye, Iba made arrangements for him to stay in a small hidden monastery in the outskirts of Lagos, Nigeria, where he was kept under close watch and was cared for by a loyal local Nigerian nun called Mother Anna. When they were growing up Effion would imitate his brother and fool everyone by pretending to be Kotubu. He still assumes the identity of his brother once in a while enjoying the attention that gives him from his family and others. The Vatican regards the Cardinal Iba as a deep, holy thinker, and in 2038 AD, no one was surprised to see the white smokestack rise from the Vatican chimney, heralding the election of the first African Pope. An African Catholic Church soon took root in America, separating itself from the traditional Catholic Church, while still acknowledging the leadership of the papacy.

The first African-American President, Barack Obama, has become very popular and has created a lot of political changes in the United States. After his terms in office he goes on to become one of the most active leaders in world affairs, following the late President Clinton's footsteps and travelling around the world to secure peace and harmony after the Resa disasters. The new African-American President, Al Cooper, came from a poor family in Brooklyn. He took the White House by storm to strengthen the nation's weak coalition of states after the third great collapse of the financial systems of the world. His appeal touched most Americans who came from disadvantaged backgrounds in the now struggling US economy. After the Comet Resa disaster, Al Cooper played an important role as a Marshall of unity among displaced Americans. As a General in the US army, he was responsible for creating order and organizing the flow of immigrants between the flooded Midwest states and the coastal states. His powerful arm reached everywhere and he became a household name, the "Iron Fist". Al Cooper organized many donation centers and he worked with many foreign governments especially China to bring in large amounts of food and medical supplies to restore some measure of order in the fragile nation. The USA needed every help it could get from the outside world.

The political power of the FDA is gone. In its place a new International organization has been created to monitor the Global food and medical supplies. Food and medicines are now freely shared by every nation on Earth on a need basis. The World Food and Health Organization has taken root and it now controls nearly all supplies. The Patents issued to Monsanto Corporation on genetically engineered corn and rice seeds now strangle the world food supply and Monsanto controls half the food supply of the world. This started back in 1995 when Monsanto started giving "free grain samples to farmers in the USA, Mexico, India, and Africa. They did this so that their genetically engineered grain would take over the farmlands. The aggressive strains caught on, replacing the natural grains and became the staple source of corn, wheat and rice worldwide. Little did anyone know that the patented grain would be a noose around the necks of farmers worldwide. Monsanto is rumored to be a Bilderberg controlled company. There are some good things that happened. Within three years of Al Power's political career, medical care became affordable for all the people. The medical profession had been stratified into a few commanders and battalions of salaried "worker doctors." Al Coopers became a symbol of the cultural reform taking place all over the world. For the first time there is now a Black Prime Minister in England. Cultural progressivism and scientific thinking flourished under Powers. With the support of the Bilderberg Group, Powers was able to become a very powerful world leader.

Today, Pope Paul the Seventh has called an urgent meeting with Father Josephus Mulcahy and Nairo Nambu, the discoverers of the "Principles of Quantum Theology." This morning, Camerlengo Rodini instructed me to reinforce a protective field around Father Mulcahy and scan him for his religious strength and devotion. It is fortunate that I no longer need Memorin to channel myself into the dimensional gateways since it weakens my body every time I got out of it. After channeling, I enter the sanctuary of a small church in Rome where Father Mulcahy is performing mass. He is a tall olive-skinned man with an imposing large white beard. Mulcahy is in the midst of the sacrament when an altar boy runs to him with the news that the Pontiff is on the phone.

"It is urgent Father," the boy whispers to him.

"Tell the Pontiff, I am in the middle of the sacrament and cannot come at this time," the Priest answers calmly. I am impressed by his devotion to his work. After mass he walks to the back entrance of the church to meet and greet his flock as he must have done for many years. The altar boy runs up to him and again bids him to come to the phone.

"The Holy Father? He is surprised and has forgotten already.

"Yes Father, it is urgent," the boy insists.

Father Mulcahy rushes to the small back office of his church and grabs the phone. He speaks to the Pontiff for a few minutes and sweats profusely. Abruptly and without explanation, he packs a bag and heads off to Rome in his MPV. Father Mulcahy meets Nambu and Rodini at the altar of St. Peter's Cathedral in Vatican City.

After the specified hour passes Rodini comes into my room and asks me to follow him to the papal offices for the meeting with the Pope. I ask him to allow Diana to come with me and he refuses. Despite my objections Diana is not allowed to be at this meeting. It must be one of these meetings when only men are allowed to bear witness to the secret sacraments of the Papacy. I follow Rodini into the altar of St. Peters and join Nambu and Mulcahy for the long walk to the Papal office. Not knowing the protocol for the meeting and since Rodini does not introduce us, I do not greet Nambu or Mulcahy. Sometimes, there are strange protocols that have to be followed depending on what the Pope wants to project to his visitors. I decide to walk behind them followed closely by the Swiss Guard who mirrors every step Rodini takes. We enter into the Papal office and Rodini motions that we wait just inside the large door. I have been in this office many times before and I know every dimensional field inside this room. The Pope is standing by the window and his silhouette is clearly visible from where we are standing in wait. Effion is sitting on a chair by a large conference table with his back turned to us with a pen in hand busy writing on a pad.

"The Pontiff is much taller and heavier than he appeared in the newspaper," Mulcahy whispers to Nambu.

It is obvious that he has never met the Pope in person and I am waiting for the surprise look on his face when he sees Effion and the Pope together. The Effion and the Pontiff are dressed in traditional priestly garments and look like regal images of one another. The Pope is not wearing his usual ceremonial headdress but I could tell him apart from his twin brother by his thin head of grey curly hair peppered by a few small black patches. Effion looks a little rougher and has a thicker head of greying hair. Effion does not rise from his chair when we enter. I am instructed by Rodini to sit at the far end of the room with Effion and to observe and say nothing until I am called upon. My job seems to be to listen and try to understand the "science" behind the fields that control my life and being and this may be my only opportunity to come face to face with the secretive scientists that understand the strange world my life revolves around.

Rodini walks over to the Pontiff and falls on his knees, kissing the Pontiff's extended ringed hand. He motions to Mulcahy and Nambu to do the same. The scientist kneels reluctantly and kisses the Pontiff's outstretched hand following the priest. The Pontiff turns to the Rodini and beckons him to follow through an open doorway into an adjoining room. They vanish for a short while and then they emerge from the room with a metal box that is a model of the famous Temple of Resa. I recognize the delicate structure cradled in the hands of the Camerlengo. I have seen this model of the temple before but could not remember where or when.

"You are Sons of this Earth, gentlemen," the Pontiff begins, taking his time to gaze upon each of us. Rodini finds a seat at the edge of the window to give space to the Pontiff's authority.

"You have found a connection between the ancient mysteries of creation and modern science. We have decided to enlist your help, for the sake of what is soon to befall the Earth."

The Pontiff motions for us to sit beside him on the large mahogany conference table. He motions to the guard to close the door and leave the room.

"Seth Malaki meet Father Josephus Mulcahy from the Vatican Labs," the Pontiff introduces the Nambu and Mulcahy.

"These are the great scientists that the Church its rightful place amidst science and mathematics," he says pointing to the two.

The reaction on the two faces says it all as they see Effion for the first time. I can see visible shock on the faces of Nambu and Mulcahy as Effion turns and faces them for the first time. Effion says nothing and for some reason, the Pontiff does not introduce him. Pontiff brings his fingers together as if he is going to pray.

"Seth Malaki, is a valued guest of the Vatican," he reports, "he is here at my request. He is a brother in our church and has assisted us with matters of the church worldwide."

I bow to the Pope's introduction surprised that I am now in the spotlight. I thought I was only to observe and learn.

"The ancient powers of the Bible have been unleashed upon mankind!" the Pontiff begins then pauses to take in the reaction of his audience. The shocked priest's mouth drops open and I know he is serious and not pretending interest. Nambu sits quietly with his hands supporting his chin.

"You will understand after I have explained," the Pontiff reassures us as he begins pacing the room with his hands folded over his chest. I start wondering what he is going to be talking about. Is he going to start talking about the Scroll? Is he going to tell them about me and my strange world of fields?

"It has come to my attention that a great and mysterious event is unfolding around the world. According to the Bible, a great prophecy speaks of an unholy priest of the ancient world who will come to the Earth to fulfill a horrible prophecy."

He pauses and starts to walk in a small circle.

"This priest is born in the middle of the most chaotic mess the world has ever seen. He has amassed great power in the world, making his way to the highest ranks of government. He disguises himself as an Islamic leader, although he is their enemy. But our Moslem friends do not yet know."

Interesting! He is going to talk about the prediction of the Scroll, but Resa is already known by the world so what more could he add? He waits as if to gather his thoughts and I decide to scan the Scientists. Their aura is pointing to slight disbelief and confusion. They are in the field of Anticipation and I see very weak fields of Hope.

"I have been informed that the world of Islam is not part of his unworldly scheme but like many others before him, such as Osama Bin-Laden, he must use Islam to disguise his devilish acts. Your work has unwittingly exposed the source of his power."

"Your Holiness is this man already among us?" the Priest asks.

"Yes! Indeed, he is already among us," the Pontiff replies regretfully shaking his head. The Pontiff is talking about the Scroll predicting the rise of Resa, but we all know that Resa is dead and so perhaps I was not correct in predicting that Resa is the Antichrist, unless some other information has come up that changes his death.

Nambu is listening with interest to the strangest story ever told by the most powerful man of the Church. I am looking out the window into the beautiful serene gardens below to calm my wild imaginings. Effion is biting his nails and I sense that he feels left out. He keeps looking at the Camerlengo as if he needs to say something but cannot. I can sense that the Camerlengo has given him strict instructions not to open his mouth until he is told to do so.

"According to the Koran," the Pontiff explains, "if the Prophet Mohammed cannot go to the mountain, then the mountain shall come to Mohammed. There is a power that built the pyramids of Egypt, and now we know that the same power lies hidden in secret, in the sands of the deserts of North Africa."

"But I do not understand your Holiness, what have such matters to do with lowly men like us?" the priest beseeches.

"I must tell you of many things before you begin to understand," the Pope assures him.

"I come from the ancient Ibo tribe in Nigeria. In Nigeria we have many customs that relate to ancient rituals that you know nothing of. When I was a young boy, I used go to church with my brother and my mother and pray each Sunday morning. But after mass, we would go to watch the Poro play their games in the outskirts of Lagos."

Now that is interesting because I also know the Poro. What on earth is the Pope talking about? Is he saying that he is Poro? When I worked in the mines, I joined this society for my own protection and to study the scroll and its predictions. I am Poro! Is the Pontiff taking a part of me and trying to pretend that he is also Poro so he can be part of all these predictions?

"What is the Poro, your Holiness?" the Priest inquires.

From the look on Nambu's face I can tell that he does not want to participate in the discussion the Pope has started and he is biding his time but also listening intently. I am also waiting for the Pontiff to answer the question posed by the priest after all I am Poro and if he is, then he should be able to describe the Poro very well. The Pontiff pauses and he looks penetratingly at me as if he is in need of help. I want to say something but something tells me that this is to be his audience unless he calls upon me with an open mouth of words.

"I expect you to keep all this information secret and confidential. The world must not know what I am about to tell you until we have succeeded in winning it back from the devil."

Father Mulcahy is now visibly shaking and frightened by the strong words of the Pope. He repeats his question,

"Who or what is the Poro?"

"The Poro is a secret society that is found only in the West African Jungle. It is a secret society, elusive as the black night, and very few outsiders have been privileged to see them. They know nothing of the outside world and the outside world knows nothing of them."

I start to think that perhaps that is what I am. I know nothing about the outside world and the outside world knows nothing about me!

"They are as invisible to you as the Holy Spirit is to the eye. But you must promise me first that you will keep all this in utmost secrecy."

"We promise to keep everything you say in confidence your Holiness," the priest answers.

Nambu nods in agreement.

"I used to go watch the Poro in Nigeria with my mother. You see, even though she was a devout Catholic she was also a "Queen Mother.""

My God, his mother was a "Queen Mother!" I would never have known that.

"Queen mothers have an obligation to keep traditional and cultural-ties despite their faith. They must attend special Poro meetings and witness events meant only for the eyes of the enlightened of their society."

The Camerlengo walks over to the Pope and hands him a glass of water, he pauses and drinks some.

"During those Poro ceremonies, there is always mention of an ancient and mysterious power hidden in the jungles of West Africa. This power is a Mystic Eye allowing the ancients to see past, present, and future events. It has tremendous mystical powers, reserved for protecting mankind during the last days of the world."

The Pontiff's hands shake uncontrollably. He looks at me as if I am causing it. He quickly places his hands on the table holding the model of the Resa Temple, to steady them. The Camerlengo notices but the other two men do not.

"I have seen the ancient writings of the Poro Society. No man sees them without eating and drinking from a secret potion. I saw them and remembered them because I was a privileged young boy whose mother is a Queen Mother. But now, I know that the ancient writings found underneath the pyramid of Cheops are identical to Poro writings. They are secret codes!" He pauses to catch his breath. Obviously he knows more than I know. I did not know this connection before.

"Two years ago, a great oval cavity bearing similarities to Poro worship sites was unearthed under the pyramid of Cheops. We have a paid American Agent in Cairo who informed us of the existence of the same type of cavity beneath the temple of Abu Simbel.

"I must try to explain in simpler terms. Abu Simbel has carvings of spheres which seem to depict a great power, the power of the Pharaohs of ancient Egypt. The same carvings and the same cavity at the sanctuary of Abu Simbel were found underneath the great pyramids! We again received this information from our American Secret Agent operating in Cairo."

"But what has that got to do with the church and us?" Nambu asks fiddling with a pen that was lying on the table.

"Well, these are messages to us from the past. Drawings in the newly found cavity under Cheops show Amman Ra holding a spherical crystal in his hands and levitating huge blocks of stones several hundred meters into the air. These rocks are estimated to be more than two hundred tons in weight. To add to all this strange turn of events, recent sonar surveys of the foundations of all the other pyramids in Egypt, including similar surveys in South American and Inca pyramids have also indicated the existence of the same massive oval cavity!"

"That is very weird indeed!" Nambu ejects.

The Pontiff walks over to a huge, flat LCD monitor hanging on a plain wall and turns it on. He attempts to execute commands on a hand-held control, but his hands shake erratically. Effion starts toward the Pope standing with outstretched hands to take the control. The symmetry of twins looking at each other is uncanny. The Camerlengo grabs Effion by the arm like a child and commands him to seat down. The Camerlengo approaches the Pope and gently releases his grip on the controls.

"I am getting too old for this," the Pontiff jests and thanks the Camerlengo. Rodini takes control and initiates the control. The large monitor on the wall responds showing a map of the world. Twenty-seven blinking dots pulsate in various parts of the world. Rodini points the controls at the monitor to enlarge the image of the map. Effion again rises from his seat and tries to take the control from Rodini's hand but Rodini revolts disregarding Effion. The Pontiff nods to Rodini to allow Effion to take the control and Rodini complies recognizing that the Pope does not want to make a scene with his brother. Effion points to the screen sheepishly smiling at us.

"These are twenty-seven sites where great cavities have been found!" He starts to explain.

Again, Rodini approaches Effion and tries to take the control from Effion but Effion denies him and continues. I watch the reaction of the two scientists. They are puzzled by what is going on. It is obvious that there is some issue between the Pope and his Autistic brother that has been kept quite from the public. Father Mulcahy approaches the screen and carefully examines the map as if he is oblivious to the battle of wills between the Camerlengo and Effion. The blinking dots are now superimposed within blue shaded regions on the map. The Pope approaches the screen and continues.

"These same areas exhibit localized global cooling, otherwise known as GC. Now gentlemen, each of these archeological sites is also the site of a massive and monumental structure, such as a pyramid. We do not yet know for sure whether the GC sites are exact matches for the archeological sites but the evidence so far indicates that this is the case."

Father Mulcahy starts pacing anxiously about the room methodically turning and looking at the monitor as if absorbing its contents. Effion looks at me as if he needs to make contact with my eyes and I deliberately avoid them. I have learnt to avoid the appearance of being a student supporting a teacher's nod since that only makes him pay more attention to me. He turns to Nambu and looks at him. I can see that he is thrilled by the presence of the great mathematician being one himself. He has a very serious look on his face but I can tell that his autism is acting up again.

"The GC effects were thought to be a natural response to the comets' impacts. It was believed to have been caused by dust and debris build-up in the atmosphere, but this is not the case!" The Pontiff explains.

"I thought that the change in angle of the Earth's orbit as a result of the impact caused the GC centers," Nambu remarks skeptically to no one in particular.

"The tangent spill is an autonoma of the cusp relations between the trajectories," Effion blurts.

Nambu looks at Effion as if amazed by what he said. Effion feels vindicated at last someone is listening to his ramblings. He continues as if to impress Nambu.

"They were beneficial offsets to the rapid global warming differential. The funny thing is that they correlate exactly with all the major archeological sites in the world," Effion states looking wildly around the room and laughing loudly.

The Pontiff walks over to Effion and places a finger on his Effion's lips. He tries to take the control from Effion but Effion again refuses to give it to him. The Pope points a laser pen at the screen and motions to Effion to change the slide with the remote control. Effion complies.

"Look at this cave drawing found at the grave of Imran, in South Lebanon. It is three thousand years old! Yet look! There is the exact likeness of the Kabba of Mecca, suspended in thin air! The ancient texts written on the underside of the rock say it is lifted by the power of a crystal held by the hands of Nebudkadnaza!" the Pope exclaims.

"So, what? That is just archeology and the myths of the ancient world," Nambu retorts not quite understanding the connections.

"It is one of the active GC centers on the screen," the Pope replies pointing to a blinking dot on the screen. I wipe off sweat off my face and I sense that it is getting a little hot in the room. Perhaps it is the projector. The priest is leaning on the left wall of the room close to the monitor. His hands dangle over the thorny branches of a rose plant and he is unaware that he is bleeding.

"And that is not all," the Pope starts again. He motions to Effion to change the slide. A large cubic rock of immense proportions appears on the monitor with millions of people surrounding it.

"The Kabba in Mecca is in the form of a massive cubic rock whose significance you will soon learn. Humanity has witnessed many civilizations that became great by using mysterious powers. In the Kabala, the Bible, the Bhagavad-Gita, and the Koran, there is mention of powerful and ancient races of giants, who had the power to lift mighty rocks and stones. The ancient race of Atlantis is also said to have possessed this power. The Arc of the Covenant, which is a gift from God, is another example of such power and so are the Kabba, the Star of David, and the Holy Grail. We believe we have found the power that God has hidden from man since the time of Adam!"

The Pope waits for a reaction. An awkward moment of silence passes and then the priest asks,

"How do we relate to all this and why is it of significance to the Church?" He is shaking his head and looking at the Pope for an answer.

"Let me explain," the Pontiff remarks. Effion turns to the mahogany desk and starts writing some mathematical formulae on a large white pad. The Pope sees an opportunity and walks over and takes the control from Effion. Effion is totally oblivious that the control is in the hands of the Pope. The Pope manipulates the image more adeptly than before and the Josephus-Nambu equations appear on the wall.

"I wrote them," Effion turns and blurts out laughing and approaching the screen and showing off the equations he just wrote on the pad. His shadow fills the screen superimposed over the projections of the equations. Nambu looks at Effion and smiles. He knows that Effion probably wrote them but he also knows that he and Mulcahy created the equations. The Pontiff walks over and holds Effion by the arm. He pulls him over to the side and sits him on a chair. He knows that there will be a terrible scene if he tries to get Effion out of the room.

"This, as you know, is the Josephus-Nambu twenty-seven-dimension cubic field equation both of you discovered," the Pontiff points to the screen with a trembling laser pointer. The image changes to a diagram of the cubic structure of the Resa Temple.

"Gentlemen, you are familiar with this cubic field," the Pontiff states looking at the two scientists.

"Yes, this is exactly the physical expression of the twenty-seven-dimensional Josephus-Nambu field we discovered . . ."

"and the Resa Temple!" the Pontiff interrupts.

Nambu jumps on his feet, startled.

"But what has this temple got to do with equations?" He grumbles at the Pontiff.

"That is precisely the point, Nambu, precisely the point," Effion interjects laughing.

The Pontiff continues,

"When I found out from Dr. John Davis of the NASA Astro-archeological survey team that there is a connection between this, the ancient sites of Egypt, the Resa Temple, and the crystals, I realized that the Poro writings on the crystals were describing the same force!"

My mind starts to race with anticipation. The name rings a bell but I cannot place it. I search my mind for answers but could not find any. I have no information that could help at the moment. Whoever he is, or whatever John Davis has, is something I cannot channel. There seems to be a physiological block in my mind, almost a symmetry block that cannot be violated. A parity blockade that reminds me of perfect parallel mirrors making an infinite number of repeating images of the same thing over and over again. Only these are mirrors in my mind. John Davis, I imagine him to be a fat old Professor with tattoos or something, perhaps one of those Indiana Jones type Archeologists.

"How could an ancient African society know these equations?" Nambu asks. The Pontiff does not answer, he continues,

"It became a matter of International security. The Pentagon did not know of my background with the Poro. They connected the two things later and we found out that Seth also had a Poro connection. These connections with the J-N fields are what we are here to understand. These equations already existed in the strange Poro writings!"

"Who else knows of this coincidence?" Nambu questions cautiously. It is obvious his interest has been peaked.

"No one knows but you two, Seth, Effion, Rodini and myself. Professor Davis was heading to President Cooper's office to tell him about his findings but then something told him to talk to the Vatican first. You see, I was troubled by the sudden interpretation of the writings as Poro writings, especially when I knew I came from a Poro background. I am puzzled."

Six persons that know something means it is no longer a secret. I am sure there are many more that know about this connection already. What about the Bilderberg group? Surely they already know about this in Nevada.

"So, what is the significance?" Nambu persists.

"How could such a primitive society, forgive me your Holiness, be connected to all these great archeological sites around the world, and especially the J-N fields?" Father Mulcahy questions following Nambu's curiosity.

"Well, for one thing, anyone with a primitive knowledge of quantum theology will know that the cubic pattern is the Josephus-Nambu field that you two have been studying. And anyone with an archeological background will know that the same pattern has been documented at the twenty-seven archeological sites."

"How do you know that the same pattern found on the surface of the globe is that of the cube?" The priest asks.

"I had a close friend of mine do the transformations before meeting with you," the Pontiff reassures. I am assuming he is referring to the only other person in the room capable of such mathematical feat, Effion.

The Pontiff hands Rodini the controls and motions him to manipulate the images. The walls of a cube unfold to a flat plane then they fold into a perfect sphere. Rodini unfolds the sphere again and makes the left wall of the cube rotate about its diagonal formed by its far upper and near lower corner points to form a diagonal plane at right angles to its original plane. Then, he did the same thing to the right wall in reverse. The two planes intercept at the center of the cube. A pattern emerges that is similar to the hexagonal arrangements of fundamental particles, now known to all scientists as the eight-fold way.

The Pontiff walks over to Effion who is busy writing mathematical formula on the pad. He pulls Effion's arm and motions him to explain what is on the screen. Effion rolls his eyes and smiles excitedly and walks to the screen and starts explaining.

"Aloquet 27, exclude the central plane and the pattern is that of quantum Spin 3/2 particles in a triangle on the inclined plane."

Nambu comes closer to the monitor, pointing to the transformations.

"Indeed" Effion adds, "Spin ½ particles seem to be located at the corners of the cube or on the primary fields of the cubic symmetry, while the spin 3/2 particles are the internal half-width points of the cubic field pattern. Now, you would expect the spin zero particles to lie on the unmoved plane of the cube."

"Now Gentlemen," the Pontiff steps in excitedly, "let us look at the spin zero symmetry of the unmoved plane."

Effion clicks the control to open up a new image. The Pontiff sighs and sits down waiting for the scientists to digest the information.

"Very interesting indeed!" Father Mulcahy shouts pointing to the monitor.

"All spin zero particles stay away from the plane and have not been rotated through any angle!"

"It is also clear that the U-quark has not been moved in any way from its original position except that its relative angle to the other particles has changed," Nambu observes.

"Now gentlemen, this is the problem," the Pontiff breaths deep.

"We have found the same cubic symmetry superimposed on all the sites where the famous crystals were found! We believe that there is a fundamental connection between the two. If Abu Simbel is destroyed as a result of the power of the Josephus-Nambu field, then mankind has unleashed a problem far greater than any natural power in the universe."

"So, what are we to do?" Nambu poses his question with a serious look. Effion again takes the stage and walks to the monitor, turns it off and returns to his seat. He must have rehearsed this over and over again.

The Pope speaks,

"I need to prevent the world from getting this knowledge. As we all know, the Josephus-Nambu field is the ultimate knowledge of creation itself, manifested here, in broad daylight. We cannot let a force that is infinitely greater than the power of the atom, fall into the hands of the Egyptians or any other government. Humanity is not ready. We need to find a way to neutralize these fields."

Tears fill the Pope's eyes as he speaks. His hands begin trembling uncontrollably.

"Why me?" He whispers, "Oh God! Why me?"

He walks to a corner table in the room and takes some paper towels from a box to wipe his face. He leans against the wall for a minute to regain his strength. He ambles weakly back to the conference table and sits down. Effion walks over to the Pontiff and places an arm around him to comfort him. He starts sobbing with the Pontiff as if they are connected by the same field. He turns to look at Rodini as if for approval. It is all so very unordinary, so very strange. Rodini does nothing to stop Effion, it seems like Rodini expects this behavior from Effion and as if he had rehearsed it all with him already.

"You and I know that this is the very field effect that is used by the Almighty to create matter out of the cubic lattice of empty space. For the very building block of spacetime is this elemental cubic field. Now, the church knows for a fact that all the universe, all matter, all forces were created when the symmetry of the nothingness cube is broken by God."

The Pontiff blows his nose. Perspiration soaks him in the cold air-conditioned meeting room. He hands a paper towel to the Effion and then hands another to the priest. He points to the wound inflicted by the thorn of the rose plant on the priest's hand. Mulcahy wipes his sore skin and thanks the Pontiff. Nambu is preoccupied with nervous speculation and he bites at his pen as he stares down the window at the distant reflections of the sun over the garden below. An old painting of the battered coast of the Mediterranean Sea engages me. It is something to look at in the room.

"When Our Lady of Fatima told us of the power of the Atom bomb, the church made the mistake of telling the American government what it is capable of. We had hoped that knowledge of its apocalyptic powers would scare the world and end the wars. But it did not. Instead, they built the bombs and destroyed millions of lives while desecrating God's creative work. This time, however, we will not fail to shepherd the world."

The Pontiff's black cavernous eyes well up with the tears of a rising, passionate tide. He surveys us penetratingly, searching for what lies within our hearts and minds. He turns and looks at me disarmed. I have never seen the Pontiff so vulnerable before.

"This morning, a young man named Seth Malaki came to me as an apparition," he points to me and continues,

"He came to me as a young man from the past and informed me to write his predictions. He predicted the catastrophe of Comet Resa in 2016 and also revealed that a fantastic force would be unleashed upon the Earth by the devil. This "final force," more powerful than the combined energies of the atom and the sun, has guided humanity since the beginning. The nature of the force is an exact match to the Josephus-Nambu force."

"But is he not Seth Malaki?" Nambu asks confused and pointing at me.

"Yes, he is older today. He is a dimensional traveler. He came from the past to this day. He was trying to predict what was to come but had arrived in my visions this morning. I was trying to figure out why he did that. Why did he not write his predictions for the Pope Gregory in 2012? Why did he come to the future only to tell us about the past as if it was to come? I have read all the Books of Seth Malaki over the years. What I did not know is that he could not have written the words until they come to pass. He had to come to me and reveal the information that has passed so that we can be warned about the future. His message is not about the past but about the future of man, about the Josephus-Nambu force."

"The, how is it a prediction if he had been writing the history all along. How could it be a prediction if he just told you about the Comet this morning?" Father Mulcahy asks. The Pontiff looks at Effion and motions him to answer.

"Symmetry! The Past and Future symmetry! Charge, Parity and Time symmetry warns us of the past today, then it violates no law. However, the mirror symmetry between past and future is not exact. By Malaki coming as a young apparition this morning, he creates a Past charge field that enters into the future field to neutralize it. Only he is in the past and we are in his future. He is then able to predict the future without violating the past. Symmetry!"

"But he is here sitting in front of us, are you not?" Nambu asks looking at me. I do not say a word. I have come and gone, and now, 28 years later, I am sitting here looking at the same man I had met before on this day. I realize now why I was isolated this morning. If I had met my own past self, I would have changed the course of history. I would have violated a fundamental law of nature. Maybe my past charge would have neutralized and exploded with my future charge and I would have died. I can tell that the Pontiff's audience is dumbfounded. I can tell that Effion had figured all this out. He knew I should be isolated. I look at him also a bit puzzled. Effion starts to band his head with his hands, uncontrollably. I look at him again and then at the Pontiff. It is a strange feeling of symmetry that I feel. Who did I really meet in the future, well in the past now? Was it Effion as Pope, or Kotubu Iba? For the first-time doubts start to fill my mind. The symmetry is uncanny. I remember the Pope banging his head when he was writing the prediction. Effion bangs his head. I have been channeling to Effion to narrate the Books of Malaki all along. If I met Effion as Pope why is Kotubu Iba Pope Paul the Seventh? Something is amiss. I realize now that I must take a harder look at Effion and must take him more seriously. My thoughts are broken by the Pope.

"Unfortunately, we made an unholy alliance with the Pharoahnics. Even as Ramses made a covenant with Yahweh to free the Israelites, so do I make a covenant with him to free Mankind. He is to be the sacrifice. The charge must neutralize what is to come."

The charge neutralized? Is the Pope saying I must die? I eye the Camerlengo. His eyes have been fixed on me all along. What is going on?

Effion walks over to the Pontiff and holds his hand in his. He assists his brother into a chair. Symmetry, again I see the symmetry! The charge must neutralize what is to come. What charge is the Pope talking about?

"The church has always thought that the Josephus-Nambu field is proof of the existence of the Creator. This is the power of the mind force. But we never suspected the devil would interfere," the Pontiff ominously whispers. The tension mounts as the they launch the Pontiff's plans into action. I am becoming weak and tired. It is as if I have been dosed with Memorin. Is this all real or my imagination? They are talking science, mathematics, and all of it. I am thinking of the fields, real fields that rule my life. Is the J-N field a dimensional field that can be understood? Is it the generator of the dimensions I have been travelling through? If so, perhaps there is hope, perhaps there is hope for me and mankind.

"We know that a sphere, which may well be one of the cubic field points, existed in Egypt before the death of Resa. We have information from very reliable sources in the CIA that a particular crystal data bank has catalogued the exact location of the sites of these strange crystals. That is what we must find and destroy before the US government finds them," the Pope continues.

"But how do we know where to access the data?" Nambu inquires.

"The data is in the possession of one Professor John Davis of the University of Maryland. It holds the only copy of the Geo-Orbiter data collected from the sites where the field is found around the Earth," the Camerlengo reveals. He is looking at me as if I am an object of study. My mind races with anticipation. Am I going to be a victim? John Davis keeps cropping up in my head, like an escape route. I feel strange and isolated.

"John Davis works with the Nevada group, I know him. So why don't we just get it from him?" Nambu asks. The Pontiff turns to me as if I should know.

"There is a problem," the Pontiff adds, "the codes we got from John Davis do not work. He has the real satellite codes in his possession and he is nowhere to be found!" The Pontiff states disappointingly. Now, weaker as if some unidentifiable force is draining him, the Pope summons up his last remaining bit of strength to speak.

"We are fighting a battle of space and time, a battle of dimensional fields, and we beg you to help us. We do not want any government to get any of these holy crystals. If they do, they will surely destroy the world. We must locate all the field points before anyone else. And I have assembled no better team on Earth to do the job."

"But what assurances do we have that these sites do indeed contain the powerful J-N crystals? And what if it is a mere coincidence that they are cold global sites?" Nambu asks, serious, but at the same time I sense that he has some doubts about the Pope's story. The Pontiff seizes the small model of the temple of Resa and pounds it on the table as if it is a gavel in a court of justice.

"Look at this temple!" He cries out hoarsely.

"Do you recognize the cubic field pattern in the temple of Resa?"

The two scientists study the temple replica feverishly going back and forth with calculations. Effion is trembling as if captivated by an autistic episode. The Camerlengo is looking at me as if my every move is an answer. It is still not coming together for me but for the two scientists it seems that the Pope's logic has been impeccable.

CHAPTER 11
The West African Jungle
The Sierra Mining Company
(June 22, 1982)

It is 3:00 a.m. and far too late for me to be driving around in my condition. I have just arrived at the campus of Fourah Bay College after a drinking spree in the city of Freetown on the West African coast of Sierra Leone. Speeding through the small winding roads that leads up through the mountainous peaks to the college, is dangerous. The poorly lit city roads of Freetown jut out over the treacherous coastline of the West African jungle, but that did not slow me down. As the incoming fog and rain blind me on the steep pass, I promise myself never to get so drunk again. Next time, I swear that I will stay in the city with my girlfriend Angie or at least take a cab. I had been celebrating my graduation from college. In a few days, I will enter the real world as an Engineer hustling for a job in the big city.

As I stagger to my dorm room, I push on the walls to support my limp drunken limbs. When I enter my room, the smell of pee fills my nostrils. My room has been ransacked and my clothes are everywhere on a dirty wet carpeted floor. My expensive bedside lamp is broken and my girlfriend's pictures are all over the floor. My books are wet and destroyed and there are blood stains all over the carpet. The window to my dorm room is wide open and it seems like rain has soaked into my bed and has also saturated the carpet. I leave the window open to let in fresh air and dry out the floor. Shocked, I start toward the corridor to see if I could find a mop to clean the mess but my limbs are too weak and my drunken state does not help me think straight. I enter into the common area of the building. It is alight with activity and has become a strange mix of metallic control consoles and strange tribal people. I must have gone mad. This is not my dorm building. My mind reels as I see strange looking people dressed in raffia garments walking about like ghost warriors. They are doing things that I cannot understand. A black warrior walks up to me and hands me a cloth and points to my head and smiles. I rub my head with the cloth and feel the pain of an open wound that is bleeding. The cloth is filled with blood, my blood. For a while I stand there looking at the man as if he will eventually fade into past dream. He does not vanish instead more of the warriors come and surround me pointing spears at my skull. The first man aims a strange looking device at my eyes and I feel pain searing through my head as if I am being burnt by a laser. I must be dreaming, I will wake up in the morning and it will be gone. The man points a calabash at me and asks me to drink its contents.

"Drink, Seth, drink," he commands.

I feel no resistance. I put the strange jug unto my lips and drink the strange contents. It is a sweet-sour white liquid. The potion enters my body and mingles with my insides. I feel it travelling into my stomach and mingle with alcohol. I feel my body going limp and I start falling into the arms of the warrior before me. He pulls me to a sofa and lays me down. Slowly, my mind vanishes into oblivion. My body is floating into a vanilla sky that

I do not recognize and I feel immortal. I am as a bird girt with wings of wisdom. I steel my mind without doubt and I feel like I have united with Lord of Love, like the blazing fire that reduces wood to ashes and knowledge to madness. My mind is free from attachment and I am free from desires to attain the supreme perfection of freedom.

I wake up on a sofa in the common room feeling like a lark. I must have been so drunk the night before and it must have all been a dream but it has had a profound effect on my future prospects of drinking alcohol. I decide to go down town to visit my friends. As I pass by the common bill board, I notice a job Ad for engineers hanging on a billboard. A new company called Sierra Mining Company has just arrived in town. The company is hiring engineers to man a diamond and gold mining operation in the small town of Sefadu, four hundred kilometers north of Freetown. I tear off the Ad and plunge it into my shirt pocket and amble off to bed staggering.

The next day I pack my belongings and leave my key at a designated location for my friend Richard. Tomorrow, Richard will pick up my belongings and store them in his father's warehouse down town. I start my BMW bike and head out of the Campus for the last time. I will go down to the city and present my resume to a popular Egyptian called Jawed Said. I am hoping that I can land a job with this new mining company.

I arrive downtown and skirt around a classy neighborhood of the city looking for the offices of Sierra Mining Company. I am imaging them to be posh offices in some massive sky scraper in the middle of the city. It turns out to be a small office in a bad neighborhood of the city. I pack my BMW bike in their garage and head off by foot into the building. As I enter the plush headquarters of the Sierra Mining Company, I could not help contemplating the enormous amount of diamonds and money that must change hands daily in this small corner of the city. I am escorted into a huge waiting room by a Mr. Jawed's secretary. She is a charming grey-haired lady with a permanent pre-meditative smile. As it turns out, I had read an article about Mr. Jawed Said from local newspapers, but I cannot recall what the papers said about him. All I remember is that Jawed is a rich and powerful Egyptian who came to Freetown to mine diamonds. I know nothing about diamonds and so I hope he will not question me about carbon stones that decorate middle fingers. As I wait my turn to meet him, I can hear and see operators chanting in telephones and sending faxes as they negotiate millions of dollars' worth of diamonds with dealers around the world.

"Blood diamonds," I think to myself. I know about the great blood diamond war that has ravaged this country in the early eighties. Now, they are at it again. That evil diamond will do it all over again. It is people like this that I despise but I do need a job and despise will get me nowhere. A young lady darts out of a room carrying several papers and what appears to be two sealed jewelry boxes. She is wearing a beautiful short blue skirt and a tight red top that exposes the cleavage of her plump breasts. Her hands are laden with gold and diamond bracelets and rings that are for sale. A man barters with her in the middle of the waiting room, just a few meters away from me. I watch them negotiate hotly for over a half an hour and I am astonished to hear the young lady haggle the buyer up from three hundred thousand dollars to a final sale of two million dollars. Then, she also tries to strike a deal for her gold nuggets. I never realized how much money precious diamonds represented before. To me they were just stones until I saw the young lady's sales techniques.

A few moments later, Mr. Jawed's secretary ushers me into a room for my meeting and interview. The office is an exquisite display of wealth and power. The red carpet, I imagine, must have been custom made to match the furniture and wall coverings. A huge elephant tusk hangs majestically over an eight-foot high doorway. The far wall of the office sports a thirty-foot-long snake skin complete with wide open eyes frozen in death. A gargantuan flat TV sprawls over the left wall while the opposite wall seems blank. As I wait for someone to enter the room, the screen entertains me with images of huge diamonds as the company's promotional video declares,

"...Sierra group will eventually venture into the real estate market. Huge town complexes will be built using the waist clay products from the mining operations. This concept has already been tested in our prototype of Low Cost Housing in Kissy Mess Mess . . ."

I scan the room self-consciously. My eyes brush over a bunch of papers on a huge mahogany office desk. I focus on a large photograph of a strange clear spherical crystalline object about the size of a soccer ball. The photograph is stamped "Top Secret." There are several notes I cannot decipher at the bottom of the picture. I focus on the image of the object and notice a strange encryption within. Several interconnected passages of the code come together to form a seemingly crystalline electronic module. I do not know what to make of it but it intrigues me. Could that be a diamond? Do they come that big, I wonder? The inner circuit reminds me of something fiber optic or photonic.

I figure that the Sierra Group must have vested interests in fiber optics or photonics. Diamonds, photons and electrons! They go together, don't they? I jest sarcastically to myself.

Suddenly a short brown-skinned, bald headed man bearing the physique of a gladiator appears. Except for the extremely expensive suit he wears he could have passed for a notorious prison inmate awaiting death row. A huge scar spans his forehead. His black, beady eyes protrude unnaturally from their sockets as if he is constantly being choked. His nose seems boneless and flat. The man turns off the TV and walks over to the huge mahogany credenza in the middle of the office where I am standing. He pulls some papers and the photograph quickly from the desk as if they should not have been there. He gives me an angry look that calls out, this desk is my control console, the symbol of my consolidated power in this world. This is my turf. How dare you look at my papers!

He keeps his glance at me as he sits down pulling at the zipper of his pant. I notice a young woman walking across the office toward the door he had used to enter the room. I realize that the wall is not sealed-off and that there is an inner sanctum within hidden that is by two separate walls spaced about two feet apart to create the optical illusion of one continual wall. So that is what the zipper is all about, I calculate, the woman. I assume the man must be Jawed, my interviewer.

"Give me the papers," Jawed blandly orders. I hand over my resume feeling insulted by his tone.

"I thought you were here to deliver the maps?" Jawed mutters accusingly in a thick foreign accent, tossing the papers back to me in disgust. The accent was a thick Egyptian or Arabic accent and it matches the face.

"No," I reply hesitantly, "I am an engineer and I came for a job interview for the mines."

I pull the Ad from my shirt pocket and clumsily lean over the wide table that separates us to hand him the papers.

"Be careful of the table my friend, it could pay your salary for the next ten years," my prospective boss is an ass.

Jawed pours over my dissertation on fluid mechanics and thermodynamics as if he has a firm grasp of the sciences, then he abruptly hands the resume back to me.

"Come back tomorrow," he decides as he places the dissertation in a desk drawer, "leave this book with me," he orders.

I leave with the anxious hope of almost having landed a job. This will be my first job ever. I did not sleep well at night. I did not drink either. The next day, I wake up early anxious to follow up on the interview. My meeting was set by the old secretary for 8:00 a.m. I must hurry. I jump on my bike and start the engine, but it would it not start. I prime it with fuel and try again with no success. For half an hour, I fiddle with the engine trying to start the darn thing. Finally, I give up and walk to the edge of the street to catch a cab. I am getting late but if the cab hurries I can make it. Finally, with a few minutes to spare, I arrive at the plush offices of Jawed again. Following a two-hour wait outside the office, I am ushered into the office by the same old secretary. I once again find myself face to face with the ugly face of my future boss. There is another middle-aged man sitting on a comfortable arm chair opposite Jawed's desk. Jawed introduces the man as Ramez Abla, an Egyptian business man. Since I have already grown accustomed to Jawed's lack of courtesy, I do not wait for an invitation to seat myself.

"Ramez and his younger brother, Ali Jamil Abla," Jawed explains, "are exploring the possibility of exploiting the great diamond mines of Sefadu, a small mining town up north."

"They have come to make life easier for Sierra Leoneans," Jawed mutters jokingly. It is not amusing to me after all they are just scavengers trying to drain of the poor African nation. The Sefadu mines in Sierra Leone are hosts to some of the grandest and most beautiful diamond deposits in the world. Some of these diamonds have found their way to the exquisite Crown Jewel collection of Queen Elizabeth of England.

I notice Ramez's flamboyant fingers are weighed down by grotesquely huge opal rings that form an array of different exquisite patterns and colors. Ramez likes to advertise his success, or perhaps he is gay. Another man enters the room and is greeted by the name of Militant Vegas. He walks into Jawed's office casually as if he is part of the mining crew. I figure from his accent that he is Brazilian or South American. He is casually dressed in blue jeans and a poncho, Mexican style. His coarse long black hair is uncombed and I imagine he looks like what a seasoned miner is supposed to be. Ramez introduces me to Militant and asks Militant to take a look at my resume. It is apparent from his tone that Ramez has a lot more say in this office than either Jawed or Militant. Militant grabs the resume, glances quickly over it and places it on the table.

"This is fast river and carries lots of sand and grabel," he remarks in a South American accent exotic to my American ears.

"We need a machine to dredge the river bottom and bring gravel to the surface, can you build one?"

"I do not know," I answer resignedly. If it is going to be like this, I may not stand a chance.

"Draw me a machine that can extract gravel from a river". Militant hands me a blank sheet of paper and a pen. It is a strange and abrupt request. What is it they expect? I wonder. I am not intimidated. I have designed machines before and I am one of the top notch conceptual artists for my machine class. Concepts are all they are, since I have never built a working model of any of my designs. After some thought, I draw a boat carrying a simple spiral auger blade within a large hose dropping into the river. I hand the simple sketch to Militant.

"What is this?" Militant questions smiling. Without waiting for an answer, Jawed takes the sketch and bellows a laugh. Militant repeats the question.

"The spiral blade rotates by air pressure and digs into the river bed. The hose extracts the loose gravel by air buoyancy," I quickly describe to make a point.

"Buoyancy?" Militant asks?

I think for a moment.

"Yes. If you use a diesel compressor and an air motor, you would not need electricity in the bush or on a barge in the river, so air pressure would be the way to power the spiral blade, and lift the gravel," I explain.

"The air exhaust from the motor that runs the blade would then mingle with the loose dirt and be transported to the barge by pressure and buoyancy". Militant correctly summarizes for me.

There is silence.

"Air-lift." Ramez mutters, "have you ever worked in mine before?" He asks me in serious tone. I respond in the negative.

"Wow, that is good," Militant states patting me on the back, "we got a good one," he states looking at Jawed.

"A very good one!" Ramez reinforces.

After a lengthy interview, I land the job with the Sierra Mines Company. I am hired to work with an elite group of engineers and miners in the town of Sefadu. The pay is good, very good.   The camp will consist of eight Brazilians, three Australians and approximately one thousand Africans workers. Militant explains the intricacies of the mining camps to me, describing the types of equipment used in the different mining operations. I spend the rest of the day with Militant, Ramez and David Greening, the mining camp manager. They end the evening drinking scotch and warm beer in a small bar in downtown Freetown while I decided to sleep soundly for my new-found adventure into the jungles.

The next day, I wake up early to meet David and Militant for the trip to Sefadu where the Sierra Mining Company is headquartered. We lunch together at Ramez's house in the city and then board the brand-new Nissan Desert Dweller, a rough terrain vehicle. I learnt that David Greening, a blonde, blue-eyed twenty-five-year-old playboy from "down under" is an invaluable member of the mining camp. He speaks English, Arabic and Portuguese fluently. He is the key liaison between the mining camps and management. David is one of those rare individuals whose energies are fired by a dangerous mix of brain and cunning. He enjoys intellectualizing as much as chasing women. But without David as a camp site manager for the Sierra Mines, the delicate balance of power between engineers and equipment operators cannot exist. He can easily relate to both types of people. He is a "Jack of all trades and master of none."

That morning, we meet at the offices and prepare for the long trip to the mines. David is preoccupied with preparations for the rough five hundred kilometer drive we will have to endure from the city to the mining camp deep in the jungle. We have packed the truck with luggage, food, equipment, supplies, and other essentials. The drive will be extremely uncomfortable because of the primitive conditions of the roads and it will take about fifteen hours for us to reach the Sefadu site.

We leave Freetown at about 8:00 a.m. for a journey that is an experience of a lifetime for me. Apart from the bumps and crevices that make the drive uncomfortable, David speeds up to make up time and literally bounces the entire vehicle a few feet up in the air when we encounter a bump or crevice, and, there are just too many of them on the road. Every fifty miles or so we encounter a large fallen tree in the middle of the road that must be removed and we would have to wait for help from another group of travelers because the logs are so very heavy to move by ourselves. It is the first trip for me to the inhospitable "Upline," as such faraway places from the city are dubbed by the natives.

David, who is in control of the vehicle, keeps playing the monotonous songs of Sting and the Police over and over again, as if it is the only music he has to play. It is nauseating. "Wake up this morning . . . found a million bottles off the shore." David sings along every song and then the album repeats again. Perhaps I think David has a syndrome of some sort, because no one has attempted to change the music or make a comment about the situation. The cassette player just keeps repeating itself and I wish I had brought some other songs of my own to play, but from the passive response these folks have to David's music, I know there is no chance of changing the situation even if I had.

We finally arrive at a dense sleeping jungle after midnight. The narrow dirt road passes through a thick jungle with no civilization in site. We are surrounded by dark thick forage that seemingly has no end and the smell of humid dust fills the air making it harder to breathe. Once-in-a-while we encounter dust storms from other vehicles ahead of us. David chases them and finds dangerous opportunities to pass and make his own dust storm in revenge. After few long hours, the monotony of the music starts to appeal to me, it has become something familiar in this dense-dark-nowhere. It is something one can listen to and understand and it has become a chant, a mantra for the journey. I am getting used to the situation and I am beginning to understand why David keeps playing the same songs over and over again. After a fifteen-hour journey into the unknown, the mantra works and it has lulled every one into a deep rest that makes us forget about the bumps, the dirt, the humidity and dangers of the deep jungle night.

We arrive at dawn in an open area, apparently, in the sleepy town of Sefadu. There are no visible houses, just cleared forests that never seem to end in the distance, wide open fields with large holes in the ground and mountains of dirt beside them. This must be the mining sites, I surmise. I imagine where we will be staying; will it be a tent, or a house? I have never camped before and even though it sounds exciting a soft bed to lie on will be most desirable now.

Finally, David turns the vehicle into a desolate flat offshoot. A huge array of machinery appears on either side of road. These are large machines, larger than I have ever seen before, trucks so large, that you can walk under them, buckets so large that a car would fit in them. Some massive stuff! Excitement slowly replaces the monotony as I start seeing this new form of civilization. The site of huge machines is an exciting change from huge city buildings.

"Sefadu," David utters, almost reading my mind. Militant wakes up from his blissful sleep, he is used to these trips. A set of buildings appear in the distance, they must be home, I think and perhaps this is not as bad as I was thinking. Soon we stop in front of a large steel building where a few people are gathered outside as if waiting for us to arrive. Two well-dressed Africans approach the vehicle and greet David and Militant with gleeful familiarity smiles. They start to off-load the vehicle and transport the supplies off into the building, while David and Militant heads off to the gathered crowd. I get off the vehicle and walk toward the building where the workers are taking the stuff. I am greeted by a burly old man who introduces himself as Bayano, he is Brazilian. He leads me into a small room in the building where he asks me to place my stuff and be comfortable. He speaks very little English, but it is clear enough for me to understand. I enter a well-appointed living room with lots of furniture, apparently for a lot of people that live in the building. I relax. This is it, my home and life for the next year or two.

I spend the rest of the evening talking to Militant, Bayano and the rest of the mining crew in broken English. I learn that David holds a special place in the hearts of all the Brazilian workers. Militant is the only Brazilian who could speak English and so David is their only link to the outside world. This is going to be interested.

CHAPTER 12
(West African Jungle)

Ten hours after the IFP leaves BWI airport, the craft lands at the Lungi International Airport in Freetown, Sierra Leone. It is still dark, and daybreak is just beginning to show its face over the horizon. The airport is an ultra-modern facility capable of accommodating ten large or twenty small aircraft at a time. It is very hot and humid and fine misty rain fills the air. Hal and I are received by the military police of Freetown who had been informed of the trip by President of Sierra Leone, John Massakoi. We are escorted to a receiving VIP room at the far side of the airport.

The recent four-way International Sector Pact between the United States of Europe, formerly the European Community, the United States of America, the United States of Middle East, and United States of Africa, has assisted small countries like Sierra Leone to be able to afford some of the comforts of modern technology. The great catastrophe of Comet Resa has benefitted the economy of most third world countries. The tremendous push to reduce pollution in the world after Comet Resa resulted in the trading of essential technologies in exchange for the green gas emission credits from third world countries. Wood stoves have been replaced by solar powered stoves and cookers. MPV technology, although wide spread in the world, lacked the technological support needed to run efficiently in third world countries. Thanks to global cooperation, old clean-fuel cars in Africa were slowly becoming obsolete and subject to high pollution taxes.

President John Massakoi adapted a Western political style. In the late 1980s, Sierra Leone was on the brink of a national disaster. Then, a war broke out over the so-called blood diamonds. Thousands of people were killed by rebels who wanted to take over the wealth of the diamond mines. They methodically took over the mining camps, selling diamonds for weapons and in the process, they destroyed the infrastructure of towns and massacred anyone in their path that owns or operates diamond mines. Thousands of children were amputated so that they could not grow up and use their hands to mine diamonds. There was an international cry to stop all traffic and sale of the so-called blood diamonds. Thousands of people also died of starvation, paramilitary activities and crimes as a result of the bloody war. Complete ecological ruin and mismanagement of the diamond and gold mining industry resulted in unprecedented poverty and suffering in the country.

Corruption and political struggles for power kept the country from developing a self-sufficient economy. Since the farmlands were blessed with diamonds, gold and other exotic minerals, this resulted in mass conversion of the fertile lands to badly managed mining pits by the rebels. The lack of organization of land resources and destruction of natural drainage systems ruined the ecology of the Northern cities and completely eliminated the possibility of using land for agricultural purposes.

Meanwhile, corrupt leadership led to smuggling riches from the mines to the Western world, to fill the Swiss bank accounts of a few. Back in the 1980s Jawed Said was in control of nearly all the mining wealth of the country. His diamond and gold empire soon became a multibillion dollar enterprise. He strangled the economy of Sierra Leone by controlling the flow of food and grain, especially rice, into the country. If Jawed did not import rice, thousands of people died of starvation. He controlled the prices of food and oil and became very wealthy indeed. With the blessings of the then President Siaka Stevens, Jawed took over factories and essential industries in the name of privatization. He purchased fishing fleets, commandeering fishing and shrimp fleets from the coast of Morocco to the shores of Liberia. Jawed's name became a household name in all of Africa.

In 1985 an uprising began in Freetown. Sierra Leoneans, sick of the chains of poverty, took to the streets and revolted in their thousands. They ousted the President and replaced him with another more conscientious man. It was rumored that Jawed fled the country with his entourage of thieves, abandoning the diamond mines in a state of chaos. However, no one really knows where he escaped to, or where all his wealth ended. Some rumored that he went to live in Egypt.

A period of progressive improvement followed until the Pact of World Resources Conservation was signed in Helsinki in 2027. Countries like Sierra Leone were excused from paying off loans to the West and allowed to participate in the world economic growth. Mining activities in Sierra Leone were rejuvenated under a plan with the USA. France took over the infrastructure of the country, rebuilding the Capital city Freetown into a modern city not unlike Washington, D.C.

MPV technology was brought in to replace the old smoke-puffing cars of the past. A thermonuclear plant was built offshore in the Atlantic Ocean to supply all of West Africa with cheap nuclear power. In exchange, the governments of these countries, including Sierra Leone, wisely accepted to follow the lead of the Western Nations. Their accounting and financial planning were strictly controlled by the International Monetary Fund. Diamonds and gold were paid for by direct technology transfer. Agriculture in all of Africa was put back on track by China. The great drop in world population has resulted in an equilibration of resources in the world. There is enough food and water left for the remaining one billion inhabitants of the earth and they are now willing to share it. By the year 2035 Africa could feed itself. With poverty eradicated and third world countries thriving as second world countries, the world could move forward and forget about wars, famine and disease. The past decade had been extremely productive. Although Africa is still far from being technologically self-sufficient, third world countries like Sierra Leone are now capable of initiating their own health and technology programs. They are becoming the new labor base for the developed world.

So, when Hal and I enter the ultra-modern airport terminal, we are not surprised by the development we see. What did surprise us is the entourage of Military Police that has come to escort us to the home of President John Massakoi. We are piled into the MPV, a state-of-the-art fifteen-sitter "soft glide version." By diverting some of the magnetic field force to the chassis of the vehicle, the MPV partially levitates on a cushion of magnetic field force, so the vehicle could ride as smooth as an Ion-Flash Propulsion jet. This version is especially useful on the rough roads of the tropics. It speeds through the island city of Lungi to a large Hoover Craft which is to take the MPV and its occupants to the city of Freetown.

Freetown is a small city captured by mountains against the West Atlantic coastline. From the Hoover Craft, the twinkling city lights nestle inside the green mountains like a glowing sky. This unique lighting effect is the result of a new lighting system called Ameri-Light developed by General Electric in the USA. By bombarding the air with ionized particles and using very high frequency lasers, photons are released by the ionized air at the frequency of visible light. The luminous glow can light up an entire city block on just one kilowatt of electricity without using any cables, light bulbs or infrastructures.

The city of Freetown is even more beautiful in daylight. The green grass and colorful flowers on the rolling mountains merge together in the distant skyline to form a scenic view that the discoverer of the land, Pedro Da Cintra, called The Land of Lions - Sierra Leo. The row of plush green mountains appears like massive lions lying in the thick of the surrounding forests. The sky is filled with small Ion-Flash propulsion jets going about the business. Small boats are everywhere, ferrying passengers from one part of the city to another. The coast is lined with several large industrial plants powered by the huge nuclear power plant visible in the Northern shores of the peninsula's waters. As the city grows closer, the appearance of polluted city slums signals their entry into the realities that still existing in the "second world". Despite all the development, there are still many problems that remain to be solved after Comet Resa. These are some of the growing pains of the second world. The craft floats into a beach driveway projecting a hundred meters into the Atlantic Ocean. The MPV's aboard the Hoover craft are all humming now, slowly aligning their motors to the new frequency of the electromagnetic fields beneath the roadways of Freetown. The Military Police on the MPV are the last to get off the Hoover Craft.

The fierce African sun is awakening, giving Hal and I a taste of the tropical heat and humidity of a post global warming climate. We can feel the slow build-up of heat in the MPV despite air-conditioning that is set at maximum. Although the roads are very narrow, the MPV speeds at more than 50 kilometers per hour through the crowded city. People walking on the roads come so close to the vehicle that we expect an accident. But this is the way it has always been in Freetown. The streets are full of little shops all peddling merchandise just a few feet away from the mayhem of fast traffic. In a few minutes we spring out of the city and into the sanctity of the suburbs. We drive for thirty minutes through the lesser commotion of Lower Freetown. As we approach the famous beach resorts of Lumley Beach town where there are no speed limits, the MPV clocks more than 165 kilometers per hour. The military police escorts stand on a single file along the aisle of the MPV, holding their loaded weapons in one hand while supporting their swaying bodies on the overhead hand rails with the other.

The lazukars in their hands could vaporize a human arm or leg in an instant with their powerful laser light. These weapons are made in Russia and widely used as a weapon of local wars in some third world countries. They are also used as a weapon of choice to control crowds at political uprisings. A 2027 pact that disallowed the use of military force between nations of Africa made armies prohibitive and so the military police are the closest substitute power-hungry Presidents could find. There is a total of five military policemen in the MPV. I am sitting beside Hal in the front row seats looking out the window. I am a little surprised by the military escort. I know that President Massakoi knew we are on a mission of peace. Did that justify using a military unit to escort us to the hotel?

As if the driver of the MPV could read my thoughts, his voice cracks over the intercom system,

"We will be at the Presidential palace shortly, Gentlemen. I hope you have enjoyed the ride."

"It has been very pleasant indeed," I reply hypocritically.

"But why the palace?" Hal asks, "I thought we are headed for the Hilton Hotel?"

"I don't know Sir, the President wants you to go to the palace as soon as you arrive," the man replies.

Why his palace? Were we not supposed to go to the Hilton in Lumley Beach?

The front gate of the palace is guarded by no less than six military guards. After checking some papers, the gate opens and the MPV drives through to the large parking lot. We are escorted to the front entrance of the palace into a huge room where we await further instructions. Except for a picture of the President and a couple of large sofas the waiting room is empty. A TV is tuned to NBBC broadcasting. It is a welcome site for us. We are apprehensive. Why were we not taken to the hotel to rest before seeing the President? My thoughts keep racing hoping to find an answer before we get into some serious trouble in this place. Hal looks relaxed he does not seem worried by all the Military presence. I comfort myself by thinking that he is an old hand at such trips. He has been to several third world countries before. I find myself wondering about what is so urgent that we must meet President Massakoi tonight? I begin to assess the situation in light of what I secretly know. Could the President have been informed of our true intentions? Do they know about the Green Peace Data? I do not think so. Even if they know about such details, they could not possibly understand the implications with so little data. There is no perceptible military significance to their mission, even with knowledge of a cubic pattern of global cooling. So, there could be no danger of a military threat.

An hour after we arrive, President Massakoi arrives in military attire followed by four other men. The CIA briefing was not complete. I recognize the three main body guards of the President, but there is one other man that I cannot recognize from the briefings. My thoughts race as the President enters the room with his four escorts/. He looks regal in his white Military suit full of gold medallions. Although I have seen his photo in a CIA briefings and Time magazine before, I never expected a six-foot, five-inch tall President with a massive beard!

"John Davis?" the President looks at us to identify the leader.

"Yes, Mr. President," I greet him extending a hand.

"This is Doctor Hal Shea," I introduce Hal to the President.

"But my friends, you must be exhausted after the long trip. Come follow me to my living room," his voice is calm and relaxed. He approaches me and embraces me with one hand while motioning for Hal to follow with the other. He leads us through a long corridor of the palace with his four companions following closely behind us. The corridor is studded with pictures of African Presidents and other notables of the United States of Africa. The President stops at the picture of the Nigerian President and points.

"This is the former President Mobako of Nigeria, my friends. He is the man that changed our continent," he remarks and then adds, "but my friends, I will not forget the good work of my American friends across the Atlantic. You see, I have their pictures hanging on another wall. Someday I will show you my collection of photographs of the most powerful people."

The grandeur of the President's living room rivals that of the White House. The furniture is late twentieth century Arabian but also features Native Sierra Leonean sculptures of "Nomoli" carvings. The parquet flooring is accented by a large Persian carpet that anchors a set of contemporary sofas and chairs. There is enough seating for two dozen people. The left and right walls are decorated with exquisite African art. A contemporary painting of Thoi Banjo the great East African child-prodigy discovered after the comets, hangs on the front wall of the room facing the entrance. Hal and I recognize the famous "Child Painting." The Thoi Banjo paintings are some of the most expensive paintings in the world. A large bar, made from pure red mahogany decorates the space beneath the paintings. A painting of Resa hangs on the adjacent wall of the room. It is the largest painting in the room. It shows the Ayatollah in a sitting posture with the symbol of Pharaohnism over his head. The President walks to the huge mahogany bar and asks us what we would like to drink.

"Please make yourselves comfortable my friends," he offers graciously.

"We will get to know each other well before you go to your quarters."

"I thought we had reserved the Hilton your Excellency?" I ask feeling it is the appropriate time.

"Oh yes, I forget." Turning to one of his men he commands in terse native words to send a message to cancel the reservation and to make sure that rooms are made ready for us in the Presidential palace. Turning to me he says,

"You men will stay here tonight. I am honored to be your host. You see, I gave my word to President Cooper that I will take good care of you, personally."

A sigh of relief exits my breadth, President Massakoi is a friendly host after all. Ever since the President of the United States paid a visit to Sierra Leone, he has grown to like Americans. President Cooper personally telephoned President Massakoi and asked him to look after us, and it is obvious that the African President has taken every step to reciprocate for the generous aid he receives each year from the American people. He is also eager to put the wasted mines in Sefadu to some good use.

After a few drinks and small conversation with the President and his men, we are escorted to our comfortable lodgings at the President's private residence. Throughout the night, the soft moaning of President Massakoi made it difficult to sleep. We later learned that the man was suffering from a rare sleep disorder brought on by an even rarer condition known as Bell's Syndrome.

CHAPTER 13
(The start of the journey)

A preoccupied President Massakoi stares impatiently at the robust Chinese Minister of Foreign Affairs. His eyes are blood-shot from lack of sleep. Having to cancel his scheduled October meeting in Beijing incenses him. It is a pity that certain developments in China would have to put the President's deal with Chun Shu Min on hold. A fat deposit in his Swiss account as a result of granting Min the privilege of mining uranium in his country, would have to wait. He rubs his grey beard with superiority as he speaks to the tele-conferencer and eyes the arsenal of elephant and rhino tusks framing the Minister. When his next visitors arrive for a meeting, he hurriedly turns the video panel out of their sight and tugs at his expensive white Rufus Suit as he rudely continues with his business conversation.

"Gentlemen, please make yourselves comfortable," his coarse voice announces as he chews on a stick.

We take turns shaking an extended left hand that reveals a wrist watch studded with diamonds. The President's right hand holds an old-fashioned telephone hand-piece, laden with gold. He turns away from us, ignoring us. The conversation appears to be heated, but neither I nor Hal could understand the Chinese dialect. When it is over, President Massakoi waves a small stick irreverently at Hal.

"What is it?" he asks curiously.

"A root that grows locally. It is great for the business," the President jests and laughs heartily with an unhealthy cough. Hal hesitates and takes the root in his hand. He sniffs its unfamiliar smell, then takes a cautious nibble.

"Wow!" he exclaims. "It sure is bitter!" he remarks, rejecting the root and placing it on the table.

"What is it?" I want to know.

"An African root called "Bitter Cola". It is a potent aphrodisiac. I use it every day. It is my Viagra," the President confides matter-of-factly as he salivates greedily on his stick.

"Yuk!" Hal jerks forward and replies, "I guess you have to acquire a taste for it!"

The President leans over giggling heartily, as if he has succeeded in playing a practical joke on us.

"You White Men should respect our jungles. This stick's taste may be unappealing but it has no side effects. It is much more practical than Viagra," Massakoi reveals haughtily.

I cackle uneasily and I begin to harbor suspicions about the President being either a sex maniac or impotent. Whatever his problem is I suppose it cannot make for stable leadership. I decide to take another nibble at the nasty root so as not to offend the President's boast.

"It tastes like quinine," I comment mildly, placing the root back gingerly on the table. The President walks over to a refrigerator and returns with a jug of a white milky liquid.

"Do you know this stuff?" he inquires arrogantly, indulging himself with such bacchanalian gulps of the substance that when he comes up for air he is out of breath.

"This is the best thing that happened to man," he decrees like a statesman uttering the creed of his country. "Here taste some!" Massakoi insists, thrusting the jug into my face.

I really do not appreciate the President's near threatening tones. I accept the jug half-heartedly not knowing whether my ability to withstand the odious liquid would offend or emasculate Massakoi. The President seems savage yet also vain and self-absorbed. How should I react? I pass the open neck of the jug under my nose. My nostrils open, then, instinctively, my blank expression breaks into a soulful grin.

"What is it?" I question the President as an interested student would a teacher.

"Something from God passed on to man," Massakoi admits dramatically as he sucks in the last remaining drops that are in his mouth.

"It is a wine made from the palm tree. It is very good for the liver," the President's eyes gleam like a drug addict after a euphoric fix as he pushes the jug to Hal.

"No thanks," Hal declines, smiling, "one sip is enough for me, sir."

"You speak Chinese?" I ask, being quick to change the subject.

"Well, yes, of course, I studied for my PHD in the Beijing School of Fine Arts," he replies straining to point at a certificate on the wall behind his desk. That is a surprise to me. I then realize that the influence of Chinese culture runs deep in this part of the world and that my failure to instantly recognize this reveals my weakness to the President. So, I am glad tactical games come to an end when an escort interrupts us to announce that breakfast is ready. The President motions us to follow the escort out of the room without joining us.

"I will see you when you return from Sefadu my friends," he promises, waving farewell at us. It is a dismissal, almost as if the President has better things to do. As we depart, we pass through a crowd of young girls waiting outside the plush Presidential offices. Judging from their short, stitched raffia skirts, the only clothing that hangs from their toned bodies, they appear to be local dancers. They look so familiar to me yet I cannot place them yet. I remember seeing a television program on the Discovery Channel about the famous Poro dancers of Africa.

"Good riddance," Hal mutters softly to me, referring to the President.

"What kind of man is that?" I whisper. "I cannot believe he heads this country. He is drinking right there in his office!"

"He is more interested in the China deal. That is why he wants us to leave."

"But that is rude and undiplomatic!"

"Well, John, you have to be prepared for that. I am sort of used to these missions. All that stuff back there is just a diversion to get us dis-interested so he can dismiss us".

"That is the bitterest piece of shit I ever put in my mouth," I swear with a mouthy expression. That is also the last time we see President Massakoi. Arrangements have already been made to transport us to the depths of the African jungles in Sefadu. Supplies for the trip have been arranged by an escort Military Police.

After breakfast, we go back to our room to pick up our luggage and leave for Sefadu. The escort respectfully waits outside the room as we enter. It strikes me as odd that our belongings are nowhere in view! Hal and I start to vigorously search the closets to no avail and then panic sets in.

"The scanners, the maps, shit,...we are in deep trouble now," Hal exclaims in a shaking voice.

"The files! The files! Did we put them in the coded cases?" I ask trying to remain calm and retrace our steps.

"I don't know John, I did not even know about them."

I think for a moment, then relief as I answer my question.

"Yes, they are in coded boxes."

"So, what happened? Do you think they know?" Hal asks with a very worried look. I know his mind is racing back to the strange behavior of the President. He seemed too "jokey" this morning. Perhaps it is an ironic twist to our final fate.

"The cases will self-destruct if they try to force them open," I explain to ease Hal's fears.

"What?"

"The files," I emphasize.

"Do you think they found out about the plan?" Hal wonders.

"I don't know," I retort softly afraid of being overheard.

"If they know we are dead. We will hang for spying or any other excuse they can find," Hal says looking at me desperately. This is it, I conclude. If we screw up now, nothing in the world can save us. God! Where is our stuff?

The big burly escort followed by a military policeman armed with a Russian Lazuka bursts into the room. I inwardly panic as the man approaches me. Hal turns toward the approaching military police. We will have to act quickly if we are to escape death. But the escort removes a set of keys from his pocket and shakes them in the air in front of me.

"The MPV is ready Sir, your luggage is already inside," he remarks apologetically to me.

I let out a sigh of relief. My God that was scary. The dangling keys echo a pleasant rhythm in my distilled brain. Hal smiles at the military guard while his adrenaline level drops. We board the MPV and search for our belongings. The cases and the bags are neatly packed at the back of the large MPV. They seem intact and the locks are undisturbed. I open one of the cases and inspect it. It is intact and I feel relieved. I nod at Hal and relax into a seat beside the driver. The trip to Sefadu is uneventful. Thankfully, the military escort who is driving the vehicle does little to entertain us. Other than brief explanations about the trip and places we must pass through, he remains pleasant enough but with few words to spare.

For miles, we see nothing but the thick of the jungle. Sometimes a small clearing would mark the entrance to a village or town along the way. Other than that, there are no signs of where we are headed. The sign-less roads are not maintained. The constant heavy rains and heavy sunshine had taken their toll on the asphalt surface of the small roads creating very large potholes. However, we are riding a special Presidential vehicle and the comfortable mag-shocks of the MPV are specially designed to glide over such rough roads. President Massakoi has the best the modern world could offer.

For ten hours we travel into the thick forests of Africa each of us seems to be absorbed in our thoughts. Unable to discuss any viable subject relating to the mission, we were mostly silent. As we enter a clearing in the mountainous rain forests of Sefadu, a heavy fog slows the MPV down to barely a crawl. This will delay the trip further. It means that we will arrive at dusk. The vehicle could take us as far as the city of Sefadu but no further. From there, we will have to force our way through a mile of thick jungle to get to our first targeted destination of Tubudu. Tomorrow we will have to travel on foot or find a gas-powered truck willing to take us to the remote village of Tubudu.

We arrive at Sefadu at dusk as I predicted. The driver heads for what he calls a famous local hotel in outskirts of the city. It is a dark area of town and the lone building stands out like a lantern in a desert. The hotel Mariana is a large building with well-appointed decorations built like a Spanish Courtyard. A large lantern is the lone guiding light in an arched entrance covered by Angelic statues. We enter the lobby where the driver asks us to wait. Abu, the military escort, introduces Chief Mobasso, a fat and tall man wearing an informal traditional African costume. He is the proprietor of the Hotel. We are ushered into a large room with two beds. The darkness of the room masks the chief's facial features. We place our luggage in the closets, this time making sure that no crucial or sensitive plans are left in the open. I stash the important papers and maps into a briefcase and put it beside me covered by a pillow.

The driver turns to me and Hal as he leaves the room and says,

"Gentlemen, please refresh yourself and I will meet you in the Lobby. We will have dinner in the city tonight and I will leave in the morning."

We take turns to shower and refresh themselves, then, with my brief case in my hand, Hal and I re-enter the Lobby and join Chief Mobasso and the driver. After further unnecessary introductions, the Chief orders a young woman to take care of our needs at all times. He informs Hal and I in perfect English that we will be considered emissaries of the President and that we will be allowed to work freely in the city of Sefadu.

Our lodgings are no different from a typical American home, modern day technology blended with traditional surroundings. We refresh ourselves with free cold beer and we waste no time in touring the famous city of diamonds in the MPV. The city of Sefadu has a population of about thirty thousand workers housed in agricultural and mining centers a few miles away from the city center. The arrival of Chinese aid in the early 2020 has changed the city from its ravaged state caused by the organized greed of Jawed's mining syndicate, to a thriving agricultural and mining center. In addition, the advent of Western aid and strict development policies imposed by the IMF has made the city an attractive place for farmers and miners alike. The once desolate wastelands created by mining machines now serve as productive farmland for growing crops such as rice, wheat and corn. The huge diesel machines of the past no longer pollute the air. Newer Ion Flash propulsion systems silently propel huge bulldozers and other excavation machines in some of the most productive uranium, gold and diamond mines in the world. Tall buildings rise into the skyline of the city powered by plenty of free electrical energy. The beauty of the shining city lights draws miners far from their homes to roam the city nightclubs, bars and discos. The streets are flooded with drug addicts sniffing and sucking on illegal drug sticks, a scene reminiscent of Japan and Hong Kong once captured by the drug trade after the comets. While the rest of the world adopted the new Paris Guidelines for Drug Enforcement and took steps to make it impossible for drugs to be sold in their cities, these "second world" cities are still plagued by the most dangerous of drugs in the market.

Back in the early 2020s, Paris became infested with "Halucigrams," a synthesized substance that produced hallucinations when the eyes are subjected to a particular frequency of light waves in the presence of "Cryogen 14". Cryogen 14 is a new substance formed by trapping "Naprene," then, a new and damaging drug in Buckminster Fullerene carbon nanotubes. The globular Buckminster Fullerene molecule stores large quantities of drugs in the body and serves as a slow, time-release capsule for administering drugs into the bloodstream, so that its victims suffer hallucinations for days. But despite the drug trade, Sefadu is still a very safe city at night.

Our tour of the city is pleasant and informative and also tiring. We sleep early and the hotel room is very comfortable indeed. The next morning, we wake up to the raw sounds of crowing cocks in the courtyard of the hotel, making music with the native women who gossip in the shade of the huge cotton trees that line the large compound. The women rhythmically sweep fallen leaves from the ground as they conversed.

Abu is already waiting for us under the shade of a large tree. He is killing time chatting with the young women while we clean up and dress in casual clothes for the hot weather. When we enter the courtyard, Abu escorts us to the Chief's meeting room in the backyard of the compound in a well-appointed tent in the middle of the yard, where Chief Mobasso waits us to join him in a local coffee ceremony.

The Chief motions us to sit down beside him on a floor mat. Then, he invites us to drink the steaming freshly brewed hot yellow lemon-grass tea. We sit on the floor and silently drink the delicious beverage. As the Chief drinks, he lifts the base of his cup with his right hand, while the left palm offers constant support. The cup rises up and down to what looks like a jungle beat as the Chief's heavy stomach rises and falls metrically with each sip. His whole-body titters from the joys of prosperity and contentment that the drinking ceremony symbolizes. We had been informed that we can only speak when spoken to and so no one speaks in the presence of Chief Mobasso. We clumsily follow his drinking patterns as we consume the beverage and eat some fried eggs. After the ceremony is over, the Chief begins conversing in flawless English.

"I must remind you that some of the areas around Sefadu city are still very primitive. I have been informed by my people that you have a mission in the interior districts. You must pay your respects to each Chief in each territory you visit. You cannot assume that I will protect you or that your President's authority will protect you. I must warn you that some of these Chiefs are extremely primitive and disapprove of the President and Western ways. How will you proceed?"

"We want to begin by investigating the occurrence of z-particles in the surrounding towns," I explain.

I wait patiently for Chief Mobasso's next question.

"Who is z-party cool?" the Chief mouths with some difficulty.

"I know no one by that name?" He admits, confused.

"These are fundamental particles we are studying for the presence of markers to establish the usefulness of the genetic program we must implement to improve farming," I reply almost believing my lie.

The Chief nods as if he understands.

"Even though we have some indications that the concentration is highest in Yamandu town, we will also search within a thirty-kilometer radius around that city. In addition, each town will also be visited and we will abide by the rules and instructions of the guide you provide us with, but we would like to start with Yamandu first," I state emphatically.

"You must be prepared to walk for several kilometers before coming to a town" the Chief warns us kindly.

"The roads to the neighboring towns to the city of Sefadu are still primitive," he adds.

Abu turns his gaze from the Chief to address us.

"MPV technology cannot operate on some of the roads that link the city to the small towns. In most cases you may have to use gas powered Fords."

"The only other alternative is to walk, you see," Mobasso reveals in a voice that is rich and strong like his jungle.

Presently, our escort suggests that we rent some Adwalks. These small, solar mopeds are capable of carrying a man and about two hundred kilograms of equipment for several miles on rough terrain. The Chief quickly orders his people to arrange for the Adwalks. Before we are to embark on our journey, we learn that it is customary for the Chief to seek the advice of his wife. The Chief summons for his wife and moments later, a beautiful white lady and young mulatto boy emerge.

"Maria, my wife, and my son Paul, will assist you in your plans my friends, as I must leave now to attend to some matters of importance," the Chief declares biding us a safe journey.

When Maria walks into the room, my heart leaps as if it has been nudged from its slumber. She looks so very familiar. I smile and greet her with a passion. It is as if I have known her all my life. Maria is in her late thirties. She has a well-tanned olive skin and long black silky hair, probably from the Middle East. She gracefully walks over to us and extends her hand in typically Western fashion. When she touched my hand, I feel a force like I have never felt before. It is as if a stream of electrons has flowed between our bodies to neutralize positrons that reside within my being. She is wearing a white sari-like gown and her wide-open shoulder straps reveal full breasts of perfect proportions. Gold necklaces and amulets laced with brilliant diamonds, pearls, opals and emeralds, weigh her hand. As she moves, her young son Paul follows closely behind. I have a strange feeling that I have met them before but cannot recall where. Paul is a young handsome man in his twenties, and I almost feel like he is my son, a spitting image of Nadia my daughter. Why am I having these strange feelings? I dismiss my feeling and shelve them for future reference.

"I have heard that you will be staying with us for a while," she begins.

"I must admit that these are not the most luxurious of surroundings when compared to America, but I will do my very best to assist you," she promises.

"Have you ever visited the USA?" Hal asks. I am hoping that she will answer in the affirmative, since that may explain why I believe I have met her before. I did not tell Hal about my feelings, but something inside me correlates to a symmetry of emotions I could not yet understand.

"No never. I have seen a lot of American movies, though," she states smiling.

"My son Paul will be your personal escort into the villages. He knows all the Chiefs around here and will assist you in any way he can. By the way, you look very much like Seth, my father!"

I do not know if it is coincidence or some inner spirit that makes us know certain things but when Maria says that I look like Seth, her father, something creepy comes to light in my being. It is nothing evil, just a nagging feeling that my soul is somehow connected to Seth, Paul and Maria. Sometime we suffer from insomnia and our minds cannot quite contemplate the obvious connections we have made in our lives and sometimes that leaves us with an empty emotional feeling that can only connect to a dreamlike state we call memories. After our episode of amnesia, Elaine and I could hardly recall if our lives had been orchestrated by either an ambitious scientific research program for the Bilderberg group, or whether we were actual victims of a virus brought on by the comets. Whatever it is, we lost time and history and we can connect all the dots and cross all the tee of our lives.

Young Paul has a smile that reminds me of Nadia. He smiles handsomely under the weight of his new responsibilities as he listens attentively as we describe our objectives.

"We will be measuring the soil nutrients and observing the distribution of minerals around the mining sites in order to determine which soils could be utilized to grow genetically engineered grain". I say smiling.

"But first we will have to "roll" the Adwalks to get to Yamandu" Paul explains. We cannot go there by truck. The excitement in his eyes tells me that he loves Adwalks. Later I learnt that he does not often get the opportunity to go see his Grandpa across the river on an Adwalk. He cannot afford to rent one for the whole day and at times he would use his father's Adwalk to go shopping at the local markets or make a short trip into the city.

"We will have to walk to a local store that rents Adwalks. If we want to get there by evening, we have to take off immediately and go to the store". He explains.

We arrive at a local rental store in the middle of the city were we rent three Adwalks. We load the Adwalks into the MPV and Abu drives us to the outskirts of the city where he drops us with our belongings to continue their journey into the jungle on our own. After a few lessons from Paul, we start to forge our way through the ancient foot walks of the dense hot forests with the high-pitched motors of the Adwalks and jungle mingling together to concoct an alien song. I feel rejuvenated by this new adventure. This is a dream come true, I rejoice. It is an adventure well worth the risks. Suppose we find the answer to a question that even the greatest scientists in the modern world could not dream of answering. Will the jungle give birth to a new vision of the world? Will we find the answers for Abu Simbel here?

My reflections turn to Paul, he so looks like my daughter Nadia, it is uncanny. I almost feel as if they are related. I have to connect the dots and cross my tees. Who is he, why do I feel so connected to him and Maria? Is he a missing component of my mind? It is obvious that he has never met me before, but he too thinks I look very much like his father, a father I have yet to meet.

The jungle holds no secrets for Paul. He was born and raised just a few miles from Yamandu Village and his great grandfather Bitter Cola is still the Chief of Yamandu village. His White grandfather "Oportho" he explains, is married to Bitter Cola's daughter and they live in the Yamandu village a few kilometers from where we are. Yamandu is under the full control of Paul's grandfather. This will mean that Paul can give us access to all we need to learn in Yamandu village. Paul assures us that the trail to Yamandu will be easy and that it holds no obstacles for our journey. I feel totally at ease with Paul, and I feel like he is feeling the same thing towards us, at least towards me. I am very curious about Paul and his mother. How did they get to become part of this African society and why do I feel so connected to them? I decide to quiz Paul about his family as we approach the banks of the Yamandu River. Perhaps this will help us strategize when we arrive in Yamandu.

"So, tell me Paul, how did your mother get to marry Chief Mobasso?" I inquire.

Smiling, Paul explains.

"A long time ago, my great grandfather "Bitter Cola" was a young Chief in Yamandu. It is said that Yamandu village was invaded by Kasila the devil of the Yamandu River. From what I have heard, Bitter Cola single handedly killed Kasila and stored him in a bottle. He buried the devil bottle deep within the Earth. By then, most of the young men of Yamandu were killed by Kasila except for a handful of villagers and a Whiteman. So, my great grandfather became the Chief of the village and my grandfather "Oportho", which means Whiteman, is the white man that survived Kasila. He married Bitter Cola's daughter and became an elder of the village. My mother is the first of three daughters of "Oportho." She was given to Chief Mobasso, my father, as a gift from one chief to another. So that is my story. Now, how about you? Tell me about your family and about America!"

As I begin to reciprocate with an anecdote from family history, a piercing cry interrupts me. Shivers run through my spine as the mysterious shriek repeats through the camouflage of the dense foliage. Paul motions to us to turn off our Adwalks as he steps off his Adwalk and puts his finger over his lips, signaling us to silence. He motions us to leave the Adwalk by the cleared path and follow him. Puzzled, we follow his lead as he quietly walks from the open foot path into the thick covering of the jungle. The rustling of the tree branches tells us that something very large is approaching at a fantastic pace. Just then, the massive head of an enormous anaconda emerges from behind us writhing into the footpath. The snake's tail towers over the trees above us and its massive head is well over a foot wide, turning from side to side as if searching for something. We watch the giant creature arch its body over the Adwalks to tower a full foot above the vehicles as it crosses the cleared path. It takes five minutes for the snake to pass by and submerge its entire body into the river. A pool of blood follows its trail and its legacy is a pile of broken tree trunks on the footpath.

"That is Ungoro, spirit of the dead warriors. It is a god worshiped by our Poro Society," Paul explains reverently. "I have only seen it twice in my lifetime."

"The biggest anaconda I have ever seen in my life!" Hal gasps.

"I cannot believe my eyes; did you see how big that thing's head is?" I exclaim.

"Shit, that was close!" Hal realizes.

Feeling that the danger is over, Paul moves back into the clearing of the footpath motioning to the frightened visitors to follow.

"What is Poro?" Hal poses his question still looking fearfully about.

"It is a magical society," Paul murmurs surreptitiously, his eyes twinkling.

"In my town we do not ask what Poro is, it is the dream of every school boy to be Poro. When you are Poro you have a great many rights in society. My great grandfather is head of the Poro. In fact, he is still a Poro Master," the boy reveals proudly.

"Is it the same as the Voodoo?" I inquire.

"No, no it is not. You see, in school we learn about voodoo too. That is a bad thing, but Poro is a good thing. Nobody dies in Poro, but people die in Voodoo."

Satisfied by the answer, I pursue the topic no further.

"How far before we get to Yamandu?" Hal asks.

"We have to cross the river first, then, it will be another five miles from here," Paul shouts, jumping back on the Adwalk.

I visualize crossing the river with a forty-foot snake in the water and I am becoming paralyzed with fright. But when I see no sign of apprehension on the face of the young boy, my tense muscles begin to relax. For Paul, the snake is merely part of the life of the jungle and as long as you do not disturb it, it will not harm you. Disturb, that is the key word. How do we know we are not disturbing the monster when we enter the river?

Soon, we arrive at a large clearing that merges with the bank of the river. Here, the river is narrower and we can see the bank on the opposite side. Even though it is still daylight it is getting dark as the sun slowly falls into the thick covering of the jungle forage.

"I hope we get there before the sun goes down, Paul," Hal admits, half laughing.

"We will be at my grandpa's house this evening, Mr. Hal," Paul reassures him.

"How do we get to cross?" I am wondering what means we are going to use to cross.

"There should be a ferry man," Paul explains, "and he should be here any time now. My grandfather controls this territory and the ferryman M'Gboko has never been late."

"Who is M'Gboko?" Hal asks.

"He is the Ferryman, the Crosser," Paul reveals. "He has a canoe for us and should be on the other side waiting."

"How do you know he is waiting for us?" I ask amazed at the certainty Paul projects.

"Well, Dr. Davis, he has been ordered to always sit there and wait for people like us, in case we come to Yamandu. He is given the responsibility of watching the river on behalf of the villages. In this way, he has sacrificed his life but it is for the good of all."

We stand in the center of a clearing by the huge river awaiting Paul to make his move. Paul walks over to a large tree that stands a few feet from the river bank. One of its massive roots stretches like an umbilical cord into the water, joining river and land. Paul instructs us to rest inside the womb of the hollowed-out tree stump. He retrieves a strange looking conch and raises it to his lips. His deep breath blows music towards the far bank of the river Yamandu and waits for an answer. After a while of waiting he expresses that he is astonished to hear no reply. Three beats pass again and again no response comes back. The cycle repeats itself over and over until finally the silence is ritualized.

CHAPTER 14
(Vatican City Scientific Laboratory)

After our meeting with the Pope, I was ordered to spend time with Nambu and Father Mulcahy to help them understand and get a personal feel for the J-N fields. We have been given unprecedented access to the most valuable scientific instruments in the Vatican laboratories. We have full access to the entire facility. This morning, we have come to the Vatican labs to start our studies of the J-N field. Nambu is standing in the middle of the laboratory holding a bag of small colored plastic spheres in his hands. I am sitting down on a far chair watching them, hoping that they will mind their business and not bother me with their questions. Father Mulcahy restlessly chews on a "Bongo," an artificial source of energy and protein that is equivalent to a day's meal. The lights from the neon bulbs on the ceiling dance impatiently on his shinny bald head.

"Where do we begin?" Nambu seems to be speaking to himself.

He walks over to the Holovision simulator and initiates a program. A three-dimensional hologram materializes in the corner of the room. He adjusts the holovision projection parameters on the simulator and steers the image across the room to the tabletop, aligning it with the surface of the table. The image, a mere set of interfering light beams between the Priest and Nambu, sits like a solid object on the table. The Priest eyes Nambu.

"Have you entered all the details?" he inquires.

"Yes, I have," Nambu replies, "but I cannot get the colors right."

"Have you tried compensating by increasing the intensity of the oscillator?"

"No, I haven't," Nambu replies surprised.

He adjusts the controls of the simulator. A brighter more defined cubic structure appears on the table. Nambu is stunned.

"Just tricks of the trade!" The Priest chuckles.

Nambu puts on a set of interactive gloves and hands the other set to the priest who continues eating.

"Where am I and why?" Nambu again mumbles as he pushes one of the spheres positioned on the left corner of the twenty-seven equally placed points of the structure. The sphere responds by shifting slightly and bouncing back to its original posture. The Priest studies the sphere as Nambu alters its image points.

"Adjust the rotational hinges," he suggests. "If you don't increase the frictional values during rotation, you could lose the planes completely," he advises.

Nambu walks over to the controls of the simulator and increases the frictional values of the image. He asks the priest to give the plane a nudge. The priest puts on a glove and holds one of the lower spheres, pulling it slightly upward. The plane of spheres on the right side of the image spins effortlessly about a diagonal of the cubic shape. I am watching intently hardly understanding what is going on. I know that they are looking at the symmetry of the J-N field model, but I am not well versed in the theory. I practice it instead.

Nambu walks back to the table and sits in front of the holostructure. He rotates the two parallel vertical planes to his left and right, forming the exact symmetry that the Pontiff had demonstrated earlier.

"Murray Gelman would be jumping in his grave," he announces jokingly.

"Well soon enough he will join us here in person if the Pope is right about the end," the priest assures.

"Look at this symmetry. Wow! How could we have missed that?" Nambu exclaims.

"Well, I am glad we did not publish the quark symmetry connection earlier, otherwise the whole world would know of it by now," the priest predicts as he finishes eating.

He walks to the sink to wash his hands.

"I think God has forbidden man from delving in such matters, but then, I also think we are fortunate to be on His side when we found the cubic field!" the priest concludes.

"I spoke to the Pontiff yesterday," the priest explains.

"He believes that during the time of Moses, the Egyptian warriors were killed by a tremendous power that shaped ancient history. The parting of the Red Sea, the plagues in Egypt, and so on. He has noted references to the power in all the historic texts like the Bible, the Koran, the Torah, the Kabala, and the Vishnu Texts of India. You see, they all talk of the great power of the gods in a crystal sphere. They all describe a tremendous force used to punish and reward man."

"It amazes me that the ancient Proro," Nambu starts.

"PORO!" a voice calls from the far end of the room.

A well-dressed man is standing by the door looking on. He enters and walks toward them.

"Your Holiness!" exclaims the priest. I look at the man and realize that it is Effion. I have been around the two brothers long enough to know the difference. I decide not to say a word about it and let the symmetry deception continue. The man stares at Nambu as if he has never seen anything like him. The two scientists stand in disbelief as the supposed Pontiff starts roughing up numbers that were on the holograph as if he was a mathematician.

"27," he points to the Holovision image.

"Aliquot sum thirteen, the first composite member of the 13-aliquot tree with sequential values, 27, 13, 1, and 0. The real Jordan algebra of self-adjunct 3 by 3 matrices of quaternions is 27-dimensional. Aliquot sum of a pair of odd discrete semi-primes 69 and 133. Collatz conjecture starting value of 27 requires 112 steps to reach unity, more than any smaller number. First base ten composite number not evenly divisible by any of its digits. Magic constant 27 for all prime reciprocal magic squares of the multiples of 1/7," Effion states, rubbing his hands vigorously as if to wipe off an invisible film layering them. The astounded scientists stare at the genius that is still spitting out mathematical magic at them.

"Your holiness, I could never have guessed that you are a mathematical genius!" Nambu states in admiration. Effion starts laughing hysterically. He comes closer to the priest and starts tugging at his robe. He tugs until the priest becomes uncomfortable.

"These are my brother's clothes," the man states almost crying. Then, he starts crying like a child.

"These are my brother's clothes, they are my brother's clothes, my brother's," he sobs as a distant memory creeps into his autistic brain. Nambu looks at the priest puzzled. A Swiss guard enters the room and heads straight to Effion. He grabs him by the arm and leads him out of the room without saying a word. Then, it dawned on the priest.

"His brother," he blurts out looking at me for confirmation.

"Whose brother?" Nambu asks. I did not respond. There is something telling me that this is to be kept quiet. He has never heard of the brother's problem, but there was a strange rumor that the Pontiff has a security double and he thought all along that Effion was just a security double. With their thoughts silent on the issue, they continue their reverie into the Field.

"Well, I am amazed that the Poro know of the cubic field and these complex scientific finds. That also reminds me of the Dogon tribe of West Africa. They had a complex calendar of the Sirius B system completely accurate to less than an arc second!" Nambu recounts.

"They even knew of the orbit of a companion star that is invisible to even the best telescopes of our day, how?" Father Mulcahy utters.

"Well, let's tie it all together," Nambu urges smiling.

"If the Pope is right, and I think he is, then all of history must have been a record of this cubic force. From the Josephus-Nambu Field equations, it is possible to deduce that the twenty-seven field cubic points could be energy absorbers of tremendous capability," the priest states nervously.

"Why so?" Nambu looked at the priest questioningly.

"Because of particle creation in the anti-gravity field around the cubic points. Take for instance what we found in Nevada. The Nevada crystal is green and absorbs heat. You see, if matter is created around the crystals, then energy must be absorbed from the surroundings to do so, thus the cooling. So, I presume that the ice ages were also a result of this powerful dynamic, which occasionally would rear its head like a sleeping virus. Then, great energy would be sucked up from the Earth at these sites, abruptly causing ice ages. Perhaps this force caused the extinction of the dinosaurs!"

"Bravo!" the old scientist teases him.

"What if the power of the cubic field is indeed alien? What if the Earth is powered by this alien field? What if the entire cosmos is powered by a similar field? What can we do to prevent people like Resa from getting hold of it?" the priest attests shaking his head.

"But Resa is dead, is he not?" Nambu remarks shocked.

"Well I don't know. The Pontiff thinks not. He is told that Resa was never found at the accident site in Abu Simbel, nor were Aziz or others. Perhaps he did not die! Perhaps it is all a game perpetuated by the likes of Mahmoud Aziz and others. After all, the Resa Temple is identical to this structure!" the priest says spinning the cubic walls on their diagonal axes.

"Perhaps it is our bosses, the Bilderberg group that is perpetuating this fallacy," Nambu states, "after all they fund the Vatican labs!"

I watch them study the structure in silence within the stark Vatican City laboratory. The plush gardens surrounding it provide a wonderful place to relax my mind in the cool shade of the inviting flowery trees that dot the summer landscape. I focus on the flowers below. I will not really learn anything new about the field. I decide that the only way I can contribute is by being silent so that they can mathematically describe my world. Nambu finally breaks the silence.

"To start with, we must identify the twenty-seven points around the world where the crystals are found. To do this we need the map to get the GPS coordinates. We need to contact the US Department of Defense liaison, one Professor John Davis. He works for the Bilderberg group also but from the US side, so why can't we just ask our contact Dr. Hal Shea for that information?" the Priest summarizes.

My heart leaps with some unexplained anticipation. The name John Davis is all too familiar to me. I cannot quite place it in space and time, but I have a strong emotional and dimensional connection to the name. I feel like they are actually taking about me! Space and time can be cheated if you know how to travel in other dimensions. Our future and our past are just fields that are controlled by the mind. The substances perform actions and the mind rules the actions. You can go back to the past and find all the actions that have occurred as real. You will see yourself moving about like a mechanical zombie of substance with no real mind. Your mind is always with you and so the past thing that looks like you are just a mechanical system that has stored actions your mind experienced as reality. When you wind the clock back or forward, the world you meet is just a mechanical world with no minds present to experience it. It is the where and when we are living in now that the mind uses to experience reality. So, in a sense, our machine bodies just store motions that are recorded as past and future and our minds experience this reality in the now. I am in the Now, and for some reason, I feel like I know John Davis as a past or future character.

"The US is unwilling to share the information until the Bilderberger group agrees to the final destination of the crystals when we capture them. They want them in Nevada," Nambu states.

"John's wife Elaine told the Senator that he is in Africa on a humanitarian mission on behalf of the US Government. Now, isn't that suspicious? A full-fledged Professor of Archaeology and Physics journeying to the jungles of West Africa on a humanitarian mission for the US Government?" the priest murmurs a question almost sarcastically.

Again, I feel like a zombie sitting here just enacting a past or future scene and that I do have a life as someone else. The name Elaine is just so very dear to me right now. Are they just calling out these names to prepare my psychological state of mind? I know how to cheat space and time and travel back and forth in emotional fields. If you imagine that you can either travel in a spacetime or travel through a shortcut through other dimensions then you will know what I mean. I listen to the two men talking as if they are talking about me, about my past and my future, but I cannot quite place myself in their minds, at least not yet.

"Why is the field showing up in Africa anyway? Do you think whoever has the crystals is using Africa as a testing dump?" Nambu asks.

"I do not know. All I know is that mathematically, if there is a field, then there is a force, and so somehow someone is manipulating a powerful force field that can destroy mankind".

"But", Nambu states, "If the field appears in any one place, then theoretically, there must be twenty-seven such fields, since the symmetry must have been broken from a vacuum state to generate even a single observed field."

"True, true. That means it must be everywhere."

The priest looks curiously at Nambu and then at me. He realizes that Nambu has hit the core of the problem they must solve. If the field exists anywhere, then there must be twenty-seven separated fields for that to happen! So, they are not dealing with a few fields, but the entire symmetry! If they can prove the existence of one field, then the others must exist. The field is like a glass sphere that is made up of twenty-seven unbreakable sections, and so, if the sphere breaks, then, there will be twenty-seven separate fragments.

"Where are the other twenty-six fields?" The Priest asks almost whispering. Nambu places his elbows on the table in front of the hologram and rests his chin on his hands, clasping his head as if it is about to weigh him down. He is thinking. He is known as the most astute mathematical genius of the century. It is said that he plays mathematics like Mozart plays music.

"Each field generates a spectrum that is unique. So, the color of each field is unique. When they all come together, you get a blinding white harmonious light that is pure. When they are separate, they will be distinct colors. So, if somehow, we can identify the color that is in the central field, we will be able to control the entire field by controlling that color," Nambu concluded confidently.

"The Pontiff's brother," the Priest pauses, "he said 27 has an Aliquot sum thirteen, the first composite member of the 13-aliquot tree with sequence (27,13,1,0)."

"That means that the self-adjunct 3 by 3 matrices of quaternion space are 27-dimensional and that is the key!" Nambu spurts out.

"The Collatz conjecture states that for a starting value of 27, 112 steps are needed to reach Unity, that means that there are 112 steps to gather the field correctly and unite them," the priest concludes.

"A genius," Nambu states, "He is a true genius!"

"The twenty-seven-dimensional spacetime is the string theory spacetime. A real manifestation of this spacetime in four dimensions will be exactly the J-N field. That means that the field could flux between four dimensional spacetime and twenty seven dimensional spacetime," the Priest explains. Nambu looks at the Priest and smiles.

"It is a hyper-craft, a powerful spacecraft that can be used to travel through spacetime to other worlds!" Nambu shouts looking at me. Does he know? I do not recall anyone telling them that I am a traveler.

"Which other worlds?" the Priest asks.

I decide to approach them and assist.

"The Spiritual world! It is a means of bringing the spiritual world to our spacetime world and a means of taking our souls through to the other side of the afterworld." I state glancing at the two men.

"It is the scientific explanation of the spiritual journey in the last days," the priest repeats.

"That does not compute!" Nambu shouts, "there is no way you can make me believe that the spiritual world is a science, it isn't".

"What is a field, Nambu? What is a field? An unseen "thing", a ghost, an essence that is everywhere?" The priest asks,

"What is a quantum field? Why do particles respond to the observer's mind? Why are there three quarks that make all particles? Why is there a trinity? Why is time imaginary and yet tangible? You keep refusing to believe in God but let me tell you that all you have done is labeled a new god called Nature. It is time you notice that all of creation is a science."

"No, it started with the Big Bang..," Nambu starts.

"Yes, the only creation event, Let there be light, the priest counters.

"Then, all particles and fields evolved from that break in symmetry, no God was involved."

"There is only one act of creation in science and in religion, the Big Bang, and let there be light. Every other so called "creation act" is actually an evolutionary tale. God said let the Earth bring forth life, meaning let the Earth evolve. Then, he said, let Us make man in our image, meaning, let us engineer man," the priest says.

I listen intently without joining in their debate. I know the priest is right and it will take time for them to come to the same conclusion. The last thing I really need is to be a Guinea pig for these two scientists.

"So, what you are telling me is that evolution is an acceptable fact?" Nambu asks.

"Yes," the Pontiff answers, "the church has no problem with evolution. There was only one creation act and that is "let there be light". In no other instance did God use the command "let there be". In every other command He asked that the waters bring froth life just as science describes that we came from the sea. Throughout the so called seven days of creation, He never used the creating command but once, just as the Big Bang is the only creation act in science! He would reiterate that Earth should bring forth living things, the firmament separate and so on, but no other creation act where he explicitly commands an event to occur without natural causes."

"Interesting, but how is the J-N field a spiritual vehicle?"

"If we can pass back and forth into higher dimensions, we can essentially travel through time and other dimensions and vanish and appear as we wish on this earth, like angels and demons. We can go to other worlds, call them spiritual, heaven, hell and so on. But yes, we can essentially transcend our restrictive four-dimensional physical world."

I decide to interject and perhaps educate them a little bit on the higher dimensions. This might be my chance of enter into an understanding of the J-N field from the very scientists who created the theory.

"So, what if since ancient times, God has used this spiritual vehicle, in fact I will call it "advanced spacetime vehicle", to bring us I say spiritual guides, you say technological guides, I say Prophets, you will say Scientists, I will say demons, you will say evil Hitler, and so on. You can see why the demons and the angels would want to get this vehicle and use it to go back and forth. You can see why God, angels and demons, are far more advanced and powerful multi-dimensional spirits that can come and go at will and control our realm of four-dimensions," I explain.

"Then, Resa has just such a vehicle, and where are the rest?" the Priest asks.

"Well, technically, each crystal can manipulate one of twenty-seven dimensions, so in a sense Resa will only have a vehicle capable of transcending four-dimensional space through a fifth," Nambu states.

"If these are the vehicles that must take us through Judgment-day into the spiritual world then controlling them means controlling the destiny of mankind and heaven and earth. It is the battle of Armageddon! That is why we must try to find them," I said emphatically.

We spend the rest of the day studying the symmetry of the cubic field. Father Mulcahy suggests locating Dr. John Davis's exact location through his personal pager system by using the Pentagon satellite tracking system. The scientists decide to turn to Senator James Owen, the friend of Pope Paul the Seventh, for help. I jump out of my chair in confusion. I have not heard of Senator Owens for many, many, years. He used to be my very close friend. In fact, he was my class mate in St. Edward's Secondary school in England. My mind is racing with confusion. Am I experiencing a Memorin journey again? Is this all real, or am I actually experiencing this reality? John Davis, Elaine, Senator Owens, Africa! It is all so very, very familiar.

The Pontiff makes arrangements for the Vatican to locate Senator Owens and ask him to help search for the information about John Davis. By the end of the day, John Davis has been pinpointed in Yamandu by the Pentagon Satellite Search System. Father Mulcahy quickly searches the Vatican files to see if the church has facilities he might use in Sierra Leone. The files indicate the existence of several Catholic Churches there. He contacts a Father Gregory O'Toole at the main Archdiocese of Freetown, the capital of Sierra Leone. Father O'Toole is given the responsibility of assisting Father Mulcahy in Sierra Leone. He will go to Africa to meet father O'Toole. A personal call from the Pope to Father O'Toole ensures that Father Mulcahy will be well cared for.

Father Mulcahy thinks it best for Nambu to stay at the Vatican University lab to further study the cubic symmetry while he pursues John Davis in the hostile jungles of Africa. As for myself, I have no say in what I do. All I know is that they must locate the Professor and I have to stay confined in the Vatican. The priest needs the break from his daily boring life in Rome. Early the next day, the Senator arranges for Father Mulcahy to be picked up by a private, unmarked Ion-Flash jet for the trip to Sierra Leone, while a special envoy arranges for a travel card to be sent to the African-American embassy. This would allow the priest to travel freely throughout the United African States.

That night, I decided to do what I must do to help the Vatican and myself. It is obvious that the crystals they are talking about are the same crystal that I saw in the mines in Africa. After the explosion in Africa, me and my colleagues were taken to a local hospital in Africa and then transported to the USA for serious medical attention. There were shatters of crystals lodged all over my body. I was operated upon and recovered slowly over a period of four years. I went on to study astronomy and astronautics while Militant and the rest of the Brazilian crew returned to Brazil. I became an Astronaut and entered into the space program with NASA. It was not until my mission in 2012 that it all started to come back and hunt me. My visions started and the strange story of the Scroll started to rear its head as a reality. I never realized that the Scroll was a powerful thing until that fateful night when Militant arrived with it. Ever since that day my life has changed. I carry the remnants of the Ancient Eye of M'Dulu in my body, the same power that killed Militant and the rest of the Brazilian mining crew. The same power that now afflicts me in a very uncanny way. I have the remnants of a fantastic alien technology embedded deep inside my body, deep inside my physical being and also inside my very soul. At least my story of the scroll is not just a dream, it is real. My travels are real and I am indeed a real astral traveler. My life in Africa had been altered by my experience with the Poro and the Eye of M'Dulu. It had been abruptly brought to an end during an accident in the mines. The accident must have made my psyche what I am today. It opened the door to my dimensional travels and now I am sort of lost in realities that alternate between many worlds. I decide to follow the priest, at least in spirit to Africa. I decide to channel myself so I can see what is happening to him. I do not know if this is part of the Malaki books I am writing any more. I do not know if I am simply making my mind narrate a story of the Scroll about Father Mulcahy and Nambu. I find myself biting my teeth again and slowly sucking my tongue as if it were a piece of candy. I am flying through dimensional gateways of my subconscious. It takes a few hours to find the priest entering an IFP.

In the Ion-Flash Propulsion jet to Africa, Father Mulcahy chooses to sit in an isolated corner of the passenger cabin, so he could be alone with his thoughts. There are three other passengers on the plane, a woman with dark blonde hair, and two men who he had been told were on a special mission for the Senator.

John Viser, the Hungarian born captain of the IFP introduces himself to his crew and then proceeds to check the spin-disc-propulsion unit and other instruments on the craft. He starts the engines and brings them to a slow stable hum, preparing the jet for takeoff.

"We are blessed today!" the Captain announces looking at Father Mulcahy, "I see we have a priest in our midst," he relays to the rest of the passengers in a mock Southern accent as he strolls up to the control cabin of the craft. Father Mulcahy nods back smiling. After the plane takes off the priest immerses himself in reading a book written by Dr. Abdul Aziz about Abu Simbel. Abdul Aziz allegedly died under strange circumstances at the Abu Simbel excavation sites. I could hear two men in front of the priest chatting about a financial deal going down in Botswana, South Africa.

A couple of hours later, the Captain set the jet on auto pilot and leaves his cockpit to offer his passengers a trolley of beverages and snacks. The two men order Martinis on the rocks. The lady takes Bourbon on the rocks. Father Mulcahy chooses carbonated water.

Night slowly filters into the craft and the soft hum of its engines lull the priest. He does a little and suddenly he is awakened by a loud bang coming from the front of the craft. The blonde woman is standing up holding a plastic bag over her hand holding a Lazukar. She is aiming the weapon at the two men in front of the priest. The two men jump from their seats startled.

"Move back", she orders pointing the weapon at them. Father Mulcahy awakes and looks around also startled. He springs up and motions to the woman not to shoot, but she points the weapon at him and orders him to quietly sit down. The Captain hears the commotion and goes to the back of the craft back to see what is going on. The female abductor releases a second weapon from between her breasts and aims it at the Ion-Flash propulsion engine.

"Do not move or I will blow up the entire craft," she threatens, motioning to the Captain and two other passengers to move to the back of the craft and join the priest.

"If anyone makes a false move, we are all dead," she threatens sweeping the guns around the craft.

"Lady, please don't .." The priest starts.

"Shut the fuck-up or I will blow your head apart!" she shouts at the priest.

"I need to set the auto pilot" the Captain explains as the other men join the priest. The captain hopes to somehow get a distress signal out to the closest airport once he is inside the cockpit. They were flying over North Africa now and would be close to Morocco soon. The woman backs her way into the cabin entrance. She examines the red auto pilot alert and sees it blinking.

"We are already on auto pilot!" she shouts.

"What do you want?" the Captain beseeches nervously.

"I want you to take us to Maron airport in Morocco right now. Reset the auto pilot or I kill us all!" she shrieks.

 Maron airport is a small private airport in Marrakesh, Morocco. The pilot has been there many times before. The choice is perfect, since there are no security personnel in Moron airport. It is a very private airport used by affluent Moroccans. When the pilot assures her, he would comply with her requests, she sets the Lazukar on stun and points it at the priest. He shuts his eyes and hurriedly turns his thoughts towards God but then she slowly turns her weapon onto one of the men sitting next to the Captain and fires at him. The priest's body twitches with the anxious amazement that he is still alive. His unsteady hand makes the sign of the cross as the other passenger's limp body slumps to the floor. The female terrorist informs the rest of her hostages that the man had simply fainted due to Lazukar's energy overheating his blood.

"That is just an example!" she boasts, "next time I will kill, so make sure you do as you are told."

The priest makes the sign of the cross again, and the Captain, now visibly frightened, scampers to the control cabin to change the vector coordinates for the flight to Maron airport. A few minutes later the plane plunges into a steep descent and finally lands in an airfield that is completely deserted. The terrorist orders the priest to open the door of the craft and two men dressed in military attire enter the craft, walk over to the priest and ask him to vacate the craft. The female then points the Lazuka at the two men beside the Captain. The laser beams sear their bodies instantly and the stench of burned flesh overpowers the jet. Moments later, the priest and the Captain are thrown down onto the hard-concrete surface of the airfield. They are ordered to lay flat on their bellies with their nostrils touching the hot pavement of the runway. The two military men bind and blindfold Father Mulcahy and John Viser. Then, the prisoners are escorted to a parked MPV a few meters from the craft and shoved into the back seat with one of the military men. The murderous woman sits in the front. As the MPV speeds from the airport, the priest wonders what has gone wrong. Why are they being kidnaped? Could these assailants know what the church is up to?

No words are spoken, but somehow the two prisoners share the same horrible reveries of perhaps never seeing their families again. Captain John Viser's eyes fill with tears. Is this my fate? He asks God. The Captain is not a religious man but if this end is the will of God, he will accept it.

"Oh God" he pleads, "let me die fast and easy." He tries to make sense of his impending doom, remembering all the good he has done for the world. John Viser has helped several students pay portions of their college tuition and believes he has been a decent family man. But ultimately, he consoles himself with the fact that his death would be blessed by the company of a priest.

After an eternity of tumultuous introspection for both captives, the MPV comes to a halt. One of the military men opens a window and speaks in an Arabian dialect to someone outside. The exchange is brief. As they are pulled out of the car, Father Mulcahy realizes that the two Arabs and three Israelis are new strangers. Their captors must be terrorists interested in the church's mission, Father Mulcahy hypothesizes. But why? What have they found out about the cube?

As they walk, the prisoners feel soft mud cushioning their weary feet and guess correctly that they were in a swamp. One of the captors cradles the priest into his arms as if he is a small boy, while another lifts the Captain. The sound of running water tells the priest that they are in a shallow river, or possibly the edge of the sea. Soon, Father Mulcahy is rocking back and forth in a small water craft. The kidnaping has been well orchestrated.

One of the Arab terrorists removes the blind fold from the priest. Father Mulcahy struggles to make out the faces of his captors in the moonless night. As his eyes adjust to the darkness he sees the woman searching for something on the horizon, and notices the Captain is missing. Moments later, the cries of his drowning comrade ripple across the surface of the water. Then, some thrashing and the sounds of ripping flesh muffle the cries of the Captain as an alligator eats him alive.

"Father Mulcahy, my friend," utters a familiar voice.

"I apologize for putting you through this, but we had to do this for the good of our people," someone explains.

Father Mulcahy holds his hand to his mouth in disbelief, he recognizes the man.

"But what is going on Reverend Canaan?" the priest blurts incredulously.

"Why am I here? And why are you a part of this scheme?"

The Reverend looks at Father Mulcahy and laughs out loud.

"In the days of Abraham, I would have called you a great man to give up the pleasures of a woman for the priesthood. But today, I call people like you, fools."

The blonde woman is nestled against the Reverend. He kisses her lips, placing his hands on her plump, firm breasts.

"Enjoy the site of a woman Father, for you are about to die!" Reverend Canaan informs the priest in a sarcastic tone. He pushes the blonde hatefully from himself as if he considers her a whore. She falls on the priest and then thrust her open breasts upon his face. There is nothing the priest could do but pray as she bears herself down upon him again and again while the terrorists huddle around them to glower and celebrate.

CHAPTER 15
(Morocco)

The priest's boat journeys to a narrow canal. It carries them beneath the greeting portals of an enormous stone mansion that graces the coast a few hundred meters inland. Father Mulcahy, still numb from last night's experience of the flesh, gapes at four gigantic columns which support the huge tower the boat now floats under.

His female assailant is still sleeping naked in the middle of the boat, her rear-end humped towards him in an irreverent posture. Canaan slaps her butt to wake her up, and startled, she looks around fearfully and jumps into her clothes. The boat passes a smiling woman dressed in Egyptian trappings standing by a concrete dock and a large closed gate blocking the canal. She opens the massive gates leading to the base of a large tower. When they disembark, she leads them into a reception room at the base of the tower and wipes her feet on a Persian doormat before stepping onto the polished marble floor.

After freeing the priest from his handcuffs, Canaan dismisses the military escorts and pats his female partner on the butt as if she is a mule, motioning her to follow. They enter a cavernous room built from massive blocks of stone. Each stone holds masterfully painted scenes depicting ancient Egyptian history. Scattered about the room are hieroglyphic tablets that narrate each scene.

The Reverend walks to the end of the room, dragging the priest by his arm. They stop at a large square reservoir that replaces the floor of the room. A strange bubbling blue liquid gurgles like a running brook. The curious priest looks down into the turbulent reservoir and wonders whether it is boiling. Canaan orders him into it and he resist.

"You will not be hurt Father that is if you don't resist," Canaan threatens him with a knife. Then, he put his right foot into the reservoir to demonstrate that the water is not scalding. He orders the priest to do the same.

"You must clean yourself before you enter the holy chamber to see my Master," Reverend Canaan commands.

"What is this all about? Who is your master?" the priest begs.

"I cannot tell you anything at the moment you fool, but if you don't wash yourself now, you will die when you see him," Canaan shouts. He thrusts the struggling priest into the water and plunges in after him. Father Mulcahy gasps as the freezing cold liquid numbs him. He wants to swim, but the initial shock has left him breathless. As he begins to sink, the strange liquid invades every orifice of his body, permeating his organs, cells, and even spirit. The priest's eyes bulge not from drowning but from a tunnel of bright light that abruptly appears to carry his spirit from his body. The priest gazes down upon his own lifeless body falling upward and in the opposite direction of his spirit. Then, he realizes his soul is descending into the depths of the reservoir towards an underground river of yellow light. Is he astral projecting to hell, he wonders?

At that moment I realize that I am very vulnerable to the effect of the fields around me. These are the J-N fields that open up my traveling ability. These are the fields that can expose my being to whoever or whatever lies beneath this lake of liquids. If I enter the lake, my spiritual side would be inverted and I will be exposed to the spacetime continuum. I decide to wait and see until the two men enter and experience their spiritual selves. I cannot risk being exposed in their dimensions.

I know what the priest is experiencing. His life is flashing before his eyes as he weathers the sensation of a near death episode and enter into the timeless realm of the inner dimensions. Just as they vanish from my view, I enter the liquid and follow. Far below, I can see that the bizarre experience has considerably weakened Father Mulcahy as his limp body seems to have no control of his limbs. I know he is overcome with sadness and fear, with emotions of all sorts that have built up as gateways waiting for his spacetime journey to end. I notice Canaan falling into the bright yellow abyss beside the priest. I can feel the absence of space and time and I know the sensation well. The priest is also feeling the absence of space and time and I know he is feeling like he has been falling for centuries. Then, abruptly I sense that their spiritual bodies are lying on solid ground in a large room beneath the reservoir. Gravity has reversed in the churning brew of yellow liquid as they helplessly watch fire droplets fall upward into the levitating timeless lake of the mind. I hoover just above the bottom of the lake. The J-N field is acting like a huge bubble beneath the lake holding the strange liquid at bay as if it is defying gravity. I have seen this before as a young man in the Mines. Like a glass dome beneath my body, I watch as the spiritual realm unfolds beneath.

Father Mulcahy lies motionless on the floor, gaping at a god-like Pharaoh whose thunderous voice commands in a foreign tongue.

"Soi toi doo Man-aka!" the deity echoes as the servants around him tremble with trepidation. Canaan is crawling on the floor toward his dreadful master. The priest manages to squint at the horrific scene before him.

"I have brought you the servant of Jesus, Oh holy Master of the Underworld, Oh master of Mars, King of this Earth and ruler of the Ancient Worlds, Oh Master Zoroaster," Canaan chants as he dares not gaze upon his god.

Zoroaster leaps arrogantly from his throne. His massive being cast shadows over the priest.

"Bring him here!" the Pharaoh commands in English.

A dozen Egyptian-like warriors run toward the priest and raise the priest from the floor. They carry him and throw him at the feet of the god of the underworld. The priest instinctively does not raise his head and his gaze anchors behind the god to focus on the feet of two immortal beings whose feet are shackled with strands of gold. Then, the priest peers at several massive yellow crystals levitating between the three gods. The crystals unite the beings with yellow pulsating rays of light. All three pulsating metallic gods move simultaneously like a trinity. It reminds me of the dream of Nebuchadnezzar - a figure made of metallic parts representing the past, present and future kingdoms. The unholy bodies levitate around the crystals rotating until a different god's face floats before Father Mulcahy.

"Rise up, feeble human soul. Welcome to the realm of the dead!" the figure utters.

The priest realizes he is in a living hell.

"Oh my God, it is Resa!" he cries. I look at the beings in front of the priest and the familiar face of Resa is standing there quite alive as if he never died!

"Yes, mortal, you know me as Mabus Ali Resa, but here in the Underworld, I am known as Ramses Son-King of the Earth. Are you not afraid of what you have done? Behold! The power of the crystals is with us. For look! Oh, mortal soul, the great power that you seek is forbidden to Mankind. You have been offered the use of the atom of matter, but we cannot allow you to take the power of the atom of creation. That is ours."

The three figures point into the corners of the room. Finally, the priest raises his head to the replica of the Resa Temple! Then I see the crystal spheres! Just as the three deities have simultaneously levitated, I can see the entire structure of the temple all at once. Twenty-five spheres sitting in a neat cubic pattern. My apprehension twists me into intense and horrific speculations. They already have twenty-five crystals in their possession, and two more will result in the greatest fusion of energy ever known to the universe! It may be too late.

"Did you not know that we are the trinity of kings that own your world?" Resa challenges the priest as he shudders to think that the unholy has invaded his mind.

"We have come back to take control from you and your puny God of the living souls on the Earth above. Your church has blasphemed against us. You have taught of the trinity in the light of your fake gods. I Zoroaster, I Aman Ra, I Ramses have waited for thousands of years to make you aware of the truth. Instead you molded the truth for your own benefit. You polluted our Garden with your science, and with your stupidity you forgot the might of our Master, Lucifer. We left you signs to see into the past by building pyramids. We left you the lion of time. We left you clues so you could understand we have journeyed from faraway worlds through the cold emptiness of the cosmos to find your world. For here, your God hid the crystal seeds of life, the power of creation! But instead, you have broken into our pyramids and soiled our holy sleeping chambers with your foul hands."

The perplexed priest does not respond.

"We know that you wish to seek our crystals. Where are the remaining two?" The devil's servant demands.

Ramses eyes blaze with anger at the mute priest whose tormented soul desperately prays to God.

"The Lord is my Shepherd and I shall not want . . . noooo," the priest cries inwardly.

The trinity of crystals light up and a stream of anti-particles jet across the room, ricochet off the walls, and painfully attack the priest's chest.

"Please Lord free my soul" the priest begs out loudly.

Zoroaster, Ramses and Aman Ra, three gods of the ancient world. Worshiped so long ago and yet alive and well today! The Pope is right. This is not death, it is a battle for souls, and it has begun!

A million souls witness the spectacle from within the cubic temple. Father Mulcahy is not about to give up. From the depths of his mind and soul, he calls out to the living God to save him. He prays to Jesus and Mary, to the Saints and all that is holy. Suddenly, he feels his heart beating fast. He places his hands on his pulsating heart to make sure he still lives. I feel my being propelled at fantastic speed through the strange lake. It feels like an exorcism is taking place. I am confused by what happens next. I find myself holding my chest and standing a few feet from the priest and captain Viser.

"Father" a soothing voice urges, "Father Mulcahy," the voice repeats, "Wake up, we have arrived."

The priest wakes up from his astral slumber to find himself panting and frantically gripping the hand of Captain Viser. He looks at the front of the craft. Everyone has de-planed except for the blonde. I think that she thrust her breasts towards him as she disembarks with a wry smile but then I realize it had all been a dream for the priest. I know what astral travelling is and I know it was as real as it gets. I follow as the priest disembarks to the sound of earthy street-traders peddling their wares near the runway of the craft is comforting. The priest clenches his right palm to confirm the sensation of being awake and is completely baffled to find a small silver rosary given to him five years ago by his late father.

I clench my teeth again and look at the rosary again. It is a bunch of scarabs tied together into a neat chain to form a rosary. I have seen it before. After the Space station incident in 2012, I became very ill and was bed ridden for a long time. During my alternate personality episodes as an Astronaut, an exorcism was performed on me in Nevada. The priest had worn the same rosary during the exorcism. It was him, it was Father Mulcahy! I find myself clenching my teeth and sucking on my tongue. I must exit now! I feel a strong need to go back to Diana! As I exit the alternate reality, I can hear Father Mulcahy swearing that it had been lost for a long time. Yes, I now know it was lost a long time ago because each of those beetles had been placed in a pendant to protect the Bilderberg Group in Nevada!

CHAPTER 16
(August 20, 1984)

After mastering the tricks of his trade, I have quickly become an invaluable member of the mining crew. My first major assignment is to build a new mining complex for the diamond processing plant that is planned for the village of Yamandu, located about twenty-five miles from the city of Sefadu. My job is challenging and satisfying but tonight I cannot sleep. The camp is not producing enough diamonds to make a profit yet management keeps justifying its existence. Even the labor force is tight-lipped about the situation.

My suspicions are fully aroused and instead of sleeping nights, I spend many hours tossing and turning and trying to come up with answers. Instead of struggling with my insomnia, this evening I decide to talk a long walk. Looking out of the back-garden door of the camp house, I notice that Militant's door open since I can see the blazing lights lighting up the yard from his room. Thinking that Militant must have left the lights on by accident, I walk into the bedroom to turn them off. I find the Brazilian sitting by the window staring into the black void of the African night. He is leaning on the window sill, his neck tilted to one side as if the weight of his head is too great for his neck. There is talk in the mining camp that underneath his thick hair is an abnormally large brain after all he is a brilliant surveyor and mathematician. Militant sometimes speaks in "mathematical terms" about even the simplest topic. To the poor uneducated labor miners, he is a genius, like Einstein.

A slight wind protesting Militant's silence blows his thick long greying black hair behind his face. The women admire his tanned muscles and concerned weather-beaten expression. Militant is something of an enigma. He is a person with the devotion of a priest but a playboy who would be privy only to the confessions of a woman in his bedroom.

I consider the paradox of the man and almost dismiss the scene before me as normal. But I know that the Brazilian's bedroom is a storehouse of precious diamonds and gold samples and it is indeed strange that the door to his bedroom is wide open exposing the huge safe beside his bed. As I enter the bedroom, Militant turns around agilely with the alertness of a man practiced to the dangerous life of diamond mining in Africa and Brazil. His face reveals nothing. In the mines, expressions can be misleading.

"It's very late Mike. Is your night off, you should be sleeping no?" He inquires, in broken English.

"I can't sleep Militant. I was working on the Plant 4 project plans. They must be completed by noon tomorrow," I lie hoping the answer will quell his curiosity. The Brazilian miner does not respond and busies himself by waving a flashlight back and forth into the dark yard.

"Plant 4 could not be completed by tomorrow and you know it!" His weary voice cracks after a long silence.

"We must complete the prospecting reports in time for the bankers. But Abu Jamil knows he is asking for the impossible, why can't we just tell him that?" Militant reasons uneasily.

"You know Abu Jamil!" I answer, "He will be mad at you. I can't allow that. Anyway, it is only a two-hour show-off for the banks and we can impress them with the working plant and hope that the mud will take care of the rest of the site."

Militant's face turns white as he faces me. He smiles for a brief moment then abruptly contours his face into ambiguity.

"Bankers love the mud Mike, they come here to see that filthy mud. They need to see stock piles of mud," he professes. "It is the mud that they care about. The plant is nothing Mike. Me you everybody else is nothing. Only that filthy mud matters, get it?"

I can tell that I have miscalculated the extent of our predicament. It is Militant's job to show that the mud samples are healthy so the bankers will remain interested in financing the mining operation.

"I know Militant, but we must not pretend that there is anything precious in the Yamandu mud. You, I, and everybody else in this mine know that the mud is useless. Over two million cubic meters already processed have yielded little more than 18 carats of diamonds," I dare to challenge.

Militant turns his head and looks out the window.

"The banks have invested well over one hundred million dollars in Plant 4 and now it is time for us to demonstrate a good return on investment. My group must demonstrate that the mining sites have enough diamonds to justify investments in Plant 4," he attests in a low voice.

"Prospecting for diamonds is not an easy job. Perhaps the strain of work is getting to you, Militant," I confront his shining visionary, ebony eyes.

"Mike," Militant rebounds with conviction, "I am not a superstitious person. But ever since that Haman man shook my hand, I have been inexplicably attracted to him. This place is weird. I keep telling them that there are no diamonds, yet they keep supplying the money and hiring more people," he finally admits.

"So why worry about it! You get paid don't you?" I respond sympathetically.

"In Brazil the Voodoo has a strange power to influence people. I am afraid. Every night ever since meeting that man, I have had strange, lifelike dreams . . . dreams of me worshiping people like ancient Egyptian Pharaohs. Haman would come at night and take me away to a faraway place. I don't know why, but I think that I really do go in the night and worship this man. I think I am going crazy."

"The dream is a psychological take on reality. You are following the lead of Harman into a world of treasure, Egypt!" I retort to try to ease his worries. I am really not surprised by Militant's sudden confession. For months I have listened to rumors of strange people visiting the mining site and I wonder if there is a connection between Haman and the visits, but my attempts to get any more information from Militant and others had been useless. Perhaps Militant is now opening up to me.

"Try to get some sleep Militant," I advise.

"Tomorrow we have a lot of work to do before the show. Perhaps you should see the doctor."

I open the back door and walks to a small bench in the middle of the garden. The night is not as cold as usual. I linger for some fresh air and begin to relax under the clear, night sky. I could spend hours just looking at the far away stars, contemplating the greatness of the Universe and what God had intended. But the thoughts of Haman, Militant and the bizarre happenings at the camp site unsettle me. Militant had described Haman as a tall, broad shouldered and quiet man who occasionally mumbles. Militant had only met him briefly, but apparently that meeting had profoundly influenced him. Ever since that meeting, Militant's behavior had become abnormal. He would wake up at night in a cold sweat from constant nightmares and mutter words of an unknown language. One morning, Militant told me in a guarded whisper that Bitter Cola, a local worker, is planning to kill him because he believes he is helping strangers find the great sacred diamond called the "Eye of M'Dulu."

Bitter Cola, is a middle-aged man with over twenty years of traditional mining experience. The word "Bitter Cola" is the name of a bitter tasting nut found in the forests of West Africa and other tropical forests. The nut is said to have special aphrodisiac qualities which Bitter Cola can attest to. The African has been promoted to the position of Chief of Labor of the Yamandu site, not by hard work but by constantly reminding everyone of his seniority and that promotion is a must for a sixty-year-old man. But Bitter Cola's appetite for power is not satiated by this promotion. When David appoints Militant foreman of the mining site, Bitter Cola declares Militant his enemy. This furthers the animosities between the African and Brazilian workers. It is Bitter Cola who makes all the superstitious miners and laborers aware of the terrible "Kirfy" by the riverside. "Kirfy" is the local name for devil or evil spirit that the locals also known as "Kassila." In fact, according to Bitter Cola, the loss of a frontend loader could be directly attributed to the great "Kirfy" of the Yamandu River. The "Kirfy," in fact, is responsible for all bizarre and unexplainable events that occur around camp such as electricity blackouts, the sudden appearance of black birds on a Loader, strange faces that show up at the campsite and mysterious lights and sounds that emanate from the river. As a result, no African worker would go down to the river to help recover the expensive sunken loader. Plus, Bitter Cola has convinced the locals that he alone knows how to overcome the Kirfy's evil powers. So Bitter Cola assures himself a constant salary in camp since only he can motivate the locals to work.

I keep speculating about the illogic of the Yamandu mining camp. How could diamonds not be found in Yamandu? Another town, Kunduma, has enough diamonds to purchase New York City. Yet we are not mining there. What is so special about Yamandu? I mull over some evidence that is heightening my suspicions about clandestine operations taking place at the mining site. Amadu Jall, a security officer at the site, quit his job without explanation. Soon after, the officer dies of unknown causes and the salaries of security personnel are doubled for no apparent reason. It does not make sense. There are also strange Ion-Flash landing pads that are being built in the middle of a useless mining field at Plant 4. Why landing pads in this remote area of the world?

My mind wanders back to a UFO experience that occurred while I was driving home from camp after a long day. The road was flooded with water from the river, so, three British trained Caterpillar mechanics and myself had to wait in the jungle until the route was passable. When the Yamandu finally drained back into its bed, we set out towards an area near Plant 4 where the rains did not flood. Just then we noticed a strange globe of light beaming down from the moonless sky. We assumed that the craft was either a helicopter or an IFP. Our car engine stalled as the lights approached, prompting one of the mechanics to become almost crazed with fear. He opened the car door and rushed to seek refuge beneath the vehicle. As the lights came closer, I was able to make out a spherical glass globe hovering just above the palm trees next to us. Then, without a sound, the craft impulsively veered off at a steep angle and darted rapidly out of sight. That was an experience I could not forget. Cold sweat swells over my forehead as the experience came back. Perhaps that is what the landing pads are all about. I have to find out some more, I am very curious now but the night is late, so I call it in and go back to my room.

CHAPTER 17
(Mining lies and Blood diamonds, August 21 1984)

It is a week since the bankers from Amsterdam inspected the site. They had made no mention of the poor performance of the mining operations. No significant diamond stash had been found for over a month. Yet no complaints have been made either. So be it. It is simply left at that. One morning, Militant and I jump into a brand-new Nissan patrol car to start our daily journey to the mining camp under a light drizzle. Five diesel trucks full of local workers complete the convoy that must traverse primitive roads that must be constantly paved by our mining machines. The way to the mining sight is Swiss-cheesed by potholes and the drive is a zig-zag that is very annoying. There are always standby Caterpillar bulldozers to drain with mud and water from the larger potholes in case they become impassable.

I have never grown accustomed to the rides and the thought of waking up and preparing for the two-hour trip nauseates me. I hate when the truck would tilt twenty or more degrees to one side, as careening water and mud soak the floor of the vehicle and our feet. My only consolation is that the thirty workers at the back of the truck would chant and sing a native song to make the time pass by more easily. Inside the cab, the endless Police songs play non-stop and it has become a ritual, a mantra!

Occasionally, we would come upon a blockade in the roads, consisting of several large logs stacked by one Tamba Tumbu, a local villager who constantly challenges Bitter Cola's chieftaincy. He will demand a road tax from each vehicle passing through "his territory." Unfortunately for him, the mining caterpillars would instantly clear the roads. Following several such encounters, Tamba took his case against the Sierra Mining Company to courts and the dispute was settled by the company by paying him two hundred dollars for the destroyed logs. But Tamba was eventually appointed Chief Security Officer at the mining camp to ensure a lasting peace with the people of lower Yamandu, who are ruled by a hostile local named Thamu Yombo.

The two-hour ordeal to camp is sometimes stretched longer by the constant stops made at the local villages to pick up more workers. Some workers who purposely miss the truck never ride again if they find diamonds that makes their fortune by the roads side. Management does not approve of a dwindling work force so Militant and I ordered a meeting to discourage any worker from walking to camp. When this strategy proves ineffective, Bitter Cola calls another meeting to explain that the devil, "Kassila," is responsible for the sudden disappearance of men who walk to the mining site. That scares the locals and ensures that they will take the trucks in the future.

We arrive at the camp and Militant drives down to the mining site to oversee activities while I go to the processing plant site to continue my preparations for the evening's meeting with the Ablas. Huge steel beams have just arrived from Freetown, for the structure of the plant I am to set up. I supervise the unloading of trucks and stack the beams in preparation of my work.

The camp is buzzing with activity, while everybody is busy or pretending to be busy. I will spend the afternoon interviewing welders to work on the new structure I am designing for Plant 4. It is not easy to try to coordinate mining site plant activities. My work space is a makeshift trailer, with a desk and a drawing table. In the trailer I keep all the tools that help me survive the harsh conditions of the mining site during the brutally hot summer. There is no air conditioning, no fans, and no refrigerators. Everybody takes all the essentials for the day to camp. Sometimes drinking water runs out by midday and a rationed supply from nearby village will have to suffice. Miners are a special breed of people and I am still learning to cope with surviving in a hostile environment by learning to drink underground water and eating unwashed fruits and vegetables.

As the roar of caterpillar bulldozers and dump trucks signal the start of the work day, I retreat into my engineering office to labor over the last phase of the Plant 4 design. I must design the entire gravel feed and water pumping system for the plant himself. Plant 4 is to be situated at a dugout location beside the Yamandu River near Upper Yamandu, near where the great underwater cave is found. I design the plant base with concrete using what experience I have gathered about concrete foundations and other engineering principles of stress and strain from my previous part time work with a builder in

Freetown. The one-hundred-and-fifty-ton plant is to sit on a flat concrete bed, placed in the dugout by the river. Most of the surveys around the site are centered on water bed levels and soil strength, in order to ensure that the plant will not sink or settle to a dangerous level below the underground water level. The Yamandu River will occasionally flood to a height of about ten feet from its base, filling its entire basin just below the plant site I chose. Adequate drain controls must be set in place in the event of heavy rains.

Next on my agenda are the interviews with some twenty "qualified" welders who have gathered outside the gates of the camp. All of them supposedly have "QBE" degrees, or were "qualified by experience," but after some questioning I realize that most of them know nothing about welding other than the fact that it is a process of joining two metals together. I decide to have some fun during the final weeding process and ask one of the men if he knows how to "ice-weld."

"Yes, of course!" the man answers, insulted.

"And how about tea welding?" I laugh.

"That also," the man quickly retorts with confidence.

Seeing how desperate the man is for a job, I decide to hire him as a welder's assistant, hoping his enthusiasm will help him learn the trade quickly. Finally, I meet the ultimate dream welder that will make my day. His name is Willy Cockson. He was introduced to me by a security guard as " Willy the son of a Cock."

"People call me the Son of a Cock!" Willy boasts.

"I can weld my way through anything in any position and at any time," he claims.

"Follow me Sir," Willy orders, his strange gait and confidence stirring my curiosity. He leads me to an old broken-down drag liner (rope and bucket digger) and climbs onto the machine signaling me to do the same. Willy pulls up the legs of his pants exposing a horribly bowed pair of legs and clamps his bowed legs around the drag liner's engine. He suspends himself in this impossible position, swinging his torso upside down and going through the motions of welding the base of the drag liner frame! Every miner knows how difficult it is to access the under frame of a drag liner, particular when welding. Willy needs to convince me no further. I hire him on the spot as a top "grease man" in the camp. I spend the rest of the day designing and measuring the plant site under the fierce African sun. When my thoughts veer off to last week's conversation with Militant, I know it is time to quit and eat and I eagerly head off to the Senior Staff dining room. It is peaceful and shady here and some engineers and operators have also come to have their lunch and gossip about projects. As I wait for the room to fill, I recall the amusing experience of having praised Bayano, the old Brazilian cook, for his culinary abilities, even though I have never enjoyed his food. Bayano had become overjoyed by my praises and to my dismay he insisted on cooking for the Senior staff for the next six weeks.

My reverie is interrupted by an operator who speaks loudly about a huge snake that he had crushed with a caterpillar bucket. Each meal time brings another colorful story about a near death fall or a close call with a deadly creature particularly a large snake. The man claims the reptile must have been well over twenty feet long and one foot in diameter. It is not uncommon to come across monsters of that size in the West African virgin rain forests, but I know little of their spiritual value to the Poro Society. On some occasions when we have to drive out of camp by night we would encounter massive snake bodies that can block the entire road. These creatures are revered by the local miners, who believe they are incarnations of the river devil, "Kassila."

As I sit eating my lunch and listening to stories, Militant, and David approach my table.

"How is it going Mike?" David asks, referring to the design work, and waving to Militant.

"Great!" I reply. "I have one more problem to solve and we will be home." Seizing the opportunity to talk to David about what is going on, I add quickly,

"David, we have to talk," hoping to get an audience from his busy schedule.

"Not now Mike, I have some things to take care of," David remarks and rudely dismisses me making his way to the kitchen.

"But it's important! I need to know a few things," I insist.

"I said I am busy now," David shouts angrily as he turns away from his original path and walks hurriedly away from the building to the Yamandu River. I do not expect David to be so abrasive. Militant follows David and jointly they head toward the river bank.

Perhaps Militant told him about last week, I think to myself. As the two men walk away conversing in fluent Portuguese, I can barely hear what they are saying but I could make out David shouting,

"Nu Purge, Militant!", or "It won't work, Militant!"

Confused about David's behavior, I continue eating my lunch and store the incident away for a future explanation. I watch the huge bulldozers and other excavators work their way into the depths of the mining pits as I eat, hoping that somehow, I can figure out what is so wrong about Yamandu. Huge dumper trucks buzz back and forth bringing huge heaps of hopefully precious gravel to the new processing plant. In the mining district of Yamandu, the face of the Earth is constantly changing as a result of our actions forging huge valleys and mountains of worthless dirt. It is still not making any sense to me and David does not even want to talk about it. Everywhere, workers go about their chores slowly transforming the once beautiful rain forests of the Yamandu River into a compelling landscape of artificial mud-piles. Why? As a result of Global Warming, the price of each acre of the rain forest has increased a hundred-fold, making it silly for anyone to invest in a worthless diamond mine. The cost of green credits for cutting down trees seems to far exceed the worth of diamonds we have been getting in the mines.

It is getting dark as the sunset spills its orange glow over the blanket of forest surrounding the mining camp. The camp is settling down into the slower evening pace and workers are assembling into small groups to board the trucks that are bound for the different villages. Suddenly, an ice-cold wind fills the camp. Everyone feels it and the voices of the workers fade into murmurs of freight. Faraway vibrations seem to be rocking the camp almost like an earthquake. Panicked workers scamper everywhere and I run over to the generator room to make sure the vibrations are not being caused by an electrical disturbance. My body hairs stand on end as if I have been exposed to a potent electric field. The generator is still running and seems fine so I shut it down.

The first indication that something dreadful is happening is when I see a group of horrified workers running toward a truck that is speeding away from the camp site. Another group of workers that were eating in the dining room just moments before have left their food on the tables and are racing into the jungle deep shouting "Kirfy, Kirfy!", meaning "Devil, Devil".

Loud thunderous sounds are coming from the Yamandu River and lightning is striking all around the camp site. Some workers on the far side bank run to a boat on the river and start off in a hurry as if escaping something terrible. I am confused. I jog to the river bank to get a closer look. Bitter Cola is at my heels following as if he is trying to tell me something. The Chief grabs my tee-shirt and pulls me back and I lose my balance. Both of us fall onto the muddy bank of the Yamandu River as the sound escalates. I now know that the sounds are coming from the opposite bank of the river.

"Boss nor go yonder, the Eye of M'Dulu is there, you go die!" Bitter Cola shouts holding onto my tee-shirt firmly with his nails tearing into my flesh from fear of the unknown. Bitter Cola's reaction scares me since he is usually calm and confident. He is the one that usually manipulates miners with the threat of the lurking "Kirfy." Now, the mighty Chieftain cowers and I am trembling with fear also.

"What is happening Bitter Cola?" I shout above the hum.

"Boss, the machines have disturbed M'Dulu's sleeping Eye, no go yonder, you go die!" the Chief pleads with me, his voice barely audible above the cacophony. David and Militant are already pulling into the churning river to make the crossing with a boat. Waves of river waters as if initiated by seismic action start crashing against the river banks almost in sync with the unidentified source of the sounds.

I pull Bitter Cola's hand free from my shirt as I steady myself before the rocking water that is now splashing our feet. Bitter Cola erupts into a tearful plea for me to abandon my quest. When I do not respond Bitter Cola flees screaming into the jungle as the river erupts into a violent vibration. The sounds thunder from the river and a cold sweat runs down my forehead and I hesitate. A strange blue glow fills the camp as if a smokeless fire is consuming it. I call out to David and Militant to wait for me, but they can no longer hear my voice in the wild mix of vibrations that have fills the camp. I am now scared and afraid that they will die and that I am going to die as well. It is the end of the world and it is happening here at. As I run toward the Jungle, a warm blue glow engulfs my body muting the sounds of the red-black camp. I am as light as a feather and the Earth is receding from my feet. I am upside down floating in dirt-filled air and like a silent fireless bomb I feel myself being propelled at a tremendous speed to an unknown oblivion. It is as if a gigantic balloon is being blown beneath me. The whole camp site disappears into thin air and I am totally engulfed in dust and darkness. My body is being propelled through the dusty atmosphere by some inexplicable force. As I follow the path of a crazy trajectory I can see the rotating campsite below me, heading towards space. At first I think I am caught in a tornado of unearthly magnitude as whole bulldozers and massive steel frames plummet around me. But then my body begins a graceful descent to the ground that could not have been possible in the eye of a tornado obeying the common laws of gravity.

As I come to within two hundred meters of the camp ground level, debris below me begins gathering into a dense sheet of matter. An alien landscape spawns in midair and I suddenly feel myself landing on the new huge and soft bubble of debris. I am lying on the surface of a huge invisible bubble of something. The bubble has formed an invisible skin layered with the debris of the mining site that seems suspended in just thin air. In the enclosed space of the bubble is nothing but clear air. As the dust clears I can see nearly two hundred meters below me. There is a huge cavity in the Earth below forming the underside of an oval bubble I am sitting on.

"This is not Earth," I think, "Earth is gone. I am dead and this is the end of the world!" I strike the balloon's surface with a soft rebound head first, but the impact of the crash is magically absorbed by the invisible, fabric-less barrier of the bubble. Debris is all around me floating and following the contour of the large bubble. Above me a mix of water, mud and mining equipment is slowly raining down on the bubble surface to form a delicate skin of multi-colored debris that is becoming a more solid foundation of the floating surface. I can feel myself floating in the melee as if it is some soft ethereal fluid. The debris is settling fast around me nudging me deeper into the collection field. I must escape to the upper layers and avoid being trapped beneath the debris on the skin of the bubble. I begin swim upward through the settling debris, treading to stay afloat. The dust, dirt and water around me is now turning to a more solid mass of mud and is quickly covering the lower half of my body making me struggle to move. I must move quickly before the debris solidifies under gravity.

"I am in the sky somewhere, afloat in hell," I rationalize.

For a long time, I remain suspended in the floating mulch of what used to be the camp site. I look for the Yamandu River but it is too dark to see the outline. I succeed in finding a spot where the debris field is minimal and the where the thick air and dust about me is condensed to a viscous fluid that would respond to my swimming motion. I can feel my body slowly being pushed to the less dense parts of the debris field. I can see the moonlight now and I know I will be safe at least for now. After what seems like an eternity, the steadily glowing moonlight encourages me to experiment in the thickening melee that has trapped me. I step out onto the new hardened ground cautiously and my feet become cushioned by something invisible. Becoming bolder now and eager to explore the nature of my new terrain, I press my toes into the invisible fabric barrier beneath my feet and feel a force field repelling them. I see a loose tree branch and use it to push deep into the invisible force field. I feel my body give as the reactive force pushes it back sending me even higher up the field. The branch shatters with an intense vibration that rushes through my body.

A powerful field has lodged itself between me and the Earth below. If there is only one thing I can be sure about it is that the force I am embedded in is neither a magnetic nor an electric force. Debris continues to settle on the surface below me causing my body to slide uncontrollably as if I am an ice cube gliding on a glassy surface.

As the moonlight becomes more intense I am able to scan some distance and I am able to see parts of machinery from the power plant I shut down moments before. Perhaps the mysterious force had somehow tapped into that Earthly power, making it its own to create the surrealistic world that has captured me. I am surprised that my mind is still very active and alive. The dirt beneath the scattered machinery parts is glowing with the blue light leaving curved imprints in the dust. I am intrigued by the light and I begin swim towards it in the still wet debris. The light's frequency pulsates about ten beats per minute and as I swim or push my way I realize that the temperature of my surroundings is dropping, and that I am getting weaker. The debris field is changing, getting less dense and lower in height. Swimming is no longer possible and I begin to walk over to the yellow glowing matter ahead of me. I stare in amazement at a thirty-ton bulldozer that hovers nearby in thin air above the debris field.

I am felling chills and it is becomes colder in my free-fall environment. I approach the bulldozer and realize that it is cold to the touch, very, very cold. It oscillates with the slightest touch of my hand as if it is freely suspended on an invisible rope from the sky above! Beside the bulldozer is a large floating pool of diesel fuel giving birth to a stream of orange liquid floating in the air. As I walk towards the liquid it flows towards me as if it has an intelligent purpose to engulf me with its warmth. I touch the titanium weld on the steel bucket of the bulldozer. It burns my fingers with a frosty coldness. I touch some levitating papers and Operation's manuals from the loader nearby and discover that their temperature is normal. I rightly conclude that the colder the temperature the denser the material.

I look down below and realize that I am standing on top of an oval, egg-like structure of about five hundred meters long and three hundred meters wide. In the middle of the space beneath the oval, a yellow pulsating sphere about the size of a soccer ball is floating. The upper shell of the oval is formed by dust, dirt and air, and the lower one is an oblong concave cavity dug deep into the mining camp below.

I recall seeing a picture of this spherical crystal in Jawed's office. As I recall the experience a man's distant voice beckons me to the far side of the steeply falling gradient of the egg toward its polar cap. I moon-walk forward with urgency as a strong current pulls me forward. Frightened of falling over the edge of the oval, I back off. I shout to the voice but there is no reply. Then I see a large bubble of water flowing towards me. It carries human bodies as it rolls and tumbles. The bubble of water is destabilized by the hydrodynamic forces of the gradient of the oval expanse and so it rushes past me. Three dead black bodies glowing blue, brush up against me. I am bleeding and I cry out in pain and lose consciousness.

CHAPTER 18
(The Poro Society, December 23 1984)

I awake from a deep sleep with a chorus of unfamiliar voices waking me up. I sit up startled, wondering how long I have slept and where the strange harmony comes from as the sound of beating drums filter down from the sky. Before I can focus my thoughts I feel my head being cradled. Then I spy the arms of a fierce looking African man wearing a skirt, knitted from papyrus. I have seen him before. I recall the dream I had on campus just before I came to Africa. His chest is covered by a primitive weave of fresh green leaves. Facial features are highlighted by delicate white chalk markings. His eyes are glowing blue from the luminescence of the strange pulsing blue crystal sphere below. The glow transfixes the African warrior's face into a fresco of patterned light. I stare at the patterns and oddly enough my sensibilities tune into the fifty black warriors that surround me with their beating drums. I understand that their song accompanies the pulses of the blue glow and notice in my trance that the warrior's spears sport paintings of cobras. Soon the patterns of light no longer dazzle me and I recognize the face as that of Bitter Cola's.

Before I could speak, Bitter Cola places his hands over my mouth. He lays my tired body down on the invisible canvas and stands up raising his hands into the air. In his left hand is a serpent like spear, and in his right is a white rag filled with my blood. I am bleeding through my nose and ears. A long wound spans my left arm.

The warriors and their leader float in the air around me.

"Oh! How beautiful these people are!" I think to myself. I am overcome with elation from my discovery of Bitter Cola as a new god-warrior that will protect me in the jungle. I dreamily recall the "Transfiguration of Christ" on the mountain, when Peter asks Jesus if he could build a tent beside Jesus, Moses and Abraham so that Peter could live at the site to worship it forever.

Suddenly Bitter Cola is no more. In his place is a beautiful warrior boasting both vigor and grace. His skin glows ivory black. His muscles have been rejuvenated by the pulsing in the blue glow of the crystal. Refined by a godly hand his skin is decorated with chalk images of the long dead great jungle beasts like the mammoth and the dinosaurs.

"Behold all you warriors of the faithful. Today is the day our ancestors have foretold," Bitter Cola sings.

"He who survives the wrath of M'Dulu is a Master in the realm of the PORO" his chant continues.

"M'Dulu has given us a new watcher. A new member of our ancient "Poro" society. Let us rejoice and enjoin the Whiteman in his joy. Let he who knows the secrets of M'Dulu be honored with a chant of Ojeh songs, for today we have found the Eye that Thamu hid from me and my people!"

The chant of the "Poro" crescendos as the African warrior's chant to the pulsating blue lights beneath my body.

"Ogbo, ogbo, ogbo," they shout in unison as if they have one voice, one mind, and one body.

"The Eye of M'Dulu has opened onto this generation as it surely did for our ancestors," Bitter Cola continues to chant in his native tongue.

"And as he gave them the power to roam the world and teach of his might, so shall he give us the power."

Bitter Cola motions to a group of men to come forward. They approach and hand him a wooden calabash bowl. M'Dulu's warrior kneels down and raises my body to a sitting posture, motioning me to drink the contents of the calabash. I cannot resist the command that seems to emanate from Bitter Cola's glowing blue eyes. I drink the sweet contents of the calabash, draining it completely. The contents are very cold. Whatever it is, it is pleasant and makes me feel very good.

"Now, your eyes are open to look into the eye of M'Dulu!" Bitter Cola joyfully decrees. Another old man adds,

"Now, therefore, Oh our M'Dulu, great, mighty, and awesome God, you who in your mercy preserves the Covenant, take into account all the disasters that have befallen us, our Kings, our Princes, our Priests, our Prophets, our Fathers, and your entire people, from the time of the Kings of Azaria."

Then Bitter Cola turns to me and declares,

"You have seen the ancient Eye of M'Dulu and lived. Now, you must be one of us. We are the selected few who are the Guardians of the Knowledge of God!"

Bitter Cola lays my head back on a cushion of air and points to the sky above. I spot a pair of doves flying in unison, through the blue, cloudless sky. Soon one dove takes the lead and flies in a circle while the other follows behind. A bright light engulfs them as they grow into angels. One angel is black, the other white. They fly to me and free my spirit from my body, pulling my soul upward into the sky.

"Let us show you the history of the Covenants," they sing together as one.

"You will now be a trusted Guardian of the power of the Covenant that creates and destroys. It has given Adam and Adama (Eve), dominion over the Earth and its elements. It controls destiny and opens the path to greatness, the Maker of Worlds, the Destroyer of worlds, the power that begat all and destroys all. Behold, Oh man, the first Covenant made at a time when time is not; at a place where space is not; before men came; before all came."

There is blackness without a void. I know I am alive but also realize nothing is real. This is before the Universe was created. In my mind I see a strange confluence of space and time, a mixture of forces that I could only describe as gravity and anti-gravity. Then I see a cubic structure form before my eyes. On each corner of the cube is a crystalline sphere from which light of a particular color emanates. Another set of crystal lights just as intense as the first set appears between every crystal on the cube. There are twenty-seven crystals in all, forming a cube. A central crystal of light is surrounded by twenty-six crystals all equidistant from each other. I know then that space is twenty-six dimensional and that the central point represents time.

"Know that God has made your universe from this symmetry. For from naught came forth one time equal to twenty-six domains of space," the angels teach me.

Then, I witness two opposing sides of the cube unfold about parallel diagonals to open the box up like a leaf. Two opposite lower corner points rotate and join the central point to form a threesome, and the two opposing upper points rotate out of the box to form points sitting outside the structure. I recognize the symmetry of the points on the diagonal plane formed by the rotated points as that of the eight-fold symmetry of fundamental particles that I had studied in college physics courses. Particles are streaming forth from them, to form the material world.

The angels take me to another faraway place. We fly past several different terrains and times as the world reels back to a time before history.

Below me is destruction from a great flood. Water is everywhere and not a living soul is in sight. Then they take me to the middle of the great ocean and show me a huge oval bulge projecting from it, pushing flood waters far into the dry lands of Earth.

"Behold, Oh man, for your eyes see the power of the first Covenant, swelling the seven seas to flood the land, even as your land is exposed by the last Covenant. For as surely as you have been saved from death so is Noah tried and redeemed from death," the Angels explain.

We fly rapidly to another place, another time. Below us is a mosaic of colors, a stream of events that I cannot not decipher, but could only understand as a movie of the history of planet Earth.

"Behold Oh man, your ancestor Moses as he crossed the seas, which you now call the Red Sea, for in his mighty hands is the power of the Second Covenant."

Below me I see a multitude of people worshiping the Second Covenant, walking across what appears to be a path within the Red Sea. Their leader holds one of the crystal spheres of the cube in his hands, and with it, cuts a path of dry land within a wall of water.

The angels sing a song of praise for God, dancing in jubilation as they continue to escort me through space and time.

We fly to a great temple in the middle of a desert. Dry arid land surrounds an oasis on which a temple stands. Its pillars are carved of silver and carry the symmetry of the great cube. I cannot comprehend such beauty.

"Behold the Temple of Solomon, for he inherited the power of the Third Covenant over the winds of the Earth and is given great knowledge of architecture. He is taught the language of the birds and animals, great and small. In his hands we entrusted the Third Covenant."

The magnificent temple fades away into rubble and I interpret that as the decay of Humankind. The angels fly me through a great expanse of space and time, as Earth memories melt into a chaotic mix of disasters, wars, famine, and great empires rising and collapsing.

We are now high above a beautiful land of great and mighty monuments, and I could see the shapes of what appear to be pyramids!

"Behold Oh man, the time of greatness, for it is the power of the Third Covenant that built the mighty monuments of the ancient worlds. See what you now call Egypt five thousand years ago, when Hamister is given the power of the Earthly force, to lift mighty rocks from the cradle of the Earth and build his city in honor of the one true God. Your species has defiled the Covenant by dedicating it to Zoroaster-Amon-Ramses, the pagan."

Below me is a grand city, the likeness of which I could not have imagined and in my view comes a mighty being, a King of great power holding the crystal sphere in his hands as he effortlessly lifts the great rocks that built his city. And they show me other great events in the history of man. My intellect is incapable of understanding it all, but I understand enough to know the significance of the crystals or the Covenants as the Angels call it.

I am puzzled as the history of the world unfolds before me at the speed of light, but I am unable to ask the angels any questions. I cannot speak. I am a dizzied traveler, feeling a great affinity for the Earth and its inhabitants. Through the love that I have for all life forms, I hold onto the greatness of God that is the oneness that permeates all His creations and the limits of time and space. I also know that the chant of the Poro is with me. I have become a Guardian of the Crystals of the Covenants between God and Man and a follower of Universal truth.

Next I am taken into a temple that stands in the middle of a mountain peak. The simple temple is decorated with images of Buddha.

"Behold Oh man, for you see that Man hath entered into a Covenant with God, and it is here that the Fourth Covenant is made with the red and yellow and the indigo blue peoples. Surely, your Karma is followed by karmic consequences of equal name. Thou shalt not reject the love of God for all his children. For it is the red, yellow and blue rays of the crystal that hath colored their skin in bondage to their mortal bodies. Even the chariot of Krishna shall not be neglected in the final word that is to come!"

I realize that the angel speaks of the great Buddhist and Hindu religions of China and India. Then we cross a great ocean to a new and different land. The angels take me to a tribe of ancient people celebrating the birth of a child.

"Behold, Oh man, the "House of the Giants and Dwarfs." For we also made the Fourth Covenant with Quetzalcoatl, the Son God of the people you now call the "Incas." Behold their greatness and glory under the law of the Covenant."

I witness the great pyramids of the Inca Empire being erected and destroyed. I taste the bittersweet fine line between prosperity and decay.

We fly to a new world where a beautiful young man preaching love and peace is surrounded by multitudes. His voice echoes the angels' song,

"Blessed are you poor; the reign of God is yours.

 Blessed are you who hunger; you shall be filled.

 Blessed are you who are weeping; you shall laugh."

Then space and time passes until I see a dying man hanging on a wooden cross.

"Oh man, we have entered into the era of Love and Joy, the era of the Highest, Jesus the Christ. For your hand has been chosen to see that which time has hidden away. Pray that you may understand these things and become a Guardian of the Covenant of Love!"

"Behold Oh man, the Fifth and last Covenant made with Man. For it is the Covenant of Hope. Let all men rejoice who listen to the words of the Holy Prophet Mohammed, for his is the power of the Covenant of Hope. It is through him that the closure of all Covenants shall happen. Behold then the history of the Fifth Covenant as it passes on to the Holy Prophet Mohammed of Islam. For it is under the power of the Fifth Covenant he is bade to speak of things past and things to come. Behold all religions of the world, for the Fifth Covenant has not yet been fulfilled in its completeness. The Fifth covenant is for Jerusalem. Hence there are wars and famine all in the name of the Fifth Covenant."

Then the angels take me to the final days of the world. I see Jesus holding the Fifth and Sixth Covenants between God and Man. The year is 2038 AD on the sixth hour of the sixth day of the sixth month. The covenants were like faces of the cube, each with a complete set of nine crystals for a total of twenty-seven. All faces had been put in place except the sixth. Jesus holds the sixth face that would close the box. I am shown all the calamities that stand before man; global disasters, floods, fires, and electromagnetic wars, nuclear disasters and all forms of possible horrors that await all life unless the last Covenant is fulfilled.

And the angels speak to me saying,

"Go guard the center of the Fifth and Sixth Covenants, for they shall save the World. Upon closure of the crystalline cube, the Lord shall once again possess the power of the Arc of the Covenant which is long hidden from Man."

At this point, time stands still and space is devoid of matter. I cannot see any more. My descent into my body is abrupt and painless.

I am a Whiteman standing naked in the radiant sky, surrounded by Black warriors of magnificent beauty. I raise my hands up in the ancient ways of the Poro and proclaim my solidarity with them and with Nature. From their lips comes forth an Ancient language hidden for a millennium in the Ancient African Jungles. I have become one mind and one body with the Poro. I am Poro. I proclaim my faith in M'Dulu, God of the Universe and his Covenants with Mankind.

The warriors walk on the now settled and solidified mountain that has blessed their heritage. Together they take the dirt and laboriously level it over the great Eye of M'Dulu to cover its blue glow from the face of the Earth. As the rays of the sun are shadowed from the Eye of M'Dulu, the blue light's intensity slowly reduces, and the oval bubble mountain eventually slowly collapsed into the flat landscape. The ancient crystal is once again buried under two hundred tons of dirt and debris. The Yamandu River has diverted its course by over one hundred meters to accommodate the bulge of the crystal. A new landscape hid its mysterious secret power from Mankind. Only Bitter Cola knows the hiding place of M'Dulu's Eye!

CHAPTER 19
(Yamandu River)

Neither myself nor Hal are accustomed to the dangers of the African jungle, let alone its inhabitants. As nightfall approaches, we become concerned about our safety and well-being. Young Paul is informative about the jungle but not diplomatic enough to ease our apprehensions. When night is setting in and there is still no reply from the far side of the Yamandu River, the young Paul throws the useless conch down on the river bank in frustration.
"They should have replied by now," he admits desperately.
"Are you sure we are at the right place?" Hal begins.
"Yes. I am sure. I have been here many times," Paul retorts emptily.
We wait for an hour as the dark jungle canopy blankets the forest.
"What are we going to do?" Hal asks me.
"I have no idea," I reply impatiently, sitting down beside Paul on a protruding stump.
Hal walks over to where the conch fell and picks it up.
"Can't we just go back?" He proposes to Paul.
"The batteries are too weak for the journey back. We will have to wait until the morning sun," Paul deduces.
Hal let out a scream as he slaps himself hard on his bare arm.
"What is it?" I ask startled.
"An effing mosquito, damn it! They bite like bees!" He exclaims angrily. The tension is mounting rapidly. The African Jungle is slowly taking its toll on us. Soon we are being bitten by larger and unforgiving mosquitoes. Paul does not seem affected by the vermin. He is searching the jungle canopy for something.
"I must find you the leaf of the henna," he resolves. "It will stop them from biting you."
"Yeh, sure," I proclaim laughing.
"How do you blow this thing?" Hal enquires, snarling at Paul and holding up the conch.
"Put your hands over the opening and just blow as hard as you can through the little opening at the back," Paul instructs from a distance.
Hal gives the conch a hard blow and fills the blackness of the jungle with its sound. A thousand birds fly from the branches of the trees above them, soaring across the wide-open space above the river.
Suddenly, a loud reply is heard from the other side. Paul stands still and listens. He walks over to Hal and takes the conch from him and hurries over to the edge of the river. This time he blows the conch harder. Again, there is a reply. He changes the note of the conch by blocking its opening with his fingers and calls musically in a way that only M'Gboko can understand.
The reply is not what he expects!
"It is not M'Gboko! Hurry, we must hide! It is the Minga warriors on the other side. They are not our friends!" Paul cries pointing to the far side of the river.

Hal and I quickly stand up and start after the young man as he runs into the jungle. "But why should we be afraid?" Hal calls out to Paul, "I thought your parents are chiefs here?" he points out.

"No, Mr. Hal! The Minga warriors are not forest people," he starts to explain.

"They are from a faraway land from the North desert. They are keepers of the Great Garden."

"What is the Great Garden?" I ask.

Paul starts his Adwalk and motions us to follow him. The Adwalk slowly hums to life as its motor rev up to maximum power. We follow suit as Paul speeds away toward the south side of the river bank.

"The Great Garden is where the Poro believe God first created man," Paul shouts above the hum of the machines. "It is said to have everything good in it."

"The Garden of Eden!" Hal interjects as we hurry toward a thick covering of trees.

"How did the Poro know about such myths?" He asks.

There is silence. I have a sinking feeling in my stomach.

Perhaps it is all true, I start imagining. The legends are real; the forces are real!

"The Great Garden is hidden in a faraway land. My grandfather believes it is real. He thinks the Snake of the gods will end at the Great Garden."

Hal and I do not ask any more questions about the Great Garden. There are other thoughts more urgent. We continue to follow Paul away from the open bank of the river into a raised hill covered by a thick blanket of trees. Paul proceeds toward an opening at the base of the hill. The opening leads to a wide cave that slopes deep beneath the jungle floor. He motions us to wait outside the entrance while he explores further. The cave is filled with clear clean water. As he walks further into the cave the water level rises until it comes up to his knees. Paul seems to know the cave well, for he proceeds without hesitation toward the center of the pool of water. Hal moves his Adwalk toward the opening and turns the lights of his vehicle on, hoping to see what Paul is doing but the illumination spans only a few meters into the dark cavity. Presently, Paul's figure approaches. His feet are covered in mud. He is panting heavily as he speaks,

"I will hide the Adwalks, so why don't you two go on ahead in front of me. We have to dive through the water to get to safety," he informs us. There is fear in his voice.

I hesitate,

"Where could we go with all that water?"

"Trust me Dr. Davis," Paul pleads, "we must hide immediately. The Minga will kill us if they know we are here!"

Hal and I walk into the pool of water until we are knee deep. It is very dark and cold. In the distance, we could hear Paul cutting branches outside the cave to cover the Adwalks. Soon Paul enters the cave and rushes into the muddy water toward us. He is closer to me than he is to Hal. He urges me to move on and take a plunge into the pool.

"Hurry, please hurry! There is a large air cavity trapped beneath the other side of the cave where we can hide. No one outside the Poro knows of this place. Go Dr. Davis go! Hal , go! Dive into the water!"

Paul scurries ahead of us struggling with the weight of the water and stirred-up mud that cling to his soles. I follow Paul closely while Hal is still a distance away and appearing as a faint shadow in the dim light. In a few moments Paul disappears into the pool of water, plunging head first. I look in the direction of Hal, but it is too dark to see him. Then, closing my eyes, I submerge my body into the dark cold water and paddle my way downward. I think I heard Hal saying something about the lights of the Adwalks, but I am already deep into the muddy water. For what seems like an eternity, I dive holding my breath with my eyes closed. I hear the sounds of Paul or Hal breaking the water surface far inside the cave. A few moments later I surfaces into the blackness of a large air cavity beneath the cave. In the darkness, I can hear either Hal or Paul swimming to the edge of the cave.

After a while, Paul calls out to see if we are okay.

"Dr. Davis, Hal , are you all right?" He wonders, gasping for air.

"Yes, Paul, I am fine," I reply.

But Hal does not answer back.

We listen intently for signals from Hal but nothing happens! Hal has not made it!

"We must have forgotten the Adwalk lights and Hal must have returned to turn them off. I think I heard him say something to that effect!" I state.

"Dr. Davis that is a stupid thing to do. Hal should not have done that," Paul shakes his head in astonishment.

"Oh God, I hope it is not true!" I cry out in disgust.

I struggle with the long, hopeless silence from Paul.

"I am sorry Dr. Davis, I think he has been taken by the Mingas," Paul finally admits.

I cannot accept this. Just moments ago, the young jungle boy spoke of magic herbal elixirs and the mysterious power of his grandfather. He divulged special places to us with exquisite beauty and hidden from the world. Surely, he has the power to rescue Hal.

"You brought us here to meet cannibals and die? Your father assured us we would be safe with you. What reason do the warriors have to kill?" I demand angrily.

My emotions are on the edge of what I can take.

"I am sorry Dr. Davis, but my life is in danger too, and my father did not want the Minga to capture me either," Paul defends himself.

We cannot see each other in the pitch blackness of the cave. In silence we wait hoping to hear Hal resurface, but after a long period, we gave up and Hal has vanished. We can neither see nor hear anything that indicates that Hal could have made it. I speculate that the cave might be too large for Hal to be heard.

"Where are you, Hal?" I shout, but no reply comes back. The cave is very dark, and I am still not accustomed to the darkness. It is slowly depressing me.

"I am going back there to look for him," my voice echoes almost stupidly against the cave walls.

"No! Dr. Davis, don't go out there now. Wait! I will come over to your side. We must not go out there now, or they will kill us also," Paul explains.

"We can't just leave him out there," I shriek.

I grapple with my sense of humanity and instincts of self-preservation. A nauseating feeling starts to strangle me as the distinct image of Hal flashes before my eyes. He is one of the most intelligent men in scientific circles in the United States and his loss will be a real problem for the mission. It would also start an unwanted political situation. I am getting tired of it all and all I want now is to find Hal and that damn crystal and go back home to Elaine. Is it possible that Hal is really dead? If Hal had been my son, would I have gone back for him, no matter what? The warm glow of being alive creates a burning hot sensation through my veins. I do not know how long it has been since I stood there hating himself. Finally, the need to survive wins out.

"Are there any surprises in here?" I ask my guide.

"Just snakes, but they won't hurt you if you don't startle them," Paul cautions slowly as if he is ready to spring upon something.

"Snakes? How big do you mean, like . . . like the one we saw before?"

My nervousness is getting the best of me. I reason that each step I take or do not take brings me closer to salvation or death.

"No, Dr. Davis. I have spent many nights in this cave with my father and know the wildlife well. The snakes will not hurt you. Don't be afraid," Paul reassures.

Paul walks along the edge of the water, holding the walls of the cave to guide himself to the other side. Soon he is close enough to hear me breathing.

"It's me John, do not be afraid," Paul says.

"I am not afraid for God's sake," I shout somewhat surprised that I am angry,

"I am disappointed about Hal," I apologize.

"Don't worry Dr. Davis, they say he is with M'Dulu, God of the Eye. He is in another world that is better than here," Paul tries to me.

"Darn M'Dulu," I counter with anger.

"Hal is dead and all you tell me is not to worry because he is in some African demon's hands?"

I am surprised by my rude outburst but then again a good life-long friend has vanished in this damn jungle and it is barely our first day.

"You don't talk like that to M'Dulu, Dr. Davis," Paul warns quietly. That night I cannot sleep. I keep thinking of the terrible fate dealt Hal. Hal, poor Hal! After a few hours, I succumb to exhaustion and begin dozing off.

"It is only the first night" I contemplate dreamily, "and already I am completely drained." Paul kneels down and starts rumbling a prayer in his native dialect to M'Dulu. He sings quietly to himself as I slowly doze off to sleep. When I wake up the sun is flooding the cave from a small opening on the ceiling. In the haze I can see Paul standing over a fire at the edge of the water. He is holding some dried branches in his hand.

"What are you doing?" I charge shielding my eyes from the bright sunlight.

"Fishing," Paul states calmly. "You have been sleeping for a long time."

I try to focus on the makeshift rod that dangles in the still water. It is difficult to be alert without a cup of coffee and a shower. I contemplate bathing in the cave pond but then that would disturb my guide's fishing.

"What are you using as bait?" I am curious.

"I went outside and got some mango branches and dried leaves. The fish in these waters like the taste of the Mango leaves. My father taught me this. They are called Mango Fish," Paul explains.

The boy hunches over and gently lifts the makeshift rod out of the water to demonstrate.

"I tie the leaves at the end of the branch and wait for a fish to come close."

Then leaning back, he raises a sharp spear he made from a tree branch.

"Then, I take aim like this and throw the spear," Paul demonstrates.

At the mention of a spear, my mind races back to Hal.

"Did you search for Hal?" I ask.

"Yes, I did," Paul lowers his head. I walk over to where Paul is sitting.

"Well?" I ask again.

"He is not anywhere," Paul whispers solemnly.

"They must have taken him."

"But where? Can't we go there and look for him?"

"No John Davis. The Minga are very bad people. They are half White and half Black people, just like me," Paul tries to explain.

"Are you telling me that they are North Africans?" Hoping that they are not Moroccans. I am ready to grab at anything that might save Hal if he is still alive. If Paul looks like the Minga, maybe there is a chance he could befriend them.

"No. The Minga are somewhat different! They are tubabs from the deserts of the North, and they are taking over the forests of our people. They started coming here after the Devil Kasila was defeated by Bitter Cola my grandfather," Paul chronicles.

"But why did your government not do something about this?"

"Well, we don't know what has been going on. My father sometimes talks about this with the elders of our town. They believe there is a conspiracy which the government is hiding from us," Paul reveals surreptitiously. Paul's ears instinctively prick back like a hunting predator as he repositions the spear in his hand for another fish kill. I think I heard someone approaching but then realize that it is the thrashing of a fish. Paul comes down upon it with the grace of a tribal dancer.

"The Government does not acknowledge their existence," Paul continues, examining his skewered prey.

"They have sent officials to investigate reports of missing and dead people in the forests. But all along, they made no effort to find them or stop the Minga. My father believes the government supports the Minga," Paul announces as he roasts a small fish on the fire.

He hands a piece of the roast meat to me. I sit beside Paul and eat it hungrily.

"Why does he think there is a conspiracy?"

"Ever since the Blood Diamond wars, people have seen some Minga warriors visit President Massakoi in his palace in Freetown," Paul reports, spearing another fish. The fish wiggles at the end of the stick trying to free itself from the pain of the wound Paul inflicts, Paul frees the fish from its agony, dropping it on the ground. Like a flash of lightning, a thought hits me. The Mingas are from the North, from Egypt! My God, I start thinking they are also looking for the crystal of the jungle! The same crystals that I am looking for. So, they probably bought off President Massakoi and are chasing the crystal in West Africa. Paul has already caught a few more fish and has hung them over the fire that is burning beside us.

"Do you know how I made the fire?"

"No!" I grin sheepishly embarrassed at my poor set of survival skills.

"But every high school student knows how to make a bush camp fire."

"What is a high school?" Paul begins to ask.

"Never mind," I shrug it off not wanting to engage in an explanation.

"These "tubabs," have you seen them before?" I question Paul.

"Yes, I have seen them before. They are enemies of my family. They have taken my cousin Malachi from Yamandu Village and killed him. The Poro say that these devils come to steal the sacred Eye of M'Dulu from my people," Paul proclaims.

The eye of M'Dulu. My God that must be the crystal! I surmise, and what about the name Malachi. His cousin is named Malachi. I have known that name before I have met it somewhere in space and time!

"Are the Adwalks all right?" I want to move on and change the subject. I am disappointed about Hal but I need to move on as soon as possible, the excitement is mounting again.

"Yes, they are fine. Hal must have returned to turn off the lights," Paul informs me delicately noting that it must be a sensitive issue for me. As I give a silent memorial to Hal, the fire crackles sporadically as if sending message from the dead.

"I am sorry about last night, Paul," I apologize sincerely.

"Don't worry, Dr. Davis. It will be all right. M'Dulu does not anger easily," Paul assures me kindly.

"No, I meant the way I treated you."

I am now positive that the Sacred Eye of M'Dulu is the crystal in Yamandu. I need these people more than ever and I need to be in terms with Paul. We eat the roasted fish with some fruits that Paul had picked earlier as I mull over some strategies that may come in handy. After we clean our muddy bodies and clothes, I begin exploring the cave to see if there is an easier way out. I find a cleared spiral path that wraps around a rock formation. The path seems to lead to the ceiling of the cave.

"Is this a possible way out of here?" John Davis asks pointing to the path.

"I do not remember, Dr. Davis. I have not been into this cave since I is a younger child."

Paul climbs up the path as far as he could, to see if it is a possible way out. When the path leads nowhere, we decide to leave the cave the way they had come in. After the dive back out, we wash our muddy clothes in the river and wait for them to dry again. Then, we start the search for Hal's body hoping at least to find some evidence of what had happened the night before. I walk to the Adwalks and clear off the branches covering them. I search for traces of blood or some clue that would confirm Hal's fate but I find nothing. I move the Adwalks to the clearing by the river bank and wait for them to charge up their solar batteries in the African sun. A few minutes later the battery charge indicator lights up green. I examine the instrument cases and supplies to see if the Minga had taken anything. Everything is intact. Then, I secretly open the side zipper of my personal carry case and release the brightly polished American Military issue "Z-2000-Lazukar." The lazar powered gun is fully charged and ready for use. I tuck the weapon into its holster and tie the holster around my chest beneath my shirt. Then, I cover my chest with my jacket to conceal the weapon from Paul who is busy with his Adwalk. Soon, the motors start humming with the soft buzz of the magnetic engines. And I set Hal 's Adwalk controls on "Auto Linkup Mode" with mine. Hal 's Adwalk will follow behind mine.

Paul is busy searching for the missing conch. He recalls throwing it on the floor beside the river when the Minga came. He walks to the edge of the river and notices the sandy conch, bobbing to the river's tide. He pulls the conch from the bank and dips it into the water to clean it. After the conch dries out, he puts it to his lips and blows the secret tune of his people. This time a reply that is M'Gboko's code. I look hopefully at Paul. Our eyes focus on the other side of the river, looking for M'Gboko's boat. A faint fog hides the opposite bank but soon Paul is jumping with joy at the sight of the familiar black boat of M'Gboko, emerging from the fog.

We wait patiently for the boat to arrive, packing everything on the Adwalk securely to ensure that nothing would fall off the rocking boat. When the boat comes within sight, Paul realizes that the man on the boat is not M'Gboko.

"Where is M'Gboko?" Paul calls out to the man in his native language. The boat man replies in the same tongue, "He is dead. He did not come home last night. I came here this morning to see what happened to him."

"M'Gboko is dead?" Paul asks in a strained voice.

"Yes, he is dead," the answer is definitive.

"What killed M'Gboko?" Paul inquires desperately.

Baba does not answer. Ignoring Paul, he walks over to the Adwalks and starts pulling them onto the ramp. The young man immediately knows that the Minga had struck again. He feels the world closing in on him as the influence of the warriors narrow the great jungle expanse.

"And who are you?" the young jungle boy inquires authoritatively.

"I am Baba, the son of Fanta, the village baker," the stranger explains. Baba extends a ramp from the boat for us and our Adwalks. He comes ashore and stands beside us. He is a well built, tall bearded fellow who is naked except for the traditional loin cloth that covers his private parts. He scrutinizes me suspiciously.

"And what brings the tubabs here?" he demands of Paul.

"He is my friend, Baba. He is here on behalf of the government to help us with rice plantations," Paul reports.

Paul is still visibly shaken by the sad news of M'Gboko.

"These come with us, don't they?" the stranger asks in his native tongue referring to the Adwalks.

"Yes, all of them go," Paul says, "but be careful of the instruments of the Whiteman. They are expensive and delicate," he advises Baba. I follow the two men into the boat, helping to push the last Adwalk in front of me to carefully guide it across the ramp. We pack the boat carefully to make sure it is properly balanced for the trip across the Yamandu River. We did not speak as we cross the river. Paul is absorbed with memories of M'Gboko. Apparently, I later learnt that M'Gboko loved Paul very much. When Paul was a baby M'Gboko would lift young Paul up in the air and play with him by the river.

Paul would spend many hours with him ferrying people back and forth the Yamandu river. M'Gboko had taught Paul to fish.

When we arrive at the other side of the bank, a flock of villagers are waiting anxiously to meet us. Several children come running to greet Paul and explore the new white face that is visiting their village.

"Paul, M'Gboko is dead!" announces one little boy running beside the Adwalk.

"I know," Paul replies sadly.

"He died because of his heart!" the little boy explicates.

Paul stops in his tracks and looks at him enquiringly.

"The Minga did not kill M'Gboko?" Paul asks confused.

"No, my mother said he died because he is old," the youngster retorts confidently. Paul does not say a word after that. I follow his lead as he silently steers the Adwalk toward the village which centers around the bank of the River Yamandu. I can see small shacks spread all about the village center. A small cleared path flanked by huge cotton trees lead to the heart of town. From a distance Paul could see his grandfather standing in front of his home.

"John Davis, that is my grandfather in front of his house," Paul excitedly points to a frail looking old Whiteman with a well-tanned olive skin and a large flowing white beard. He is standing in front of a small wooden structure they call home. The Whiteman is dressed in a long traditional African frock standing at the entrance of a mud house about a hundred yards away. The presence of a lone white man in a small African village in the middle of the jungle strikes me as strange. I wonder what made this white man leave the modern world for the difficult life in Yamandu. I think of Tarzan of the jungle. Maybe Edgar Rice Burroughs is right about Tarzan. There is more to this place than meets the eye.

As we approach the house, the Whiteman walks towards us confidently. His loin cloth sways in the warm wind as he leans on a wooden walking stick in the shadow of a huge cotton tree. I notice that the frame of the house is constructed from large tree branches. The roof is slightly slanted on all sides and is made from a thick layer of dried bamboo and Palm branches. Several men stand by watching from the center of the village. Paul runs ahead of us and approaches his grandfather with wide open arms. I am about twenty meters behind. A cold breeze starts to blow across my cheeks and the sky is becoming vanilla with dust mixing colors with the morning sun to produce an eerie effect, a kaleidoscope of colors all mingled with the aroma of humidity that precedes a storm. In the distance, the grandfather is shouting something at Paul with extended arms. My vision is becoming blurry as rain starts to pour as if a huge shower has been turned on over my head. I look around me and notice that the rain is isolated around me and does not seem to be falling anywhere else. I rev up the Adwalk to get to the safety of the house as fast as I can. When I arrive at the front of the house, I see Paul standing there with an old Whiteman. His face is horribly scarred and his eyes are dimmed by skin patches that have healed over time. Despite all the scaring, his deep blue eyes project a beauty that I cannot describe easily. He must have been in a horrible accident in the past. He contours a smile with what is left of his face and looks at me with the curiosity of a child. He hurries and takes the Adwalks into a roofed patio and motions me to hurry inside the house.

"A dangerous storm is about to come, quick, come inside the house," he says with a crackling voice. I can tell that he is perplexed by something but I do not know what. I look around curiously for signs that could give a clue as to what their plan is but all seems to be the way I expect. I move to grab his hand but he retreats quickly as if my hands are dirty. He turns away from me and looks far away as if I am nowhere in front of him. For a moment I feel unwelcome and I start contemplating that I should not have come. Finally, after a few moments of being ignored and treated like an unwelcome stranger he looks at me again with a strange curiosity as if I had done something wrong. Something tells me that I am out of place and unwelcome here. My mind starts to race with emotions as if a tap of feelings has been turned on by the man. I feel great warmth that is like the glow from an old friend flowing from the man, yet he is treating me like an intruder. His calm brown eyes and his ruffled face seem to be in complete control of my surroundings and I feel totally resigned to the power that emanates from his being. He is something of an enigma, and I cannot quite understand what it is that makes me feel that way.

"Pleased to meet you, again Khepera," he professes as if I am some other person that he knows very well. My hands start to tremble as if they have been shocked by his words and I move back instinctively. Did I know this man? Who is he, and why does he call me Khepera? He stares at me as if he is examining every part of my being. The funny thing is I do not feel like he is intruding into my personal space or being, instead I feel like I am intruding into his. After an awkward silence with a long stare he says,

"It is a pleasure to meet you, welcome to Yamandu," in perfect English.

"You speak perfect English! Where are you from? Have we met before?" I ask curiously.

"From around here," he answers in a voice that is almost familiar and then he abruptly says,

"Let us get out of the rain and make you comfortable, my friend," as he leads me into the comfortable coolness of the mud house. As we walk, I scrutinize the Whiteman and his grandson. Where did Michael come from? Why live in such a primitive jungle? Why does he call me Khepera? Khepera is the Beetle god of Egypt and that was the name I called myself when I was giving Elaine the pendant! How did he know that name? Only Elaine knows that I used that name, but even so, I only used it a few times during our fantasy games!

Michael's flowing white beard reminds me of the old prophets, and something about him also reminds me of me. He has an air of holiness about him that makes me feel safe in his presence. A sense of belonging, like the feeling one gets when one visits an old familiar place.

"In Africa we use all-natural material to protect ourselves from the fierce heat, Dr. Davis. But in the outside world you use air conditioners which pollute and cause strange diseases. Please be comfortable," the Whiteman insists. I am perplexed. The Whiteman called me by my title, Doctor! No one had mentioned that title in Africa. How did he know?

We store the Adwalks in a small room of the house and return to the main living room which is a wide-open space in the center of the house. We sit down on large wooden chairs around a table in the center of the room. The musty smell from recent repairs on the mud walls permeates the house. After we are seated, a young girl enters the house from the backyard carrying a clay mug of water. She kneels in front of Michael and kisses his hand. Michael introduces her as Paul's cousin, Mary. She offers everyone a drink and then disappears. I am holding an old ceramic jug the girl had given him. It is decorated with some American football team's logo. The logo is worn off and barely visible. I inspect it closer and recognize it as the logo of the 1989 Miami Dolphins. An antic of tremendous value I estimate as I guzzle the miraculously sweet water.

"Very good natural water" I comments out loud, "unlike the over recycled water of the city."

"But of course, this is the virtue of the hidden world. This is life giving water, John, from our Father M'Dulu's garden," Michael reveals smiling. An old black woman enters the room from the backyard. She is introduced as Paul's grandmother. The old woman sits by Paul embracing and kissing him.

"You have grown so much Paul. I am glad you came sooner. Your great grandfather Bitter Cola has been talking about you constantly. He would like to spend time with you as soon as possible," she informs him affectionately.

"My friend Doctor John Davis and his colleague Hal were sent by the US government to assist us with the agricultural program in Yamandu. But Hal the other man is taken by the Mingas at the great cave. We do not know if he is still alive, Grandpa. We must try to help them."

"By the Minga?" Michael asks surprised.

"Yes Grandpa! They attacked us last night by the river. I called M'Gboko several times, but got no answer. Then the Minga answered our call and we ran into the cave. You know, the cave we used to hide things in when I is a child."

 Paul has not finished speaking when John Davis adds,

"But Hal never showed up. We think he is dead. There's no trace of him!"

"But how did you two escape those dreadful spirits?" Michael asks alarmed.

"We escaped into the great cave by the river," Paul repeats, "but for Hal, it was too late and he did not make it."

"We must find him as soon as possible. Sometimes they take hostages for various reasons, Doctor Davis. I assure you we will do our best for your friend," Michael promises in a soft voice.

"My government will be furious when they found out about Hal's disappearance. If there is anything within their power to bring him back, they will do it," John Davis adds.

"And how do they propose to help us with our rice problem?" Michael inquires. The question was abrupt and John Davis was not expecting it so soon.

"John Davis believes that the problem of low rice production might be solved by using new farming techniques," Paul explains to his grandparents.

"Let the Whiteman explain himself!" Michael rebukes Paul.

"Tell me, John Davis, how can we help you?" Michael offers.

"Well, it is a complicated issue," John Davis begins. "First we must either try to find Hal or at least know if he is dead, and then second Hal and I can concentrate on implementing the agricultural program."

"What do you propose?" Michael probes.

"We have found that a new strain of genetically engineered rice can proliferate here. The mining pits have destroyed the rice farms in this part of the country. My colleague and I came here on behalf of the US government to take some measurements. Hopefully, you and your people can help me accomplish this."

"Of course, we will assist you in any way we can." Michael says smiling.

"You are now a part of our family, and as Paul tells me, you have the blessing of Chief Mobasso. We will do all we can to find Hal," the old man assures me knowing that I can do little in their plush garden paradise.

CHAPTER 20
(The Poro Chief)

Bitter Cola is the oldest man alive in Yamandu Village. The villagers call him "Papa," a respectful name for the Chieftain of a village. Ever since he was a little boy, Paul has admired his great grandfather. He had heard of his wonderful deeds from the village folk. When Paul was born, Bitter Cola anointed him into the Poro Society of Yamandu in a secret ceremony. At that time, Paul was born to my daughter Maria and I did not fit into the norm of Yamandu because I am a Whiteman and I have passed some of my genes to Paul. So, Bitter Cola, and I having no intentions of leaving my Mulatto grandchild to lead a difficult life, did the only thing that would guarantee Paul and his children a firm footing in Yamandu. Unbeknownst to the child's parents, Bitter Cola and I bestowed all powers of the Poro blessings on the young child. Paul himself did not even know that his destiny lies as the leader of the great village movement.

After Paul left for Sefadu with his mother, Bitter Cola and I kept the significance of the secret ceremony to ourselves. Only we know of its importance. Bitter Cola had already decided that the sacred Eye of M'Dulu would guide and protect Paul from all evil. Now that the Chieftain is an old man, the time is right for restoring Paul to his rightful place as head of the ancient society before the great allure of the modern city tempts the young man. He will not go back to Sefadu, never again!

I watch Paul as he helps guide the frail, trembling hands of his aging great grandfather who is trying to drink a cup of Palm wine.

"Who is the Whiteman?" Bitter Cola asks in a hoarse alcohol ridden voice after he takes a long sip. He looks at me and clenches his teeth in anger. I know he is very mad at Paul for bringing this stranger into the unchanging past world of Yamandu.

"He is a friend of ours," Paul answers.

The old man comes close to Paul and pulls his ear close to his lips whispering,

"All Whiteman of the outside world are evil, Son. Do not trust them. They must not find the Eye of M'Dulu!"

Paul rubs his roughened ear and does not say a word for fear of upsetting his great grandfather. He pulls a chair beside Bitter Cola and they begin converse in our native language, Wolof. Then, the voice of the old man grows louder as if he is angry about something. I ask Paul to tell John Davis to wait outside the house. Bitter Cola looks at me and contours his face in a snarl. He does that when he is angry.

"You and I know that this is not right. It is not yet time to reconstruct the Covenants. I promise you we will suffer because of this," he states disappointingly.

I know that Bitter Cola is talking about the covenants. As humans, we are only required to know as little as possible about the covenants, enough to get us through our daily lives.

"I know it is not yet time, but that is the way it has happened. He is here and fate will take its course. What must happen, will happen. I did not expect it this way either," I explain, knowing that Bitter will understand. In the jungle, there is no psychologist. Nature is the psychologist that God has placed in his garden and my hope of every knowing my past will vanish if the Covenants were threatened. Bitter Cola has made solemn promise that M'Dulu will bring my soul back and that I will be made whole again. Paul walks back into the house and Bitter Cola puts a finger on his lips to caution me that Paul is back. The child should not know the afflictions that have held me hostage for newly thirty years.

"Why is the Whiteman here, it is early?" Bitter Cola queries Paul.

"He is here to assist our people to grow rice," he repeats innocently.

The old man shrugs, looking away from me disapprovingly.

"Help us? Nonsense! We have the Garden! We do not need his help. We must watch him closely and we cannot allow him to find the Sacred Eye of M'Dulu. The Minga have already invaded us twice for it," he reminds him.

"Michael, you are the Blessed one from M'Dulu's house. What do you think?" Bitter Cola poses to me.

"The ceremony of the Eye must take place as scheduled tonight. We cannot stop the ceremony because of a stranger, Papa. How do we keep him from the site? That is the question," I tap Bitter Cola's wisdom.

"I don't know. Blame your stupid Son-in-law Chief Mobasso, for sending him here without first informing us. Who knows, maybe he already knows about the ceremony tonight and he has come to spy upon us. Maybe that is why he is here, to learn of our secrets, like that Jack Cousteau used to do when I was a kid," Bitter Cola grumbles as the potent wine meanders into his blood stream.

"That is not so, Great Grandfather," Paul calls out in the Professor's defense.

"The Minga killed his friend. He certainly knows nothing about the ceremony, but I do not yet know his true purpose. I cannot vouch for his integrity. Then again, we know he is not with the Minga. They could have killed him too had it not been for the hidden cave," Paul points out.

"Look! My eyes are already gone with age," Bitter Cola muses.

"The only eye I have now is the sacred Eye of M'Dulu, which only you, Michael, and I, are allowed to see. Keep your eye on the Whiteman!" He advises as he searches the basin of wine for his cup. I decide to walk outside the house and see what John Davis is up to. He is sitting outside the house on a wooden bench watching the children of the village playing on a pile of sand. He must be enjoying the chicken, dogs, ducks, goat and sheep, that are everywhere. Unlike the great cities of his hometown, I know that this is a different and a natural place. I can sense that he is uncomfortable and that he is wondering what is going on inside the house. Perhaps he suspects that we know what he is up to. Perhaps he already believes we have the holy eye and that we can see his thoughts. After all, he knows a mysterious force lurks in Yamandu.

I watch John Davis as he gets up and walk toward a clearing in the center of the village. He looks like a younger version of me when I first came to this village. He is searching for signs of the past in an effort to understand the present. He is an enigma for me and it seems like I have known him all my life. His deep blue eyes and his well contoured face remind me of what I used to be before the accident in Yamandu. His walk, his gait, his looks, his every mannerism reminds me of a fractured past that could have been myself. I watch him as he examines some old mining machinery half buried in the ground. There is old scrap machines scattered all around the village having been left over from the early mining days in Yamandu. The rusted loader bucket I used to enjoy driving about in the mines lies intact in the middle of the field of grass we now call home. If only he knows that he is intruding into his own private life, into my private life. If only he knows that we do not want the outside world to mingle with us. He is a curious man and I know he must be imagining what it must have been like fifty years ago when these monstrous machines stripped away the land in pursuit of diamonds and gold while the fumes from its engines ate away our beautiful atmosphere. I can tell that his mind is examining the evil of the past when old diesel smokers would pollute the world with their fossil energy. I can tell that he is marveling at how much times have changed, how much he has changed. I know he is not here to mine diamonds since they now have artificial diamonds and natural diamonds are no longer aesthetically desired but are now only used only in certain industrial applications.

A few kids dressed in torn and ragged clothing come running to him, teasing and taunting him. He is amused but pays little attention to them. He is still deep in thought. The accident report said all the people in the mining camp of Yamandu died after the accident. What it did not say was that I survived and ended up here in Yamandu village. It did not state that I was rescued by the Poro. I do not recall all of what happened but the past is gone and opening it up again will only destroy my mind and make me remember a life that is gone and past. I do not want to remember the past. I only journey into the past when I drink the potion. Only the potion can take me there and open the dimensional doors of M'Dulu's Eye. Would he be able to find evidence of the accident and the Eye? Would he still be able to locate the force that wreaked so much havoc, so long ago?

I hear Paul calling John Davis from a distance. He had passed right in front of me without me noticing him! John Davis walks back to the house and follows Paul inside. Khepera is here dining with us! The scarab king is here to push the sun and hide it from view. Dinner has been prepared for us. I walk back into the house following Paul and John Davis. I watch his feet as he walks. His footsteps echo everyone I take as if we are time stepped in the same continuum. We sit at the small round table and start eating a fish stew delicacy that my wife has prepared. It is a silent and arduous task. Sitting here eating with a Whiteman that I could have been, had I gone back to the outside world after the accident. I look at the ring on his finger as he swallows the sweet water from our sacred soil. He is wearing the scarab, the scarab of Khepera. He is sending me a message that he knows what we are. He knows where we are and what I have become in this little village. It is like two mirrors facing each other, only one mirror is clean and the other rough and the images they form alternate each other like a series of time stepped fuzzy and clear oddities of the mind. The more I examine myself and my mind, the more I fear John Davis.

After dinner, I ask him to join the family at a celebration of the "Ojeh dance." The Ojeh dance is a splendid performance of aerobics by our local boys posing and dressed like demons. They entertain and scare the children with their antics and strange voices and a few young men and women dance to the music of several drummers, coiling their bodies like the jungle snakes in the ecstasy of wine. As the sun disappears, the celebrating becomes frenzied and the drums beat faster and faster. I watch him as he delicately sips the potent wine, the fruit of the palm tree. I watch as he slowly begins to feel the ecstasy within his being. Soon he will know that it is different here in the jungle, soon Khepera will be one with the jungle. It is surely different here.

He is asleep before the ceremony ends. We lift his body and place him in a comfortable bed where he should spend the rest of the night. He cannot be allowed to multiplex the field of the Eye of M'Dulu. He cannot make the Eye visions of our physical beings multiplexed into one soul. The Poro did not have time to clear the physical site of the accident quickly. They took my spiritual vessel and reconstructed me from what was left of David's body. The rest of my body still had some small shreds of the shattered crystal in them. Then, the helicopters came and cleared the sight and they took all the Brazilians and the rest of me with them into realm of the living. There they made him what he is, a mere shadow of existence scattered into a multitude of realities that even he does not understand. The effects of the crystals slowed time for him and made it possible for him to experience a new life separate from his inner self. Like a new born Zombie, they reconstructed him and what was left behind of me, of my mind. How can he know that I am his ego, his alter ego, his future? I am Poro.

The palm wine has made him forget about Hal, his wife and children. Here in the domain of M'Dulu, life passes at a different pace from the outside world. Here in the deep breast of the jungle, nature is one with Man. Even the dark night does not want to worry about him and his soul will slowly slip into a world of dreams. The reality that has been built for him of a family back home will tugs away at his subconscious mind putting a strain on the security of Mankind. We will make all this right before it is too late. We will unite the covenants for the sake of man. A deep sleep slowly consumes him as he enters the dark dense world of the Poro.

Bitter Cola enters the room and breaks my reverie.

"I prepared the spell" he states.

"He will dream tonight like no other night. He will see Hal living and real. He will run from strange men dressed in ancient Egyptian trappings. He will die in an abyss and he will never know what is happening at the ancient altar of the Great Eye of M'Dulu".

CHAPTER 21
(The Potion of Truth)

Tonight, the cry of the "midnight beast" will be heard again by the children of Yamandu. It is the most dreaded sound in the mystic village of Yamandu. Each year, on the eve of the ceremony of the "Eye of M'Dulu" the children of Yamandu crawl into their parent's bed for fear of the cry of "AREYOGBO," the devil. At exactly midnight the devil will parade through the town to scourge it of its evil. Each household will have await their turn, hoping and praying the devil will Passover them. Very few claim they have actually witnessed the Areyogbo. Bitter Cola had fooled the villagers that the devil is sent by M'Dulu to seek out the enemies of the Poro. The Areyogbo is said to carry a huge and mighty hammer to break open little rocks and crevices of earthen huts that are the hiding places of devils, witches and evil spirits. No one could hide from the Areyogbo. The Areyogbo could pass through walls and swim in the rough currents of the Yamandu River. He is a powerful spirit capable of reading human souls with the Eye of M'Dulu.

According to the legend we have created, the Areyogbo is given to the Poro by M'Dulu to protect us from corruption. As guardians of the sacred Eye, the Poro has to be pure and chaste. People who did not obey the will of the Poro will be destroyed by the Areyogbo. Thus, the integrity of the Poro Society has been maintained for generations since M'Dulu first walked the Earth over ten thousand years ago in the form of a man with the face of a man and the limbs and body of a lion.

On the night of the devil's journey through Yamandu, the cries of a family in evil's path rip through a normally tranquil night. This is the night when all the Poro will come out into the jungle celebrating M'Dulu's Eye, far from the harmful hammer of the Areyogbo. All first born who do not belong to the Poro must crawl into the safe arms of their mothers in vain, consumed by the might of the Areyogbo!

After the Whiteman goes to sleep, Bitter Cola calls upon us and the Poro members to attend prayers at his house before M'Dulu's messenger arrives at the village. Sitting in front of a fire, he places his hands on me, his children, grandchildren and great grandchildren, blessing us from the persecution of the Areyogbo. The old man walks into his bedroom and pulls out an old chest that lay under his grass bed. From within the chest, he removes eight huge diamond pendants strung on solid chains of twenty-four-carat gold. He walks back into the circle of five families that surround Paul and me. He hands a pendant to each of us and to each first born and asks us to cover our eyes. We place the pendants over our heads while Bitter Cola prays to M'Dulu to pass over our homes as he had seen his own grandfathers do in their day.

The pendants would make M'Dulu's messenger aware that they are Poro, keepers of the Great Eye. Then, he drinks from a large bowl of liquid extracted from the leaves of the desert cactus which only grows in the cold dark chamber where the eye of M'Dulu sleeps. We pass the bowl from person to person, taking deep sips. The potion spread its magic to each cell of our bodies, rejuvenating young and old, alike. In an instant Bitter Cola is a young man. He lifts his hands in prayer to M'Dulu, God of the Jungle, God of the Poro, Keeper of the laws.

After the prayers, Bitter Cola, Paul, myself and the others, completely transformed by the magic potion, leave the house for the forbidden astral journey to the ancient hiding place of M'Dulu's Eye. We travel for one hour chanting the song of the ancient Poro en route, until we come to the Yamandu River. Then, Bitter Cola leads the way. No one could remember the way but him. He is the gifted one. After his death, Paul will begin remembering the way. The treacherous path is covered knee deep in elephant grass, biting our ankles and legs as we tear a path through. Armed with machetes, we cut our way through the jungle, with Bitter Cola in the lead. From the pendants on our chests comes a bright blue light that lights the way through the moonless night. At one hour before midnight, Bitter Cola stops in his tracks.

"This is the sacred hiding place of the Eye of M'Dulu!" the old man declares, raising his machete in the air. We fall on our knees, placing our foreheads on the ground to prevent glimpsing the sacred eye. Bitter Cola walks to a clearing in the jungle. The jungle is uncannily silent. All life fears the hiding place of M'Dulu's Eye. He asks us to form a circle around a clearing. Then, with his bare hands he starts to dig a hole in the center. As he digs, a bright flash of blue light fills the clearing, illuminating the forest for a hundred meters beyond. Only Bitter Cola is allowed to see the source of the light. No other should see M'Dulu's Eye. Then, in a wave-like motion, we throw our bodies back on the floor of the jungle, gazing at the sky above with our knees fixed to the ground. The pendants on our chests glow with the light from the Eye of M'Dulu, giving our souls a brief, sweet taste of its glory. We form a circle of glowing pendants about the peeping Eye of mighty M'Dulu.

Bitter Cola cries out as the devil appears clad in bright blue robes with a rectangular metallic headdress bursting with fire. Bitter Cola prostrates himself in front of the Areyogbo, bawling like a little child, making known his fear of the wrath of M'Dulu. The Areyogbo and its blue glow begins circling us.

"Oh, Master of the Heavens," Bitter Cola shouts.

"Oh M'Dulu, my eyes are filled with tears for fear of your mighty Eye. Yours is the glory of the universe carried within your palms. Yours is the fire of the stars carried on your head. We come to praise you and to make our vows with you again. My old body rejuvenated with your blood, and my old eyes opened by the light of your Eye."

Then we chant,

"Open the gate to M'Dulu's hiding place;
So, we can bathe in the light of His Eye;
So, we can be cleansed by His sacred blood;
So, we can smell the perfume of His body;
So, we can fill our lungs with His breath;
So, we can gain the wisdom of His mind."

The Areyogbo stands erect in front of Bitter Cola and raises its hammer into the night sky, stretching its elastic hands far above the forest trees. The hammer rams down on the ground beside Bitter Cola in the center of the circle striking the spot where M'Dulu peeps. The ground flies in all directions. The Areyogbo spits on the hole to fill it with a mysterious blue liquid. Then, with a frightening cry the Areyogbo disappears into a flash of light that lights the skies of Yamandu Village. Bitter Cola immediately stands up and enters the pool of blue liquid, calling the rest us to follow him. We gasp with ecstasy as the freezing cold liquid enters every orifice of our bodies. A tunnel of bright light appears before us as we witness our spirits departing from our bodies. Suddenly, up is down, and down becomes up. Our souls fall into the depths of the blue light that comes from a deep reservoir.

We travel to the central chamber that houses the crystal Eye of M'Dulu. We are exhausted by ecstasy of the journey. Bitter Cola crawls toward the Eye of M'Dulu, to touch its red, pulsating core. Caught in the very fibers of spacetime, our souls are free to roam the universe and instantly understand the secrets of the cosmos. We are ecstatic gods, embracing that power which rules the universe. The pulsating light of creation has opened the door of knowledge to the minds of the Poro. This is the trap that holds the Poro together. This is the secret of our forefathers. This is the ultimate drug that lures us away from civilization when all around us the world indulges in material pleasures.

CHAPTER 22
(Devils and Angels)

I must be dreaming of Hal, but it is so very real I cannot tell if I am dreaming or actually living a true reality. This time, Hal is in charge of a rescue mission with the help of Yamandu and President Massakoi's militia. They are now entering the infamous cave where Hal had disappeared. The water is deeper and darker than before. The President's men are apathetic to our mission and I could hear the wailing of village people as the President's Militia invades their homes to search for Hal. Suddenly, the President's men burst through the mud walls of my hut and I am propelled to another reality as I feel awake. I am trapped in the arms of a strange looking creature! The creature lifts me up into the air but then just as quickly, sets me gently back down on the bed. Our eyes lock as the intense blue flame from the creature's metallic head gear illuminates the room with a blue glow. The creature is staring at me like I am a zombie or something it has never seen before.

"Resa!" it shouts, pointing at my steely grey eyes. I am puzzled. Then, I remember that the Ayatollah Resa also had steely grey eyes. He must have associated Ayatollah Resa with my grey eyes. How strange! As the creature kneels before me, a shadow dressed like an Egyptian warrior enters the room aiming a spear at the creature. I rapidly shove the creature out of the way of the sailing weapon.

"Areyogbo!" the Minga warrior cries out running. The Areyogbo turns on its attacker crying, "Minga! Minga!" Its mighty hammer strikes the warrior in the face, killing him instantly as his blood splashes all around the little room. The room is swarming with Minga warriors poised to kill with snakelike spears. They are after the creature it seems. I quickly reach for the case under my bed and take out the Lazukar. One single-minded Minga closes in on me as the Areyogbo swings his hammer at three other Minga predators and pulverizes one of their faces.

The fair skinned Minga moving in on me aims a spear at my heart. But I fire the Lazukar and vaporize his brain. The smell of burnt flesh fills the room pumping fear into the remaining Minga. The Areyogbo taking advantage of the distraction swings his mighty hammer at their skulls, scattering blood, bones and brains all over the room. The Areyogbo acknowledges the awesome power of my weapon with a nod of respect and lays his huge hammer down at my feet. A silent truce forms between us in the deep virgin jungles of Africa. I have forged an alliance with the devil against the invading Minga warriors from the North. From the room, we could hear the cries of the captive women from deep within the village. I set the Lazukar to maximum power and motion to the Areyogbo to stay by his side and protect my back. It is a fortunate coincidence that the light from the creature's flaming headdress continually charges the Lazukar's solar batteries. Swinging its hammer in all directions, the Areyogbo terrorizes the surprised Mingas with its agility while I evaporate their brains and bodies into thin air. Before long, our alliance has massacred many of the Minga warriors and we are able to stave off the attack from the center of the village.

As the battle wages on I suspect that Bitter Cola and his family are too drunk to participate. They must be enjoying their dreams of the delights of the mystic wine as they lye drugged from the cactus potion in their beds. I run back to then house to wake them up and hope that they can help stave off the mysterious warriors that have invaded the village. I pushed on Bitter Cola's body and then run to Paul and Michael. After a long effort, they finally wake up as the flaming light of the Areyogbo enters the room. Bitter Cola's family starts to scatter in all directions as they realize that the Minga are invading their village. They run and arm themselves to help in the battle, but Bitter Cola himself is too weak to join them. The battle, however, is almost done, almost won. A stray Minga fleeing from the village meets Paul's spear as the young man curses and looks around the plundered remains of Yamandu. Another retreating Minga takes advantage of the ailing Bitter Cola tearing the old man's chest open with a throw of his spear. The old Chief still has enough left in him to throw a large hunting knife at his attacker's throat, killing him instantly.

Michael is the first to arrive at Bitter Cola's side as he collapses to the ground. He holds the old man's head up off the ground. Bitter Cola is bleeding badly from the deadly wound. Bitter Cola motions Michael to lean in closer and lend him an ear. I run toward them and stand guard looking for any approaching Mingas.

"Michael, it is the end for me," he mumbles, blood flowing from between his half-opened lips as the poison of the spear spread through his veins.

"I am going to M'Dulu now," he whispers in a weak voice. "Take care of Paul."

He pauses, looking into Michael's eyes.

"You will not die," Michael blubbers the words and Bitter Cola smiles.

"We have come full circle, Michael," he mumbles.

Michael smiles as tears fill his eyes.

"Remember when the Eye of M'Dulu nearly took you?" he reminisces.

"I held you in my arms and nursed you back to life just as I am dying in your arms now."

He pauses and takes a deep breath, struggling to inhale as his lungs become weaker and blood spews from his open lips.

"Paul is a chosen Apostle of the Earth. He will be head of the Poro from now on. Guard against Thamu Yombo. As for the Whiteman, may M'Dulu bless him forever. He has saved our people, as you did a long time ago."

Laying his head back into the embrace of M'Dulu, the old man closes his eyes for the last time. When Paul sees his grandfather running to his great grandfather's side, he also hastens to see what has happened, but by the time he arrives the old man is dead. He kneels by the bleeding dead body and cradles it in his arms. He shakes the body violently and drenches it with his tears.

"Grandpa, Grandpa, stay with us," he cried out. Bitter Cola is dead. Raising his spear, Paul curses at the Minga and starts running toward the jungle. Michael intercepts him gripping Paul's spear firmly.

"This is the fate of your Grandpa, Paul, life changes hands and you must be strong and brave."

After consoling Paul, they take Bitter Cola's body back to his home, and lay his body on a ceremonial mat. Areyogbo and I venture outside the house to take in the aftermath of battle. Some villagers are rejoicing our victory while others cry for their dead loved ones. Some have lost their first-born males to the Areyogbo. In his desire to preserve the power of the Poro, the Areyogbo has killed the first born and those children with the potential to take over the power of the Poro from the Bitter Cola sect. It is what Heron did to protect his crown from Jesus. As daylight breaks, the Areyogbo curiously examines my weapon as if it is some magical thing. I notice that the flames on the creature's head gear have been replaced by puffs of black sooty smoke and that the air carries the smell of kerosene fuel, incense and sulfur. The morning light reveals an ordinary man in a blue menacing cloak that was the devil called the Areyogbo.

As the excitement of the battle wears off, the villagers start to gather around the creature and me. The disappearance of the flames of the Areyogbo piques the curiosity of the children and the women. One by one, the angry women gather with their children slowly forming a wall of bodies around the creature and myself. They jeer at the man who appears not to be a devil after all. A female leader with a face partly shrouded in a gold-threaded scarf, approaches us. She is covered in gold trinkets and bracelets of diamonds.

She cautiously approaches the Areyogbo and touches its shoulder quickly withdrawing her hand in fear. Some bold children start to tug at the Areyogbo's red cloak with fear and hatred. As more and more children gather around the creature, I find that I am not their target after all and so I reluctantly place my weapon in its holster as a precaution for the children. They start tearing the Areyogbo's clothes. A crowd of young women with naked breasts and scanty skirts of raffia lead me away to the village center while another crowd of angry women corner the Areyogbo, the imposter, cursing and throwing stones and sticks at him. With his head gear knocked off, they discover that he is a young man of about thirty years. A woman in the crowd recognizes him and starts to call the creature by his name.

"Poro Bobo, it is Bobo!" she shouts.

A couple of the more daring women take the head gear from the floor and raise it high in the air dancing and running around the village square with elation. Children take to the streets singing songs of victory as they hurl insults and stones at the man they used to fear and call the Areyogbo. The mystery is gone and new era has begun for Yamandu. Several women strip the Areyogbo naked and start stoning him until he starts to bleed like an ordinary man. He cries in pain as he is dragged with his ankles and wrists bound around the village center while the villagers chant victory over the despised man.

I on the other hand find myself showered with flowers and fruits. Three beautiful virgins lead me away from the village and several children follow behind us beating drums, sticks and pans. I am the hero for saving the village from the Areyogbo and from the Minga invasion. They tell me that I will partake in the celebration of victory at the home of the "Queen Mother." I am escorted to a secluded portion of a small house at the end of the village. When we arrive at the Queen Mother's home, a large pool has been filled with perfumed waters and lined with fresh fruits and vegetables. The women start to perform a strange ritual inside the pool de-robing themselves one by one and entering into the pool. I am de-robbed dragged into the pool of water with all the women. They start to bathe and massage me and scatter the privileged waters over my body with their small graceful hands.

Following the victory bath, they hand me a long white robe and a pair of white leather sandals made from dried snake skin. They escort me to a woman in a crowded room as if I am a bridegroom on my wedding day. She presents me with a Scroll written in a strange language. I recognize it as the language on the Spheres of Abu Simbel!

"My God," I cry out in disbelief. The jeweled woman proceeds with the introductions.

"What is your name, young man?" she asks.

"John Davis" I answer.

"You are a victor, and according to our traditions, we must make you a hero in Yamandu," she explains.

"Who are you?" I ask.

"I am Queen Mother, Yabu Yombo," she answers proudly, lifting her head high.

"I am the next in line in the royal family of Yamandu. I am now ruler of the Poro of Yombo!" She decrees.

"What is this scroll?"

"It is handed to us by generations of Poro leaders. It comes from the Garden of Life, a place created by M'Dulu for us a long time ago," the Queen Mother reveals.

I examine the material of the scroll. It is made of Papyrus, no different from those used in ancient Egypt. The symbols on the bottom depict the ceremony of Zoroaster's Victory Bath and wedding to a Virgin, a ceremony common to Ancient Egyptians! This "Bath of Victory," is the very ceremony I am experiencing. At last, here is complete proof of the theory I have harbored for so long. For here, in the middle of the African jungle is knowledge of the ceremony of the Ancient Egyptian religion of Zoroaster, practiced more than three thousand years ago in Ancient Egypt! Here I am with the opportunity to find the secret of the crystals and possibly a chance to decipher its strange writings. I am closer to solving the greatest mystery on Earth!

I know this ceremony well from my archeological studies. I start to participate as if I am an adept. I perform the part of the ceremony I had learned in my studies. I lift the scroll and close my left eye, symbolizing the setting sun while opening my right eye as wide as I can to symbolize the rising Sun's birth. Shocked at my knowledge of secret Poro rites, the Queen Mother beckons the other women to fall on their knees in adoration. For an encore, I performs a ceremony written in the "Second Book of the Dead," discovered in the ruins of Abu Simbel in 2015 A.D. An extremely beautiful female priestess enters the room, to join us. She wears revealing white robes that expose her naked body in the center when she lies on the floor and raises her hands to the sky.

I repeat the ancient words from Zoroaster's fertility dance.

"Soi Toi Doo Manaka!" I say while I sway my head back and forth. The women at once fall to their knees and crawl towards me to worship me. The women prostrate themselves in readiness in honor of their new fertility leader, Zoroaster, high priest to their god. I realize that playing my part well is now crucial to unlocking the crystal's power.

"I have found my bride, Oh Holy Master of the Underworld, Oh Master of Orion, King of the Earth and ruler of the Ancient Worlds, Oh Master Zoroaster," I recite from my memory of the Ancient Egyptian manuscripts. I stand up to tower over the naked priestess before me. Her skin is shrouded in gold-threaded linen and her hair is adorned with delicate shiny silver beads and diamonds. A large sparkling diamond gleams from the middle of the priestess's temple. To me, she is the personification of the Crystal.

"Bring her here!" some woman orders.

A dozen other women crawl toward the priestess and raise her from the floor. They present her to me eulogizing in a language that is a mix of Wolof and Ancient Egyptian.

"This is the priestess of mercy, the woman chosen by Zoroaster. Take her and make her your own."

The high priestess leads me into a larger bedroom of the house and closes the door behind us. She announces in flawless English,

"I am your scarab now my lover, you have saved my people from certain death and I am your gift from the underworld."

She strips herself naked and falls voraciously upon me. I lay on the bed as if I am a zombie. I let her do what she wants and pretend innocence. I am too tired and too engulfed in thoughts of the crystal to consummate a marriage. Soon it would be mine for the taking. Soon!

CHAPTER 23
(The Poro Court)

Michael Sparks is standing in front of the courtroom of Yabu Yombo, the "Queen Mother" of Yamandu Village. Beside him is Paul, his grandson, and the badly beaten man who wears the costume of the defamed "Areyogbo" of the Bitter Cola Poro sect. Several other members of the Bitter Cola Poro sect have been captured and brought in front of the court of the "Queen Mother." After the death of Bitter Cola and the exposure of the Areyogbo's mystery, Yabu Yombo, the high priestess, immediately establishes herself as the rightful heir to the original Poro throne. With the support of nearly all the women of Yamandu, the Queen Mother quickly sets up a complete chain of command and demands the arrest of the existing Bitter Cola Poro sect. The arrests are not resisted by the sect, who in their ecstatic ethereal vigil the night before, and devastated by the sudden death of their leader, had no energy to fight back.

I am seated beside Yabu Yombo who is now "Savior of Yamandu and Servant of the Poro of M'Dulu." Just as Bitter Cola had brought a Whiteman into the Poro Society, so too would Yombo do the same. I will be the symbolic truss with the Eye of M'Dulu, a sign for the commoners to know she also is a chosen leader. Beside me is the young and beautiful priestess who is now my bride according to Ancient Egyptian law. She is no longer a virgin, we spent the night making love and learning all we could about each other. Her name is "Zola," which means "gift" in Poro and she is a member of the Queen Mother's one and only holy family tree in Yamandu Village.

We are now in the courtroom and Zola's task is to prepare me for my inevitable position as a major member of Yombo's Poro Society. The elderly Queen Mother sits in a raised chair of the courtroom, dressed in a delicate, black silk decorated by a diamond studded emblem. The emblem is an image of a symbolic bald eagle with a knife stuck through the center of its breast. The eagle wears a crown of jewels on its human-like head. Its bulging eyes betray no pain from the stabbing knife. In its massive claws is the dead body of a huge black snake. The snake bleeds from the deep wounds inflicted by the gripping claws of the majestic bald eagle. The eagle appears to be in flight.

The Queen mother also wears a wig of pure gold hair, made from finely woven strands of twenty-four-karat gold thread. Her Poro crown is made up of the greatest assortment of diamonds, sapphire and emeralds, each stone set to produce the effect of a jeweled snake wrapped around her head of gold hair. Her wrists and arms are covered with a continuous bracelet of solid gold and diamonds pendants that extends over her shoulders like armor. Paul, Michael and their family members are seated in front of the Queen Mother facing the female elders. The man who used to be Areyogbo, now known as Abu Bobo, has his hands and legs tied to a pole placed in the middle of the court house. A lot of family members are sobbing quietly as they await their family member's fete.

I know that the Bitter Cola sect is on trial for the misuse of the great power that was given to them by M'Dulu, the keeper of the eye. I learn from Zola that for fifty years, the Bitter Cola Poro sect had transformed itself into a society of terror, capable of imposing its will on the people of Yamandu. The introduction of the Areyogbo to the people of Yamandu started only after the great accident of 1984 when the young Bitter Cola was the first to wear the menacing suit.

In the 1980s, the Poro were divided over the political control of Yamandu Village. Before the great tragedy that destroyed an entire mining colony in Yamandu in 1984, the Poro were a peaceful people, coexisting despite their differences, under the leadership of an old man called Thamu Yombo. Thamu is the last of an ancient line of holy men and women called "Guardians," who ruled the Poro of Yamandu. Back then, everyone in the village was regarded as a Poro member and the Poro were the Guardians of the great Eye of M'Dulu, whose secret had been kept from even the local governments. The secret hiding place of M'Dulu's Eye was unknown to all but Thamu Yombo. Bitter Cola was then a young man of great standing in the village. In 1960, Yamandu was invaded by the brown skinned Moroccan sect of the Poro called the Mingas. They came to Yamandu using ancient routes through the forests. It was rumored the Moroccans had journeyed from their land to Yamandu via a winding, underground cave route called the "Snake of the Gods." The leader of the invaders, Idra Contifilli was believed to be a devil spirit of tremendous power. He was also known as "Kassila," ruler of the Yamandu River since ancient times. After three years of fighting against the invaders from the North, Bitter Cola organized a surprise attack on the invaders and drove the enemy out of Yamandu with the help of other villagers. Bitter Cola became acclaimed as great leader of the people of Yamandu.

Bitter Cola was no ordinary man. He was born on a day when the sun, moon and the rain saw one other for two continuous days. In Poro tradition, Bitter Cola is important, since he had a birth mark of the "white elephant" of the Yamandu forest. After the invaders were defeated, Thamu Yombo outlawed the ambitious and powerful Bitter Cola from taking part in Poro ceremonies in an effort to protect his sacred lineage. This offended Bitter Cola and his family very much. Bitter Cola decided to leave Yamandu with some of his faithful followers and set up a new village just a few kilometers North of Yamandu beside the banks of the great river. There, they lived for five years, terrorizing people from Yamandu village whenever they came close to the river banks. Yamandu village was completely cut off from the outside world during this reign of terror. No one from the village could leave by way of the river, nor could outsiders enter Yamandu. Children were abducted from Yamandu, and indoctrinated into the ways of Bitter Cola's Poro sect. The Yombo sect blamed a strange tribe from the North called the Minga for their misfortune. They had broken a pact with the Yombo sect to rid the Bitter Cola sect of the villages nearby. The Bitter Cola sect grew rapidly and became a strong force in the region. Soon Bitter Cola became a recognized leader in the rest of the forest. In other circles, stories abounded of Bitter Cola's alliance with the devil Kasila.

Although the Bitter Cola sect did not know where the Eye of M'Dulu was kept, they remained committed to finding it. Then the diamond rush of the 1980's in Sierra Leone exploded just before the blood diamond wars. The government allowed foreign corporations the rights to mine in Sefadu and many other areas, including Yamandu. At that time, Bitter Cola recognized an opportunity to use the extensive mining activities to his personal advantage in the search for the Eye of M'Dulu. The machines can uncover the Eye faster than hands. He went to work for a large mining company called Sierra Mining Company. Bitter Cola soon became a foreman of the mines, and this gave him the opportunity to employ a lot of his own people to work the mines. The new Bitter Cola sect grew rapidly as fresh recruits in the mining camp site joined his cause. He prospered and won the respect of some of the people of Yamandu and surrounding villages with his new-found influence in the mining camps.

Using money from the mines, Bitter Cola and his sect purchased swords and spears from nearby villages such as Kunduma and Kerang. He built an arsenal of arms and a powerful army. Meanwhile, Thamu Yombo died leaving his young daughter Yabu Yombo to take the throne as Queen Mother of Yamandu Village.

When Yabu Yombo was crowned Queen Mother, Yamandu village was already weakened by Bitter Cola's barricade of the village. The Bitter Cola embargo had converted Yamandu from the most significant trading community on the basin of the Yamandu River, to a society that was merely able to feed itself. This weakened Yamandu and the popularity of Yabu Yombo, while Bitter Cola prospered by virtue of his influence in the mining community.

When the great mining catastrophe occurred, Bitter Cola and Michael Sparks survived and were able to find the hiding place of the Eye of M'Dulu. The catastrophe was so severe that the landscape was changed completely by the huge earthquake that shook the territory. The Eye became buried under a new landscape and Yabu Yombo lost the power of the Eye and soon lost control of Yamandu. Only Bitter Cola knew the new hiding place of M'Dulu's Eye. He launched a terrorist campaign in the forests of Yamandu and made life miserable for Yabu Yombo and her sect.

The mysterious presence of Michael Sparks, a Whiteman in the Bitter Cola sect, had baffled the Yombo sect. With the help of his family and Michael, Bitter Cola united his

Poro sect and created the terrible Areyogbo to control the minds of the people of Yamandu. For over fifty years, the Areyogbo had become a symbol of the might of the Bitter Cola Poro Society. But ironically, it took another Whiteman, me, to topple the power of the Areyogbo and the Bitter Colo Poro sect. The people of Yamandu, though loyal to Yabu Yombo and his holy family tree, pretended that they approved of the Bitter Cola Poro sect. But with Bitter Cola out of the way and the Areyogbo discredited, the people decided to revolt and take over the "Eye of M'Dulu."

CHAPTER 24
(Yamandu Village)

The surreal dream Father Mulcahy has just experienced dazes him. He saunters off the Ion-Flash propulsion Jet and through the exit door of the immigration office believing there is a fine line between reality and the supernatural. Did he cross the line? He has almost convinced himself that he had. A few parishioners he knows have confessed to witnessing the transcendental. One even claimed to possess a power that willed solid objects to move. Perhaps his faith is in question. Outside the gates of the airport, a man called Imran embraces the confused priest and loads his belongings into Saint Anthony's Parish MPV limo. As they make their way back to Freetown, Father Mulcahy wastes no time explaining his mission to the Vatican aid.

"I am here on a very important mission, Imran. The Holy Father sent me to investigate a strange and serious case of tremendous interest to the Church," he reveals urgently.

Imran tunes into the note of alarm in the priest's voice. He nervously glances up at the driver to make sure he is still focusing on the roads. The driver's posture remains stiff with attentiveness. A mild unseasonable rain begins tittering against the windshield.

"And why has the Holy Father chosen our Parish to do this?" Imran inquires.

"He did not ask me to come here," Father Mulcahy admits.

"I came of my own accord to seek out a Dr. John Davis. We got word that he is here on behalf of the United States Government, on a secret Mission," the priest lowers his voice steadily. I know that he is in a very tight spot because I recall from my mining experiences how difficult it was to get any type of information from the government.

"I have not heard of this man," Imran booms back excitedly with his strong Irish accent.

"But what about this Davis man? What about the US Government?" Imran asks.

"He holds the key to a puzzle that has long troubled the Vatican. I cannot explain all of it at the moment, but we need to find out where he is immediately. The GPS satellite system traced him to a small town called Yamandu, in the northern part of this country," Father Mulcahy explains.

"Yamandu! Land of the Poro!" Imran exclaims. Of course, my immediate reaction is to flinch if you can imagine a spirit aura flinching. I worked in Yamandu before the accident and I find myself wondering if anything exists there now.

"You know about the Poro?" Father Mulcahy is surprised.

The priest snaps to attention. He waves his hand in a circular motion signaling he wants to know more.

"Everyone in Africa knows about the Poro," Imran answers, waving his hand and smiling.

"They fought off the church for fifty years in the North. We have attempted to build a church in Sefadu, and we met a lot of resistance from the Poro. They terrorized and killed many people out there," he chronicles.

"What else do you know about the Poro?" Father Mulcahy beseeches him enthusiastically.

The superstitious Imran looks around maliciously as if to scare away evil spirits that might have been eavesdropping on the conversation. The rain is coming down steadily now and Father Mulcahy rolls up his window to avoid being drenched. The driver leans into his steering wheel.

"They worship an ancient god-king, said to come from the Northern part of the African continent," Imran whispers quickly.

"No one has been able to infiltrate their culture but there are rumors that a Whiteman lives among them. But what has the Poro to do with this?"

Father Mulcahy takes a deep breath. I can tell that there is a tension building up inside him that signals the beginning of a long, spiritual sojourn.

"The Holy Father has known the Poro for a lifetime and has finally discovered a link between the Ancient Poro and the Famous Josephus-Nambu theory."

Imran's Earth-green eyes flash with a heavenly glow of hope.

"We believe that the Poro may hold one of the keys to the mystery of the Lost Arc of the Covenant!" Father Mulcahy fuels him.

The priest's charged proclamation galvanizes the flustered driver and the MPV comes to a screeching halt to avoid running a red light.

"The Lost Arc? Josephus-Nambu theory?" Imran asks almost disbelieving.

"Yes, the Lost Arc, but it is not so simple," Father Mulcahy cautions. The priest rests his shaking hand on the seat to steady himself. Small converging gusts of wind choreograph several sheets of rain into a heated dance.

"You remember the Ayatollah Resa?" He probes Imran's memory.

"Yes, the great Egyptian prince."

"Well, when he died in Egypt, there was a lot of mystery surrounding him and the Abu Simbel Temples. We found out that the Poro temples and the famous Resa Temple are indeed connected to the ancient Poro teachings, as well as to the Josephus-Nambu theory."

"I have heard of this famous theory, but frankly I don't understand it," Imran bites his lip in embarrassment.

"The Holy Father believes that the Poro somehow obtained the secret of the Lost Arc and that Resa and his men are trying to recapture it!"

Imran stares at Father Mulcahy like an ignorant child as the MPV plows through the old, impoverished sector of the city. He looks around as if the buildings are listening to the priest.

"But you must be joking, Father!" Imran loosens his grip on the console and brings his two hands together in a gesture that is half applause, half prayer.

"Surely," he continues, "the lost Arc is something of the past. Are you sure you know what you are talking about?" Imran challenges the holy man's theories with a nervous giggle.

"But this is a serious matter!" Father Mulcahy counters. "Why would the Vatican joke about such a serious matter?"

"But Resa is dead!" Imran interjects.

Mulcahy takes in the whole of Imran. Imran is a more modern priest, with a propensity to silk ties and a hint of expensive aftershave. He is not the conventional priest and must have little regard for esoteric traditional dogma. He is the type that encourages guitar players in the Church.

"Oh! So, do many believe," insists Imran concentrating on his own echo.

"No one has found his body as yet. There is no doubt in my mind that he is in Ethiopia, or Morocco, somewhere there, trying to put the Arc back together!" Mulcahy states, trying to convince the priest. Imran gives Father Mulcahy a funny look as he turns away to take in the market women who are posed statuesquely under the frayed green awnings of the sidewalk stores waiting for the rain to subside. He turns the radio on and starts scanning for a news station.

"I can't get the damn thing to work properly!" Imran blows out to his driver.

"It is always full of static," he adds.

"Get a satellite radio for God's sake," the driver retorts. I sense a good long relationship that is friendly. Turning to Father Mulcahy the driver complains,

"This man always says he will get a satellite radio and at least a CD player for the car, but he never does. Tell him it is not his money!"

Imran disregards his long-time driver and starts thinking about what Mulcahy is saying. He is uncomfortable with what the priest's words suggests and is trying to shake the moment by busying himself with other things. The radio still does not respond to his efforts. He settles and keeps his focus outside the car looking far away.

"You do not believe me, do you?" the priest intonates softly.

I can sense an enveloping aura of the rising tide of anger in the priest. He has no patience for this kind of speculative talk.

"I believe you are insane, Monseigneur, if you believe such a story!" Imran states in a very serious manner.

"It is the Pontiff, is it not?" Mulcahy almost dreads saying it.

"What do you mean by that?" Imran shouts, whirling his attention from the radio back to his fellow priest.

"I mean, it is the same old prejudices that have broken this church apart. You do not believe me because you do not approve of a Black Pontiff," Father Mulcahy's voice rises with frustration.

"Far from it, Father," Imran defends himself. I know that Imran's aura is lying.

"I am not a Separatist and you should know that from my files. If someone else were speaking of the Lost Arc in such a manner, I would again consider them insane."

The young priest sits back into the comfort of his plush seat and folds his arms with the repugnance of a wronged king. He looks at the driver and reassuringly smiles. The driver smiles back and nods approval. Mulcahy breaks the silence.

"Look Imran, I know this is hard to swallow, but it is the truth. I have come here to work on behalf of his Holiness and the Church and I need you to help me."

Imran remains steadfast in his beliefs. He looks outside the window for some inspiration and finds thirsty ground greedily accepting a rain that mellows back down to a drizzle. Father Mulcahy presses a lever on his arm rest to bring in some fresh air through the window.

"What proof do you have that you are for real? What if you are just another crackpot, proposing a new theory to romance with the world?"

Father Mulcahy is taken aback by Imran's choice of words.

"I am not a crackpot, nor am I here to romance with the world. But I know that something serious is happening and I am here on behalf of the Church and Christ. Only faith can prove me right so you have the right to remove yourself from this all right now. But as a servant of Christ, I implore you to listen to what I have to say and believe. It is all I ask of you."

Imran believes that the priest's fervor is sincere but he is still unsure of the truth.

"So, what do you want me to do for you in Africa to stop this John Davis?" He shrugs his shoulders compliantly.

"Well, first of all, it is not John Davis that must be stopped but Resa and his followers. John Davis has the secret files that the Vatican needs to find the Arc."

"The Arc! The Arc!" the young priest throws his ringed hands up in the air.

"Surely Father, the Arc is long gone, a myth, a story for lessons. And Resa is dead, so your problems are over!" Imran jests but more seriously this time. Father Mulcahy is silent for a long while trying to piece together his strategy. Imran is studying the driver's white gloves when the priest finally speaks.

"I must first find Doctor John Davis in Yamandu. Could we at least start there?"

"The Poro!" Imran sighs,

"You are crazy man. I will never go there, count me out buddy."

As the MPV speeds toward downtown Freetown, Father Mulcahy begins reading some documents he borrowed from the Vatican library on the Poro Societies of Africa. Despite success at penetrating most parts of the West African coast with religion, Sierra Leone had remained hostile to foreigners. In the jungles surrounding the country known as the "Whiteman's grave," many missionaries were kidnaped and cannibalized. The Poro, who are believed to be able to repel the Whiteman's weapons, and are the most feared tribal sect in West Africa, are at the heart of these terrors. Although the local chiefs and tribesmen do not publicly acknowledge the existence of the Poro, the Church knows many are secret members who participate in the mass murders and sacrifices performed in the name of the Poro god, M'Dulu. Imran breaks the silence.

"Look, Father, I cannot go with you. But I will be glad to help you any way I can. I must be here in the diocese of Freetown to do some important work."

Imran is making excuses and he carefully avoids the eyes of his fellow priest. Mulcahy gazes back down at his documents and Imran begins eyeing him suspiciously.

"My health is failing and I tire quickly," he breaks down.

"When I was younger, I mounted a bold expedition into Poro land. But that is a long time ago. I have not been there since. I will find you a good aide who can help you better than I can." Imran pledges.

Father Mulcahy continues to read silently, deciding not to respond. When they arrive at the large Parish house in Freetown, Imran decides to call on Bumba, a local native of Sefadu who works at the mission in town.

"Bumba was born in Sefadu," he explains to Father Mulcahy.

"He is well accustomed to the ways of the people of Yamandu. He should serve you well, Father."

Imran walks over to the telephone and dials a number. Father Mulcahy could hear the Vatican aide urging Bumba to join them quickly.

"What makes you think," Imran resumes "that the Arc is here, and how does the Holy Father know that it is the Arc?"

Father Mulcahy will try to convince him once more.

"We know that the Arc is the most powerful object in the world. It is described in one form or another by almost every religion on Earth. In the Poro Society of Nigeria, they have traditions dating back to antiquity about the Arc and its powers. The Bible talks of the destruction of the Earth by a great Flood. The Arc has power to lift huge rocks high into the air. It has a mysterious glow. We know that the same cubic pattern as that was found in the Josephus-Nambu theory is used to describe it. We know that the Josephus-Nambu field equations describe z-particle formation. We know that the global cooling associated with z-particles has been detected at twenty-seven points around the Earth. We know that these points form a symmetry that is unique to the cubic field transformations of the Josephus-Nambu field. I do not have to explain what this field is. It is a complex subject."

"Try to explain in simpler terms!" Imran insists.

Father Mulcahy ignores him, his monolog is for his own benefit, not for the skeptical Imran.

"We know that Resa is involved in Abu Simbel. I had a terrible dream on my way here," he confesses.

Father Mulcahy glares at Imran with a piercing seriousness the aide had never before witnessed in a priest. There is turmoil within the Vatican priest and suddenly the bristles of his graying beard, the arch of his frown, are features and gestures of an individual, a common man, outside the wing of the Church.

"This is now personal for me," Mulcahy continues.

"I saw Resa with two other Egyptian Pharaohs in a temple that is structured around twenty-seven crystals, like the cubic field. In the temple were twenty-five crystals. The two missing crystals might be the Abu Simbel crystal and the Yamandu crystal. We believe that Dr. John Davis is here searching for one!"

Scattered rays of pristine sun from a window speckle Imran's dumbfounded face as the priest leans towards him.

"I believe that the Arc is scattered around the Earth," Mulcahy speaks with his hands.

"When God hid it from both Man and the devil. He hid it deep within the crust of the Earth until Judgment Day when He will put it together to return the Universe back to its fundamental-particle beginning, its spiritual form. I believe that the final battle for the Arc of the Covenant has already begun! Imran! Armageddon is here!" The priest announces excitedly.

Mulcahy's charged proclamation galvanizes Imran.

"Look Father, I am not a Physicist," Imran retorts almost breathlessly.

"I know nothing of the fields you are talking about. I have a very sensitive political position in this place. I do not wish to be associated with such a wild and unconfirmed theory."

He begins pacing nervously around the room.

"The theory is not crazy!" Mulcahy debates. "It is very well established and accepted by the Vatican. Surely you must have studied its basic principles in Theology school?"

"Well, I for one never thought it had any particular practical significance for religion. It just surprised me that such a highly mathematical theory has anything to do with a savage native tribe such as the Poro. Now that is weird!" Imran declares shaking his head to put an end to the discussion.

The priest rekindles the fire.

"I know this theory very well and as a priest, I cherish its implications. It is a real theory of the creation saga. It is equivalent to the church's theory of evolution," the priest protests.

"How do you know all this so well. This is just a theory for scientists. You are a priest. Are you so familiar with the Josephus-Nambu Theory?" Imran pursues him.

"I am Josephus Mulcahy, the priest who discovered the theory with Nambu!" he answers gently.

"You! You are Josephus Mulcahy? The famous Vatican Mathematician?"

Imran takes a step back from the priest as if he is being attacked.

"Yes Imran, I thought you knew that already. I am the man you have heard of," the priest smiles humbly. The pages of Imran's scholarly readings flash before him. He is flushes from the revelation.

"My God! Einstein is in our midst. I did not know. I thought you were just another crackpot!"

"Yes, I am the Father Josephus Mulcahy of the Vatican School of Science and Theology," the priest attests to it smoothly. Imran finally acknowledges the severity of Mulcahy's words. Here is a famous priest and scientist sent by the Holy Father to search for the Arc of the Covenant. Perhaps it is all true and the end of the world is near. There is a knock on the door. A seven-foot, three-hundred-pound man enters.

"Bumba! How are you my friend?" Imran greets him with a handshake.

"Meet Father Mulcahy, a famous man sent to us directly by His Holiness himself!"
Bumba smiles at Father Mulcahy, exposing an ugly set of broken brown teeth. His mammoth hands grasp the hand of the priest.
"Hello Bumba," Father Mulcahy reciprocates quietly.
Imran wastes no time. He explains the situation to Bumba. The priest relaxes his tense shoulders. At last, Imran is taking him seriously.
"The Poro!" his new acquaintance begins,
"Ah yes, the Poro! People of the Forest. They are difficult people my friend," Bumba admits in broken English.
Imran and Bumba are suddenly members of Father Mulcahy's quest. They hurriedly pack suitcases while the priest waits on the patio drinking hot coffee and watching the afternoon sun spill its beautiful yellow light on the garden below. There is little he could do without John Davis.
His thoughts wander back to the strange cubic hologram Nambu had created earlier in the Vatican labs. How could such a force become part of a Poro ritual and what is the Egyptian connection?
"Bumba is ready Father Mulcahy," Imran interrupts.
"We should proceed to the bay to catch a boat for River Yamandu. The last boat usually leaves at midday. We must hurry."
He hands Father Mulcahy an atlas of the area then escorts him outside to the waiting cab. Bumba is already sitting in the front seat.
"The Poro! Ah yes!" is all he says, banging his hand against the side panel of the vehicle's door. Imran speaks to the cab driver in a native dialect as the vehicle speeds away from the house. The bay is choppy under a slate-colored sky when the taxi pulls into the dock. The group charters a French tourist boat captained by a silver-bearded, stubby man whose arms are covered by tattoos.
"The land of the Poro! That is a dangerous place my friend. What are you going to do there?" the man is intensely curious.
"We are searching for a friend," Imran tells him.
"Well, I will take you as far as the River Yamandu. I will not go down river to the towns. That is too dangerous."
"That will be just fine," Imran comforts him. The waters of the Freetown peninsula are now calm. It is a hot and humid despite the clouds. Father Mulcahy even sweats inside the air-conditioned cabin. Bumba is being lulled by the swaying surf against the boat. The priest summons him over.
"So, Father, where are you from?" Bumba is interested.
"Rome," Father Mulcahy replies, "from the heart of the Vatican city."
"What brings you here to look for the Poro?" Bumba continues.
The priest explains why it is so important he finds Dr. John Davis.
"Don't worry Father, Bumba is a trusted servant of the Lord," Imran assures Father Mulcahy.
Bumba needs little persuasion. He is a devout convert to the Catholic faith and would help Father in his holy mission at any cost. After the great tragedy at the mines, a missionary had saved him and his starving mother. Before that, Bumba had been a child laborer in the Sierra Mining Company in Sefadu. Although he suffered from Kwashiorkor, an illness that swells the stomach, he worked as a digger in the mining pits under the cruel Bitter Cola who lashed workers with a whip to increase productivity. After huge machines cleared the top soil of the mining pits, it is twelve-year old Bumba's job to fill sacs with diamond dirt and lift the heavy load for transport to the waiting dump trucks. Minutes before the great explosion at the mining site, Michael Sparks, one of the "tubab engineers," had asked Bumba to run a gallon of diesel to a stranded Scrapper operator who had been repairing the roads three miles away. When the operator refueled

his vehicle, he refused to offer Bumba a ride back to camp and so, he is saved from the great blast that destroyed the camp site. But with no mine left to work in, he is out of a job. The young boy roamed the city of Sefadu looking for work for days until sickness and hunger overtook him. Finally, Catholic missionaries took him in, and made him an altar boy. Bumba is thankful to the Church but he is also curious about this new adventure the priest had offered him that so suddenly stirred up his boring life in the parish service.

"Father Mulcahy, I heard what you and Imran were talking about in the Parish. Is it true that we are saving the world?" He questions fervently in broken English.

"Yes Bumba, you are the one who will also help save the world," the priest inspires him.

"How did this all begin?" Imran asks. The priest thinks for a while. How is he going to explain a complex mathematical theory and its relationship to the Poro?

"We believe that a long, long time ago, long before the time of Moses, the Arc of the Covenant existed as a single piece in the form of a cubic crystal."

"Like a sugar cube?" Bumba asks.

"Yes, like a sugar cube. But this sugar cube had twenty-seven spherical balls or what we call atoms. We now believe that the points of the Cubic Arc were a great force known as anti-gravity and anti-spacetime points."

Bumba listens with great interest.

"At these points all matter may be created out of nothing like the great creation story of Genesis describes."

"Yes, Genesis," Bumba repeats softly looking far out into the horizon.

"Somehow, the Arc became torn apart and scattered into twenty- seven points all around the Earth!" Father Mulcahy explains. Bumba listens intently as the priest explains, but his concentration is waning. His eyes become dull from lack of sleep. Slowly his eyes role into a deep sleep and the back of his head leans on the hard bench.

"Bell's Syndrome", Father Mulcahy states to himself.

"My God, this is the second person I have seen with this syndrome," he states to no one but himself.

CHAPTER 25
(Michael the Archangel)

Michael Sparks knows that his days are numbered. In his hand is the diamond studded, four-cornered star his dear friend Bitter Cola had given him. Paul is chained to another courtroom post where prisoners are held for questioning in the court house of Yamandu. During the old order, Michael was a judge who sat beside Bitter Cola. Now he is reduced to a prisoner of the new order of Yabu Yombo! The Queen Mother of Yamandu has no pity for the renegades of the old order. When her father was dethroned, she was but a little girl, barely able to comprehend the sudden changes that had taken place in Yamandu. Michael Sparks had pleaded with Bitter Cola for her father's life to be spared and now he wished he hadn't. The proceedings begin with a wave of the Queen's hand. The courtroom is very silent. I am permitted to sit in the back of the Courtroom for my own protection. Whatever I have started in this little village can no longer be reversed. I am now an integral part of the future history of Yamandu.

"Who holds the secret of the Eye of M'Dulu now?" Yombo asks triumphantly shaking her fist at Michael. I feel sorry for Michael. If I were in his shoes, I would be very worried about my life, but that is me, for he appears to be very calm

"Only Paul has been given the knowledge of the Eye," he quips.

"Then show me the hiding place of M'Dulu's Eye. I want to go to the Garden of M'Dulu," the Queen Mother demands bitterly.

The court is hushed. There is fear that the Queen Mother's wrath will spread to her own followers, no one knows who is loyal to the Bitter Cola sect.

"No man can go to the Eye of M'Dulu without the blessing of the Eye. You will die if you see the Eye of M'Dulu!" Paul threatens, hoping to scare the woman. The Queen looks down upon him as if he is a bug to be squashed.

"I have eaten of the fruit of the Garden of M'Dulu when I was a child. I remember the sweetness of the Garden with my father and my forefathers. I have the right to see the Eye of M'Dulu," she protests.

"I am blessed Queen Mother of Yamandu. The whole world knows of my father, Thamu Yombo, the great Poro Chieftain. Show me the Garden of M'Dulu now or you will die like all the others before you!"

I can sense that Michael and Paul have nothing more to say. If they show the Queen Mother the hiding place of M'Dulu, they will surely die and if they do not, they will have a chance to live until they do. Paul's face is bent low and looking at the floor, keeping his eyes away from Yabu Yombo. The Queen repeats her question as she approaches him. Michael raises his head and bravely sings out the ancient song of the Poro.

"Now, therefore Oh our M'Dulu,
Great, Mighty, and Awesome M'Dulu,
You who in your mercy preserve the Eye,
Take into account all the disasters that have befallen us,
Our kings, our princes, our priests, our prophets, our fathers, and your entire people, from the time of the Kings of Yamandu."

Suddenly there is a commotion at the back of the courtroom. Several Poro warriors enter the room pushing two white men and a huge African ahead of them. The Queen Mother stands up to look at the intruders.

"Who are these men?" she demands, her voice shaking and booming like a cannon.

One of the Warriors walks up to the front of the courtroom, abruptly falls on his knees before her and starts explaining.

"Oh, Merciful Queen of my people, Oh gracious one of all time, I stand before you in self-pity. I bring before you strangers who were found prowling around the river of our ancestors. "

"What were they doing in our forest?" the Queen commands.

"We found them on the shores, fearful of your wrath, Oh holy Queen. They were alone, abandoned by their boatmen," the warrior reports.

"Bring them here!" the Queen orders.

A young man, an older man both dressed in priestly robes and an African are pulled by their arms and pushed down on the ground before the Queen Mother's throne.

"Book of Nehemiah, Chapter 9, verse 32," one of the priests communicates in the Queen Mother's language and points to the Bible in his hand. All eyes in the courtroom rivet on the priest and there is silence. The Queen mother's eyes wander across the room and rest on the talking priest. He is younger. The older priest looks familiar to me but I cannot place him as yet.

"What is the meaning of your words, stranger, and who are you?" The Queen Mother begins the inquisition.

The younger priest attempts to move toward her, but a warrior pushes him back to the floor. The African, probably a guide, angrily breaks free from the grip of his captor and pounces on the warrior who has thrown the older priest down. He lifts the warrior high in the air and lets his body drop to the ground with hard thump.

"Bumba, stop!" The older priest orders.

Bumba grunts and relaxes. The Queen Mother motions to her men to release the older priest. He walks towards her and the elders of the court.

"What is the meaning of your visit to our humble village?" the Queen Mother asks in broken English.

"We come in peace. My name is Father Josephus Mulcahy, and this man here is Bumba, our aide. My fellow priest is called Father Imran of the Archdiocese of Freetown," Father Mulcahy introduces. My heart leaps with anticipation. The great mathematician Father Josephus Mulcahy! I met him and the great Magnus Nambu during a conference in Italy. Several of his papers on the J-N force have circulated NASA and I am somewhat knowledgeable about his theoretical work. But what the hell is he doing here? What is the discoverer of the J-N field doing here in Yamandu? Surely he knows about the crystal field and his presence confirms for me that the Crystal is indeed hidden in Yamandu!

The arrogant Queen keeps her gaze steady on the priest.

"We have some very important things to discuss with the white man speaking in front of your court. So, we have come by way of the river to find him," Imran elucidates.

"And what are the words you speak of?" the Queen Mother asks pointing to the black Bible in the priest's hands.

Father Mulcahy opens the Bible and starts reading.

"Now, therefore Oh our God,

Great, mighty, and awesome God,

You who in your mercy preserve the Covenant,

Take into account all the disasters that have befallen us,

Our kings, our princes, our priests, our prophets, our fathers, and your entire people, from the time of the kings of Assyria."

"Those are the words of M'Dulu," the Queen mother exclaims in amazement.

"And the black book you carry in your hands, it has the words of M'Dulu? Who wrote it?" the Queen Mother inquires.

"It is the Book of life. It is the Book of truth, the word of the Lord Jesus Christ. It is the great book that has all knowledge in it," the priest maintains.

"Show it to me!" she commands.

The priest walks steadily up to the Queen mother and offers her the Bible. She inspects it carefully, turning the pages slowly. Then, she stands up again and commands,

"We will not be disturbed by this incident. We shall continue!" She orders. The crowd is still gawking at the priest, fascinated by his knowledge and the words that have been sung. For me, it is sort of very strange to hear the great Mulcahy spitting out biblical verses that exactly match what is already known in the Jungle. I start wondering what is really going on. How does M'Dulu tie in with Jesus? How does the bible tie in with the Poro? I know they have exactly the same ceremonies as the Ancient Egyptians but where does the Bible enter the picture. There is something grand happening in this village and I am getting really excited but worried. If the great power of the crystal is hidden in this jungle, and it indeed relates to the J-N field, then it is all true and this is all part of biblical Prophecy. I must learn all I can from them. The Queen Mother grabs a spear from one of the warriors and aims it at the priest and then turn to Paul. Paul is unmoved, but the priest is trembling with fear.

"Where is the Eye of M'Dulu?" she threatens.

There is tension in the room as she raises the spear like a well-trained Amazonian warrior. Father Mulcahy interrupts her quickly fearing that the ruthless woman will strike the young man and kill him.

"I also have the knowledge of the Eye. No mere mortal can see it in life. It will destroy you."

I am curious so I walk over to the priest and look him straight in the eye. It is my chance to get some answers.

"What do you know of the Eye, and what does the Church have to do with it?" I demand.

"It is as bright as the sun, yet it is cold. It is the Eye of knowledge that only the Good God may behold. It is the Arc of the Covenant that has made this world a living world," Father Mulcahy willingly replies. I have read his papers on quantum theology and he is a true thinker. What makes this great man believe that the crystal in Yamandu is the Arc of the Covenant? I look at him in the eye and smile,

"Surely father it is a mathematical unlikelyhood that a mere diamond in the jungle can have such great powers as claimed by the arc of the covenant!" I say hoping to dig out more information from him. The queen mother must feel insulted by the sudden intrusion of the priests into her domain of power.

"The Eye belongs to M'Dulu, our God," she shouts "and no Whiteman shall take it from us!"

"How then do I know the words of your God M'Dulu?" the priest challenges. The queen mother pushes me back and points to seat for me to sit on. I comply. Her jewels jangle furiously as she raises an accusing hand and points at the two priests, Michael and Paul. The restless crowd murmurs as she tosses the bible back to the priest snarling in anger.

"You are a fraud, a liar. You know nothing of M'Dulu's Eye."

"Yes, we do," Imran steps anxiously. In a matter of seconds, the priests are surrounded by a multitude of the Queen's followers, each awaiting her command. Father Mulcahy remains calm but Imran is visibly shaken.

"Wait!" I shout. I need to delay any incident until I learn more. I need to know the nature of the priests' mission in Yamandu.

"We are all of the same opinion here," I point out to the Queen mother.

"The God of these people is our God also and M'Dulu is the same God. They worship the same God as you and me, Oh Queen Mother. Perhaps we can learn more from them about where the Eye is kept!"

I succeed in drawing out the priest, perhaps he sees hope in my words. He too must be searching for the Eye. My rhetoric must have struck a nerve.

"You have spoken the words of our Holy Bible which we hold in our hands. We come in peace, not to destroy but to save us all from the grave danger of the unbelievers. A great calamity is about to befall the entire Earth. We have come to secure the Eye even as you seek to do, so that it may not fall into the wrong hands," Father Mulcahy shouts, raising the Bible into the air. Some of the superstitious natives standing near the priest scatter in fear of the black book.

"And who do we need to protect the sacred Eye from?" the Queen Mother jeers, clenching her teeth like a wild dog.

"From the people of the North!" Father Mulcahy calls out desperately.

"They will soon come to claim it and make it their own. Then a great disaster will befall Mankind as this man has decreed in the words of the Poro!" he points at Michael. The Queen Mother slowly moves back to her throne. She seats for a while absorbing all the information that has just been thrust upon her and her new-found power. Then she rises to her feet with a shimmering of colors as her jewels scatter sunlight into a million rays upon her people. She is becoming annoyed by Mulcahy's dramatic oration, yet she is fascinated. He is stealing the show. Michael and Paul must be fearful that that pandemonium will be unleashed by the followers.

"I have no faith in you white men of the outer world. I am inclined to believe that you come here to take our Holy Eye," the Queen Mother accuses.

"That is not the case, your highness," Imran insists.

"We also have been given the responsibility of the Eye by God, who you call M'Dulu. We must work together or else we will all perish. There is no time to waist and if the men from the North get the Eye before we do, they will have all the power of the Eye. Please help us," he pleads.

Bumba is ready to pounce on a warrior that is nudging Imran with a spear to move back.

"I cannot allow you to roam our village freely. The Eye belongs to my people. I must find the Eye now, so I can secure it from outsiders."

Then turning to the guards, she orders them.

"Take the white men and their aide to the village prison. Tomorrow we will set out to look for the Eye of M'Dulu."

A dozen warriors surround Bumba and the two white men, pointing their razor-sharp spears at their chests. Other guards surround Michael and Paul, pushing them out of the court room toward a small prison at the end of the village. When they arrived, Michael and Paul are locked inside and two guards are posted outside.

I remain seated in the courtroom, waiting for the excitement to subside. I hold Zola's hand, affirming my commitment to the cause of the Poro. The Queen Mother stands up and orders the Areyogbo to be executed by hanging. In an instant four guards lift the poor man and carry him out of the courtroom. As the Queen walks past me and Zola, we rise and bow our heads to respect the new leader of the Poro of Yamandu, then follow her out of the building and head back to her house.

I have heard all I need to hear about the Eye of M'Dulu. Now, I have to mull over everything I have learned. The entire picture is now slowly coming together in my mind. The spheres are certainly those depicted in archeological finds all over the world. They are memories left of a strange and powerful reality far in the past of man's history. But the story does not end there. The crystals are only part of a grander global picture. There are twenty-seven archeological sites in the world where the spheres have been found. The priests have referred to them as parts from the Arc of the Covenant? Wow!

I am filled with questions. What has the end of the world got to do with the archeological finds at Abu Simbel, Yamandu, and other sites, scattered all over the world? Is the Eye of M'Dulu natural or alien? Why would a Whiteman, an Engineer such as Michael, forget about civilization and give up everything to live here and protect the Eye of M'Dulu? A picture of the puzzle starts to form in my mind but then it becomes fuzzy. Somewhere in my mind, a Poro vein blocks the solution. I feel a strange loyalty I cannot explain to the Queen Mother of Yamandu but I am still driven by questions.

As night falls, I find myself wired up with anxiety. I cannot sleep and relax despite the loving attention that Zola tries to give me in our nuptial bed. One thing I know for sure is that I have proof beyond a doubt that the Abu Simbel mystery is similar to the mystery of the great calamity that befell Yamandu over fifty years ago. Now, I know that the force of this crystal is indeed real. I know that global cooling power and the intense z-particle emissions are caused by the Eye of M'Dulu. I would speculate that the tremendous energy of the crystals can be harnessed for man's benefit, but from the lessons we have learnt from atomic energy, it may be best to hide this from the powers that will use it to their selfish ends. Perhaps this is the driving power that creates gravity. Perhaps it is the great power that destroyed Sodom and Gomorrah. I have to move fast before the wrong people harness its power. I have always considered the existence of a similar eye in Abu Simbel although no crystal was found there. Maybe that is why the temperature dropped, following the Resa accident. The presence of the crystal sphere in Yamandu might explain why satellite readings continue to log the temperature drops in the village at still well over fifteen degrees Fare height. I am convinced of a powerful force at work in Yamandu. It is urgent I update President Cooper as soon as the secret security channel satellite aligns at midnight.

CHAPTER 26
(Particle Generator in the Jungle)

Commander Philip Martin stands in front of the five thousand specially trained Special Forces assembled in the "Briefing Room" of the in Andrews Air force base in Maryland. In his hands is a document titled "Project X." He speaks in a soft quiet voice, reading the instructions that he claims have been handed down to him from the White House.

"The purpose of this mission is to capture a super-secret Particle Wave Generator in the Jungles of West Africa. Ever since the end of the great Resa catastrophe, our satellites indicate that the Russians or some other military has developed a new and powerful weapon capable of far more destruction than the nuclear bomb. It is a new crystalline structure capable of generating a powerful wave on a particular frequency that could trigger a mass hypnosis similar to Bell's Hypnosis. Its operations are based on the predictions of the Josephus-Nambu field theory."

Commander Martin takes a moment to wipe the sweat off his brow with a monogrammed handkerchief.

"What you do not know is that our government has been attempting to build a wave generator for a long time. But the destruction of our research facilities in Arizona by Comet Resa over twenty years ago put us far behind the Russians. We have, however, been monitoring their developments through our contacts in Egypt, Europe and Africa. We do not want a repeat of Einstein and the A-bomb. We must now take steps to terminate all development of the new and strange sub-nuclear weapon before it starts a new Cold War. We all know that history generally repeats itself. But this time, we shall not let that happen!"

The military speaker takes a step back to survey the crowd. The disciplined multitudes hold a steadfast, silent gaze that gives the Commander the confidence to carry on.

"We have received word that the first test of the Wave generator has been successfully completed by, we believe, the Russians in Cairo. Other tests of this great weapon have recently surfaced in Africa and Asia.

"Our mission is to capture the most pronounced wave generator ever detected by our intelligence. We are headed for the African Jungle with success as our goal!"

After the briefing, five thousand marines walk out of the building and pack their gear for the long trip to the African Jungles. Little does he know that another army of IFPs are about to take off on the real mission of saving the Earth. He and his men are sacrificial lambs for the military build-up around the world that has been organized by the Arians. In a separate secluded area of a desert base in Area 51, one thousand crafts are preparing to take off unbeknownst to Commander Phillips and his men. Each craft is armed with two tons of liquid sulfur dioxide. The gas is to be dispersed into the atmosphere around the world. It is the only gas that will absorb the rays of the crystal.

In the village John Davis cannot sleep. Tossing and turning in bed, he anxiously waits for the appointed time after midnight to call President Cooper at the White House. At exactly 1:02 a.m. local time, the secret communication satellite will be aligned with the latitude and longitude of his location. Then, he can activate the codes that will allow him to clandestinely communicate directly with the White House. When the time comes, he makes sure that Zola is sound asleep and proceeds to the small living room of the mud-house. He punches a coded sequence on his wrist watch communicator and waits for the three flashes of the liquid crystal display. The three flashes appear, clearing the way for confidential voice communication with President Cooper.

"Sir, I have confirmed the importance of great forces that destroyed Abu Simbel," he reports.

The President replies by asking a question.

"Is force of Military significance?"

The answer comes back,

"Yes. Code 1 measures must be taken to secure the area. This could be connected to the Vatican's claim of an Arc of the Covenant. Egyptians could also have invaded here. There's a priest and a Vatican representative present."

"That is strange. What is their mission?" the President asks.

A sudden noise distracts John Davis. He leans back away from the open door and peeps in to see the source of the noise. He sees no one. He pulls the Lazukar from its holster. He had not parted with it. Things happen too suddenly here. Then slowly, with the Lazukar in full power, he walks across the room to look outside the open hut window in anticipation of a presence. A large rodent runs past him and hastily makes its way out the hut window. He breathes a sigh of relief. The President calls out a few times to draw his attention. He is becoming concerned with the silence at the other end of the line.

"Sir, it is strange and uncanny," John Davis continues realizing that the President is awaiting a response.

"They talk of Arc of the Covenant. The locals call it Eye of M'Dulu. It may be the Josephus-Nambu Technology. I must locate the crystal and confirm that Abu Simbel was destroyed by massive z-particle nucleation. I have already been accepted as member of the local society. Request activation of Forces Project X for operation Take-Over. This thing is more than nuclear. I will activate my personal GPS for easy detection of this location."

The answer comes back from the White House,

"Confirmed receipt. All codes received. Code 1 measures activated. Will act as soon as possible. Expect arrival in twenty four hours. Will find. Good luck!" The President had not waited for the code from the professor. He had already ordered the Special Forces into action. It is out of the hands of John Davis.

CHAPTER 27
(Prison)

The priest is sitting on the cold prison floor apparently resigned to his fate in the feared land of the Poro. Bumba and Imran are sitting a few feet away from him busily talking. Outside the small prison, I can hear two guards speak of the terrible death of the Areyogbo in our native tongue. I already know that Imran can understand our language. He is listening intently leaning an ear toward the door. I have heard of Imran before from the locals in our village. He had lived in this part of the North Country for five years during the great Apostasy of the Church in Poro Territory.

"The Poor man in the Courtroom," Bumba laments to the priest, "is hanging upside down by his legs on a tree over the banks of the Yamandu River!"

The priest listens intently.

"His body is so close to the waters of the Yamandu River, that the alligators are able to tear chunks of meat from it while he is still alive," Bumba continues.

"Boss, we are in big, big trouble now," he mumbles, shaking his head dejectedly. "By late evening, his body has been completely consumed to the bone by other animals. His remaining skeleton still hangs over the river."

My grandson Paul is sitting close to me at the other end of the prison. We keep as far away from the priests and Bumba as possible, for fear of associating ourselves with them. After all, we are still part of the Poro society and we are in enough trouble already.

"Where do you come from Michael?" Imran calls out to me in our native tongue. I neither answer nor look up. Imran is studying me intently as if an object and I can tell that he is searching his memory for something. After he tries a few more times I shrug and turn my body away from him. He persists and walks over to Paul extending his hand in greeting. Paul shakes his hand but I do not. He kneels down beside me and looks at me straight in the eyes. I let my eyes focus at infinity.

"His pupils are overly dilated; he appears to be in a trance-like state!" he says looking at Father Mulcahy. He smiles at me and I do not smile back. I feel insulted.

"Why don't you leave us alone?" Paul shouts at Imran.

"A classic case of Bell's Hypnosis!" Mulcahy diagnoses me from the other corner of the room. Imran walks back across the room to sit near Mulcahy.

"Bell's hypnosis?" Imran asks.

"Yes. The J-N field," Mulcahy responds.

"I thought that vanished a long time ago? Is it a drug perhaps?" Imran asks Mulcahy. The priest ponders a little, then answers,

"Bell's hypnosis is not drug induced, it is caused by a particularly rare combination of fundamental particles bombarding neurons in the brain."

"It could be the crystal emissions," Imran suggests.

"They will produce just such a combination of rays. After all, we know that z-particles bombarding human tissue such as the eye could induce neutral currents in the synaptic structure of the forebrain. Perhaps the crystal's light could induce it." Mulcahy states as if I am not in the room.

"Do you know of a simple test?" Imran asks.

"No."

They meditate on my supposed hypnotic state for a while.

"Ever since the catastrophe of Comet Resa, Bell's hypnosis has increased a hundred fold around the world."

Mulcahy states as a matter of fact. I know that back in 2018 officials from the government of Sierra Leone had sent emissaries to villages including Yamandu to study the phenomenon. We never really learnt what happened after that so I listen intently to understand what this syndrome is about. I start to wonder if they are right. Perhaps I do have this syndrome.

"For some reason, masses of people in China, India, Russia and Europe would spontaneously congregate at the same time and the same place for no apparent reason." Mulcahy tries to explain.

"The 2028 massacres!" Imran responds,

"Three hundred people were massacred in China as a result of one such "spontaneous gathering" called The Bell Massacres. Is that what caused it?" Imran asks.

"Yes, they had spontaneously gathered at a military base in China that was regarded as a top secret military site." Mulcahy explains.

"So how did it spread? Is it a virus?" Imran asks again.

"It was rumored that the Chinese army was doing special experiments on hypnotic rays. Perhaps, they were experimenting with crystals. Military personnel were sent to the site to disband the people. But the hypnosis was so strong that they had to be shot dead to secure the site. Husbands and wives who could not understand what drew their spouses to the site still talk of the incident on television. Perhaps Michael and all the followers of the Poro sects are being drawn by the mysterious force of the crystal's rays" Mulcahy states.

"Perhaps M'Dulu will kill you all if you delve into His power", I state openly as a warning. Imran and Mulcahy ignore my warning continuing their conversation.

"What if Bell's hypnosis is the force that will gather the whole world on the day-of-judgment?" Mulcahy speculates.

"My God! Could this be the way the entire world will be judged?" Imran is surprised by his own question.

"If the end of the world is near, then it makes sense that the Arc of the Covenant could be present. It will be the Arc that will be used to transport the faithful!" Father Mulcahy reasons.

Imran's revelations make my heart race. My breath becomes short, spasmodic gasps and I feel like I am fainting.

"Maybe God has built the natural world to judge the world by natural law. All this Nature business, perhaps if the crystal cube is put together, all humanity will be drawn toward it by Bell's Hypnosis!" Imran concludes.

"That is why the Holy Father tells us the crystal cube is the Arc of the Covenant!" Father Mulcahy agrees.

"In His great and mysterious way, God has ensured that even the final forces that will pull mankind together on the day-of-judgment will obey the laws of physics!" Imran manages to utter. They continue talking for a long time, trying to figure out why the field is what it is. Why they are here and what they have to do. I continue to listen but play it along as if I am disregarding them. As if I am not paying any attention to them.

Nightfall comes and we can no longer see one another in the darkness. My grandson and I continue to segregate ourselves from the holy men. The noises of victory have subsided and Paul and I know that unless we find a way to escape from the grip of Yabu Yombo we will all die sooner or later.

CHAPTER 28
(M'Dulu's Eye)

Whatever happened to the body of Bitter Cola following the arrest of his sect is a mystery to everyone outside the new Holy Order. The only thing the village people know for sure is that the Queen Mother had ordered her guards to dispose of the body in a remote part of the Yamandu River.

Dawn has barely arrived when Zola and I are escorted to the front seats of the courthouse. I scan the determined eyes of the icy Queen Mother and then notice Paul and Michael nearby. I try to make my way to them but the guards form a barricade. Paul and Michael are rapidly ushered in front of the Queen's throne, while Father Mulcahy, Imran and Bumba are secured by four guards and I stand as a free man. Yabu Yombo studies the prisoners before her. She gazes suspiciously at Paul with the confident look of a mother who knows a mischievous son is hiding something.

"Where is M'Dulu's Eye?" her voice pitches with confidence.

"I cannot tell you Queen Mother," Paul booms back disappointedly. The Queen Mother remains calm and keeps on.

"Who keeps the secret of the potion?" She utters this slowly and with a great curiosity.

"I have been given the secret of the potion by Bitter Cola," Paul explodes. The goodness of Paul's soul has overpowered him. The corners of the Queen Mother's thin, red lips curl upwards. She knows she has the boy now.

"Then make me the potion!" She orders.

"I should not make you the potion it is the secret of my fathers and forefathers. It should not be shared by anyone outside my family," Paul stammers.

I know it is just a matter of time before the Queen will drain the young boy of his will.

"I must find the eye of M'Dulu, or you all die!" she persists.

The room is dead silent except for the involuntary wheezing of an old, asthmatic man. The five prisoners stand ready for the Queen's next command. Imran and the priest cling to their Bibles, praying fervently within.

"If you cannot tell me where the Eye is then I have nothing more to say to you," the Queen Mother barks back at Paul. She flings herself in the direction of the guards and screams bitterly,

"Take them to the river and make sure they suffer before they die!"

Imran winced. He seems concerned about the order. Paul is a brave boy too young to die.

"We have no part in this Queen Mother. We are messengers for the good of mankind. Please do not kill us," he pleads.

I leap out of my seat like a pouncing jaguar, sending a few of the villagers in front of me fleeing to the corners of the courthouse.

"Queen Mother," I plead with authority, "I can show you where the Eye is hidden, but you must spare these men first."

The Queen's guards rush towards me and everyone in the room pivots in my direction.

"How do you know where the Eye is hidden?"

The Queen is outraged that she is confused by my statement apparently puzzled by how an outsider can crack the secret of the M'Dulu Eye.

"I have been given the knowledge of the Poro. I am also appointed by M'Dulu," I lie.

She ponders my words a while then she must have remembered the strange battles of the previous night and the role that I played. She relaxes and commands,

"Then take us to M'Dulu' Eye!" she say shaking a pointing finger at me.

"I need to get my instruments from the house," I contrive.

Although the proud Queen Mother is unsure of the specific purpose of my instruments in finding the Eye of M'Dulu, or whether, in fact, it is really necessary that I use them, she does not want to show her ignorance of my world.

"So be it, we shall follow you," she decides easily.

A group of four strong guards leap forward carrying a high chair for the queen. The queen steps into the high chair and holds on tight as the four guards slowly and carefully lift her over their shoulders. The guards and the Queen Mother proceed to follow me out of the courthouse while Paul, Michael and the others follow behind. There is great excitement in the village at the sight of a royal procession accompanied by strange white men. The bizarre parade stops at the house where I stayed the night before and I enter to gather my instrument cases. I bet that no one will accompany me inside and I am right. I deftly tuck a Lazuka under my belt and pull my shirt out of my pants to conceal the weapon. I rejoin the group and holding Zola's hand we continue on toward the river. By now, most of the village is following behind us including some boisterous children whose playful cries distract the Queen Mother. She irritably orders them back to the village. I have been isolated from the priests and Bumba so that we cannot easily or secretly communicate with one another. I search my pocket for an ear aid and place it in my ear. I could hear them talking in the distance without having to be near them.

I wait until we reach the thick of the jungle then I activate the Z-particle detector and begin to probe. As soon as the instrument calibrates itself a faint signal of z-particle emission comes up. The beeping sound generates curiosity around the procession and the queen mother demands that I explain the sound. I explained it to her using the concept of a radio. I taught that the instrument is receiving a station that is composed from z-particle emitters. This satisfies her. I follow the signal to the edge of the river and find that the source seems to be coming from across the river. I point across the river to the direction of the source and the queen mother quickly requests a boat from the guard. I am almost hoping that the strange Mingas who had supposedly murdered Hal will decide to attack again. Two guards trudge back slowly hauling a boat along the uneven banks of the muddy river. The procession finally boards the boat and the guards feverishly row towards the other bank of the river. I can smell the sweet musty smell of the drying African jungle as we approach the bank. I am perspiring profusely and sweat is running down my body like a rain shower. The sun is bright and the jungle heat is rising with a vengeance and I can barely read the light indicator on the z-particle detector against the sun's glory.

The path through the forest is filled with bushy thorns surrounded by large armies of busy ant colonies. My face contours to a smile and I nearly laugh as I start thinking of the Queen Mother as the Queen Ant. We walk into the forest for about two hours until we come upon a clearing where the intensity of the z-particle emission is suddenly very high. The life sounds of the forest have died out completely and apparently we have stumbled upon the place where the sacred Eye is hiding. It is uncannily silent except for the constant buzz of the detector circuit as it strains from the intense flow of z-particles that bombard it circuitry. I stop to address the Queen Mother,

"The Eye is definitely hidden somewhere here," I inform her with a smile. By the time they remove this Eye, the Special Forces will be here, I hope. The Queen Mother orders the guards to stay back while she steps down and accompanies me on the final stretch of the search. We walk for another hundred meters in a circular path until we come to a sudden halt at the entrance of a small cave covered by weeds and shrubs. The entrance to the cave is barely large enough for a single person to pass through. I squeeze my body into the narrow passage and peep into the cave. There is a manmade path inside and it is clear and clean as if it has just been used. Could this be the cave through which the Minga travel? Is this the Snake of the Gods? I stop in my tracks with fear and stare at the Queen Mother.

"Go inside!" she orders.

"What if the Minga come. Maybe it is their cave?" I reason.

The Queen Mother turns back to the group of guards standing beside the priests and the others. She orders one of them to come to her. The man runs toward her in a hurry. He is well chosen and appears to be the strongest in the bunch. The Queen Mother smiles to the young guard and pats his back and then she asks him to come closer to her. She stoops and whispers something into his ears. The young guard smiles as if seduced by the Queen Mother. He shyly turns away from her as she moves closer to the man and affectionately strokes his face. Then she orders him to enter the dark cave. The man hesitates and his eyes meet mine. The smile has disappeared from his face and in its place is pure freight. Slowly he enters the cave until he vanishes. He returns moments later motioning to the Queen Mother and I to enter the cave. The Queen Mother orders him to go back to the others. I hesitate and look back at Michael a distance away. Imran turns to Michael and his voice crackles in my ear with a feverish pitch,

"We will surely die after she finds that Eye!" He says in English.

"Michael, what do you know of this strange Eye?" Father Mulcahy asks. I wait to listen intently to the men speak. Perhaps I should know more before I dare into the cave of M'Dulu.

"Father, go back to Sefadu and forget about this place," Michael states smiling.

"Michael, you must realize that the Eye has importance to us all. It is sacred to the Church," the priest assures him. Michael throws a curious look at the priests. My thoughts are confused and I am beginning to worry. I deliberately ignore the snarling Queen mother who wants to press on. I can tell that Michael means the priests no harm. He must feel a strong loyalty to the priests after they attempted to plead for his grandchild, but at the same time, I surmise that his Poro instincts are not be easily suppressed by the notion of freedom from Yabu Yombo's grip. Paul's introspections are making him nervous, and the last thing he needs is for the world to know of the secret of the Eye. If Paul and Michael loose the power and status that the Eye had given to their family, they will surely die. They had only one mission in life and that is to guard the sacred Eye of M'Dulu.

"Who are these people from the city?" Paul asks Bumba in their native tongue. Bumba grins but does not answer. Imran understands the question. I pretend as if I am sorting out the particle detector and open my briefcase and lay it on the ground. This will give me more time to sort out what is going on. The Queen mother watches me intently and curious as to what I am up to.

"I need to calibrate the instrument so that it does not burn us alive," I lie. She nods patiently.

"We are from the Church," Mulcahy states,

"We are not here to harm the Sacred Eye but to protect it from harm. We mean you and your family no harm," he assures Paul.

"The man is right!" Paul exclaims to his grandfather. "If Yabu Yombo finds the Eye she will kill us all!"

Michael nods in acknowledgement of Paul's theory. I can tell that a slow bond is developing between the five men as they stand in the midst of the armed African guards. Each is worrying about the fate of the precious crystal in a different way. Each has their own intentions and unveiling the mystery of the Eye will send them on separate missions, but for now, they can benefit from each other's help. The priest is here to save man from definite destruction. Paul and Michael are determined to save their sacred status as the keepers of the Eye of M'Dulu, and the guards' duties lay with safeguarding their Queen Mother. Father Mulcahy is visibly restless. He is close to the most holy Christian relic ever. He cannot bear the thought of an African Poro Queen possessing it.

The guard who first entered the cave returns to the group, grinning and enjoying his new found status since he was chosen for an important job by the most powerful woman in Yamandu. As I enter the cave, my heart rate accelerates. At last, I will see the strange power that has carved the history of humanity for the last twelve thousand years! Here, in this primitive African cave lies a perfect spherical crystal down to the atomic level of perfection that had been the driving power behind the pyramids of Egypt, Abu-Simbel, the great flood of Noah, the Bible stories, and now global cooling! Here is a new power that could change the face of the Earth, and again make the United States of America the most powerful nation on Earth. Here is a force a million times greater than nuclear power and merely a few hundred meters away.

The flashing rate of the z-particle detector increases with each step we take into the depths of the dark cave. Nearing the end of our journey, the Queen Mother becomes excited but fearful and starts blabbing at me.

"Now John Davis, I will take back the power that my father had over Yamandu. Our family will be in power for generations, if not forever. Never again will I lose the Eye of M'Dulu!" she mutters excitedly. We come to a sharp turn in the cave and the cave expands to a huge oval cavity filled with a dull but intense blue glow. The z-particle emission has become so intense, that the detector's indicator flutters wildly. The blue dull glow is oscillating slowly, as if it is breathing. A creepy feeling flows through my spine as the rate of the pulse slowly but perceptibly increases.

I conjecture that the dull blue glow that fills the cave is caused by the z-particles ionizing the stale humid air in the cave. We approach the intense source of radiation in the middle of the cave, and I point the detector away from it to reduce the flux of z-particles bombarding its receptor. The color of the cave interior has changed to a bright blue pulse that seems to emanate from the cave walls themselves. The cave is getting colder as the z-particles consume the thermal energy of our bodies. For the first time ever, I am able to feel and verify the great scientific predictions of the Josephus-Nambu theory! At last, I will actually see the powerful force at work. I am so close to the most powerful object ever to be conjured up by mathematical theory. I expect Russians or those strange Minga to drop in any time now but I am no longer afraid, I am excited. The scientist in me has taken over.

We are now one hundred meters from the source of the rays and the z-particle receiver can no longer give a proper count of z-particle concentration even when pointed away from the source. I turn the detector off and stop to inspect the source ahead of them. The visibly frightened Queen Mother is following behind me closely, holding my arm as if it will protect her. The source starts to pulse at a higher frequency as if it has suddenly become aware of our presence. My mind races methodically through the steps that my mission calls for. I look at my wrist watch excitedly to see if I am on time for the Project X team. If all goes according to plan, I will become very famous.

"Just five more minutes," I think to myself.

"Why did we stop?" the Queen Mother interrupts.

"I must make sure that the source is still harmless since you did not take the potion," I explain convincingly.

"But we must go to the Eye!" the Queen Mother commands.

I do not answer the primitive woman who has no capacity to understand the true intensity of M'Dulu. As our eyes become accustomed to the relentless, blue light, we notice a large opening that houses its source.

My God! I find myself thinking, the intense light we are seeing is not even in this room! The Eye must be superbly brilliant! I am becoming almost paralyzed with the realization of what I am about to do. I feel my scientific logic melting into oblivion and I feel like I have lost some capacity to reason with sanity. Do I realize that I am going to look upon the light of perhaps God himself? But if I am fortunate to do so, then I must not fear and I must press on.

"That must be where the Eye is hidden," I point to the opening in the wall with the realization that it is now or never. As we approach the gateway to the crystal, a loud roar bursts through the walls of the cave piercing my amplified ear with a searing pain. I remove the hearing aid. The sound is a ferociously foreign sound that drives the Queen and I back a few meters in fear. Suddenly, a strange creature standing tall as a large dinosaur and resembling a hybrid of a massive cross between an elephant and giraffe stands transfixed at the cave entrance to the crystal. I estimate the creature of the crystal to be at least forty feet high, thirty feet long and at least a full fifteen feet in diameter. It has short elephant-like legs. The Queen Mother, now a good distance away from me cries out in fear,

"M'Goro!" she shouts. "My father is right. At last, I have seen M'Goro. Oh! Now I know I am blessed, for the guardian stands in the midst of the Eye of M'Dulu just as my father had foreseen!"

I have become deaf to her cries. The creature's alien motion, as if it is not accustomed to the gravity of Earth, engrosses me. It paces laboriously back and forth, plowing through an invisible force field within the glow of the crystal. When the giant creature begins to bark like a wild dog I stand stunned while the Queen Mother breaks towards the cave's entrance in terror.

"What the hell is M'Goro?" I gasp to catch up with the Queen Mother.

"M'Goro is the ancient Guardian of the Eye," she huffs back desperately. She is running in great leaps and strides and her straight silk scarf floats behind her like a flying carpet.

"It was put here by my ancestors to look after the Eye of M'Dulu!" she spurts out breathing heavily as she continues to run. I turn to look again at the scene. The creature is still pacing back and forth following the spherical field of the crystal as if trapped in its rays. It lashes its huge tail on the invisible wall of its prison causing the cave to vibrate with a tremendous rumble. Outside the cave, the cries of the creature can be heard by the waiting men and their captors.

"What is that?" the priest asks one of the guards. The guard stares back blankly. I realize that I am frantically running away from the very thing I so fervently sought. I catch my breath and collect my thoughts. The first duty of a scientist is keen observation. I must regroup myself and then proceed with caution. I wait a little while and catch my breath. The Queen mother comes out panting like a beaten athlete. I inch my way back into the barking sounds like a preying panther, it is almost impossible for me to distinguish the beastly echoes from my own loudly beating heart. The Queen mother has followed back into the cave. Then I realize that there are sounds of beating drums and human voices bubbling out from within the cave. Almost as if rising out of my subconscious, the voices of thousands of men billowing in the air away from the echoes of the crystal's creature fill my head. The hot breath of the Queen Mother grazes the back of my neck and I motion her with a half wave of my hand to be still. They let the rising tide of the human voices on the far side of the crystal's cave caress them until they seem alarmingly close.

"It is the cry of the Minga!" the Queen Mother begins to panic.

I swing around to silence not the cold-hearted icon I got to know in the courtroom but a frightened, vulnerable girl whose face is furrowed with despair. In the darkness she looks like a frail child.

"We must go back out now, or they will kill us all!" she warns.

I take the Lazuka from its holster and point it into the interior of the cave just as a Minga warrior emerges from the deep. He wears Ancient Egyptian war trappings and carries a spear in one hand and a sword in the other. The Queen Mother and I have been so distracted by the new sounds of the warrior tribe that we barely realize how close to the beast we actually are.

"Chebab! Chebab!" the Minga announces bravely in a weird Arabic tongue, although the creature now lurks placidly between himself and me only a few feet away. The Minga deftly strides to the other side of the outer cave, careful not to get too close to the opening of the inner one, where the Guardian of the Crystal lay. Other Minga warriors follow their leader's path ceremoniously. It is obvious to me that the tribe knows how to elude the guardian beast. The powerful crystal pulses at a faster rate than before. The strange creature paces at the same rate, slowly freeing itself from the melting fluid field around the crystal.

Certain transformations are taking place that I cannot explain. Each time a warrior moves towards the crystal, its pulsation increases. It is as if the proximity of other life forces fuels the energy of the crystal. In turn, each time the rate of pulses shoot up, I am magnetically drawn toward it. Even the Queen Mother becomes completely overwhelmed by the mysterious pull of the gem and I have to physically hold her back until her urge subsides.

When I realize that the Minga are getting close to us, I arm the Lazuka and move backward, pushing the Queen behind me. From the dark, the chants of thousands of approaching Mingas and their echoes explode off the cave walls like an erupting volcano. "We better get out of here, those Minga are coming!"

I shout readying the Queen Mother excitedly as she nods in agreement. We turn on our heels and start running toward the entrance of the cave. A Minga spear comes flying past us barely missing the head of the Queen Mother. I turn and fire the Lazuka directly into the cave and continue running. As we flee, we hear the war cry of tens of warriors in pursuit, mingling with the roar of the creature deep inside the cave. The Queen Mother is frantic. Balancing the weight of my instrument case, I pull the stumbling Queen to her feet. There is no time to lose and running faster is the only hope that will save us from sure death. Soon we can almost touch the light of day through the opening of the cave that is guiding us out like a flashlight in the dark. I am calculating that I will have to drag the Queen through the hole, but getting down on all fours to crawl through proves an easy task for her. Once outside we run to the waiting party excitedly. When the priests witness my exhausted self and the terrorized Queen Mother racing towards them, their faces twitch with fear.

"What happened?" Zola shouts.

"The Minga, the Minga, run!" I blurt out, but by the time we reach the rest of the group, the Minga have retreated back into the cave. The entire expedition quickly turns on heels and starts to run toward the village.

"What is inside the cave?" Zola asks again.

I take a deep breath as streams of dirty sweat stain off my face.

"There is a strange creature guarding the Eye of M'Dulu," I explode. The Minga seem to be coming from the other side of the cave and perhaps this is indeed the Snake of the Gods. But there must be thousands of them in there!"

"M'Goro!" Zola whispers and she grabs my hand tightly.

"Who is M'Goro?" the priest asks.

"It is the most feared devil of the River Yamandu. Only the Poro have power over him. He is M'Dulu's slave and messenger," she reports.

When the guards hear the name of M'Goro, they tremble and nervously look over their shoulders at the entrance to the cave, now barely visible in the distance.

"We must go back and see for ourselves!" Father Mulcahy insists, but no one, including the guards responsible for the two prisoners, stops running.

"Better safe than sorry, let's go back to the village before we get killed!" Imran comforts Father Mulcahy.

"Yes father, we must go back now or they will kill us here," Bumba beseeches him.

"I have no intention of going back to prison," the priest argues. "Besides, the Lord has sent me to protect his beloved and most Holy Arc, and we must do so even in the face of death," he insists.

"But those people in there, they will surely kill us. We need to get some more help before we can go back in there, Father!"

Imran is truly disturbed at the words of the Father and looks at him as if he is silently questioning his sanity.

"We must go into the cave to secure the sacred Arc of the Covenant. You must have faith in the Lord, Imran. The Lord said we can move mountains with faith, and I believe, don't you?" Father Mulcahy challenges.

Imran is mute. From the entrance of the cave the cries of what seem to him to be a Million Minga, play havoc on our brains. For a moment I think I am trapped between two destinies, one as John Davis and the other as someone else who is here but does not why. Paul and Michael turn back toward the village following me. Several hundred Mingas have materialized at the entrance to the cave. They are shaking their spears while the fantastic roar of the strange creature invades the jungle. A distance is now separating us all apart and I pull Zola and decide to follow closely behind Michael, Paul, the priests and Bumba. The five of us attempt to camouflage ourselves by retreating behind a thick covering of trees and shrub. But the group of Mingas is not alarmed by the encroaching roars of the creature and they keep coming unconcerned. From the cave, luminescent rays flow rhythmically past us like vengeful fingers of light trying to pry their way into the sanctuary of our minds.

"They have disturbed M'Goro in his slumber, and have not taken the sacred potion. They will all die," Michael prophesies ominously as the Minga, now streaming like ants, bear down from the entrance of the cave. The light of the crystal is now very-very intense and I can tell that something is happening to the entire Minga population in the distance. Almost magically, as if operating under one central brain, the Minga abruptly stop and change direction. Back at the entrance of the cave they all stand as if hypnotized by the pulsing and increasingly intense light of the crystal. I gape at the paralyzed Minga who are completely bathed by the light force of the mighty crystal of the Covenant. Almost unbelievably, the movement of the Mingas automatically becomes synchronized as the spell of the mighty crystal comes upon them.

CHAPTER 29
(Minga Warriors)

President Cooper wastes no time deploying his best fighters. After receiving instructions from me, he affirms the orders with the Pentagon to continue with "Project X," dedicated to locating and retrieving myself and the crystal from the African jungle. The team of five thousand forces is already assembled at the Andrews air force base in Maryland. Within five hours, the covert operation is underway. Ten Military class heavy duty Ion-Flash Propulsion jets lift into the air and head towards West Africa.

My communicator receives the signal of the approaching fighting force when the IFPs are only twenty kilometers away from Yamandu Village. As I scurry toward the village, the first IFP lands in the village center, creating confusion amongst the villagers. The bewildered villagers having never seen an IFP before and they scamper into their homes like wounded animals, locking the doors behind them. The nine other IFPs hover far above the skies of Yamandu awaiting orders. Commander Phillip Martin zooms in on my detector array and it starts to beep giving me the signal that they have locked on my coordinates. I am only fifty meters from the craft that landed in the center of the clearing in Yamandu village. My communicator sends a signal back and a sharp bleeping sound starts to follow. I hurry toward the IFP with Zola following closely behind. When we close in on the craft it has already landed and a fat man dressed in military fatigues appears on the portals.

"John Davis?" he affirms, pulling me onto the ramp of the craft.

"Yes," I answer gleeful and happy that I am getting out of the mess that is building up in the village.

"Commander Philip Martin," the Soldier identifies himself.

"Where is the target?" he asks.

"Two kilometers away from the village. I have mapped its route in my communicator. I must warn you, Captain, there is a strange prehistoric creature protecting the target, probably a dinosaur of some sort!"

"What are you talking about?" Commander Martin replies unsteadily. I take a deep breath and look behind at Zola who is raising a hand to reach me. The craft is slowly humming and my attention is with the questioning general.

"I said there is a dinosaur down there! If you want to be eaten alive just go down there!" he shouts.

"We are on a special mission to take possession of a new mineral crystal wave-generator that you were to locate. We do not have any time to waste. Could you please be more specific?"

The businesslike Commander is making every attempt to keep calm and go by the book.

"I cannot explain, but you must believe me. We must proceed immediately, before the Minga get here!" I am becoming hysterical.

"What are the Minga? Could you be more specific, Sir?"

The Commander studies me for a moment. It is obvious he will not be able to digest my bizarre claims when time is of the essence. "Do you have the coordinates of the crystal?" Commander Phillip Martin inquires.

"Yes!" I gasp giving up the information gladly as if I will instantly be dismissed of all responsibility. The Commander takes the chip away from me and starts running to the control console to enter the coordinates of the cave into his craft. The craft transmits the codes to the other crafts, soaring high above Yamandu. I turn to pay attention to Zola and she appears at a window of the IFP trying to peer into the craft searching for me. She is hitting the glass window of the craft to attract my attention. Commander Phillips spots her.

"What does that villager want?" he asks.

"There's no time to explain. Just let her in," I plead.

"Are you crazy?" the Captain counters. "What the hell do you think we need her for?"

"I cannot explain now, please let her in," I repeat.

"Sorry, professor Davis, my orders were to receive you and capture the crystal, nothing else!" he shouts turning to the pilot of the craft.

"Take us Zulu1 and pass the word to the other units," he orders.

For a moment I cannot believe the blind cold-heartedness of this single-minded Military man, I am up against a career officer.

"Commander, you must let her in!" I shout.

The craft ascends slowly into the air, humming with the strain of its ion propulsion engines. I peer out of the small window to find Zola.

"Dr. Davis, please strap in and forget about that piece of ass!" the Commander warns as he springs into position in his seat. I ignore him hopping from port to port and scanning for Zola in the village mass below. The craft rises higher and the village slowly vanishes into a mass of trees and human dots. Until this moment, I have not realized how much I have been connected to the priestess. Leaving her behind is like tearing off a part of me and I now have a Poro vein that is holding me hostage to the village below. I run to the Commander and grab him by the throat.

"You bastard!" I shout, "I will kill you if anything happens to her. Turn this craft around now!" I order as if it will make a difference to the General's orders. The assistant flight commander sitting beside Commander Philip Martin quickly removes his straps, and in an instant, he restrains me to a seat.

"For a villager? What has come over you, man? You will have to explain this Dr. Davis," the Commander shouts,

"I will get you for this," he swears as he massages his sore throat.

The flight commander wrestles me into a seat and straps me down with double restraints. Tears stream down my face as I recall the tender moments I have shared with the priestess, my surrogate wife. As if overtaken by a magical spell, I start imagining her and looking over the window for her. I start feeling intense cravings for her. Part of me is down there in the melee and confusion of it all. Part of me is lost to the scattered dots running around below like ants following their queen mother. I am a coward and I have deserted her. Below, I can see a mass of antlike people gathering on the empty spot where the craft had landed. I can see the children playing letting the dirt run through their hands as if it is magical or holy. Some will gather it in urns, while others stare hopelessly at the sky as if a god had visited them. This is the way legends and myths are born.

As the crafts flies toward the cave, little did I suspect that the cry of a million Minga is already filling the jungles of Yamandu.

The pressure of the moment opens my mind into a panorama of thoughts. A flowing field of visions fills my head. A history that is born from a strange tale in the jungles of Africa. My Poro veins start to visualize the reality of what is happening down below as if my mind has been opened to a new dimension. I am reeling with feelings and emotions that I have not known before. It is as if my eyes have become one with the Poro and my mind is opening up to a history that has been recorded in my veins.

Inside the cave, M'Goro is becoming restless. It paces back and forth between the confines of the crystal light as the tribal warriors mingle further into the spacetime field of the crystal, increasing its intensity. For sixty million years, M'Goro had been trapped in the almost eternal spacetime loop created by the crystal field and yet from its own point of view, time had virtually stood still. For M'Goro, time had been curled into a slow, tight spiral in which a mortal millennia is but a mere year. Bathed in the super crystal's eternal life-giving power, the creature had stood guard for an eternity, trapped at the foot of the massive cave where so long ago, its mother had brought the crystal into the nest she had built for M'Gboko's egg. While M'Gboko's mother was grazing in the then green open plains of the super continent, the last of twenty-seven crystals had fallen from the sky into the green fields of Pandora's African plains, a tropical paradise.

M'Gboko's mother picked the massive crystal up in its mouth and brought it into the only cave that had kept its egg safe from predators. She had lost all her other eggs, except for M'Goro. The crystal activated neurons that were dormant in the mother's synaptic nerves. Intelligence that would later position its evolution to an intelligent two legged creature that will roam the Earth. A massive cataclysm had succeeded the arrival of the twenty seven power crystals as the Earth's crust is bombarded by the remains of the gigantic spaceship from deep intergalactic space. Most of Earth's creatures were killed including M'Gboko's mother and her great relatives, leaving a handful of smaller creatures behind. Then the craft exploded and vaporized into the atmosphere, poisoning the planet with an alien fuel. The mammoth fallout of the twenty seven crystals into the Earth's crust caused a huge amount of life energy to be sucked up by the particle power plant made from the crystals. A massive Ice Age ensued covering most essential grazing fields for M'Gboko's kind with a thick blanket of dust and ice. Absorbing energy from the surroundings, the crystals created matter in the form of z-particles, ejecting them far into the upper atmosphere. The matter slowly engulfed the crystals to hide them from the face of the Earth and thus preserve the secret of their power.

As millennia sped by, time moved slowly for the M'Goro egg. The crystal slowly curled spacetime in an ever diminishing sphere of energy entrapping the massive dinosaur egg within its spacetime shell and separating the egg completely from the material world. Only tiny flaws in the thin spacetime shell were visible as light energy penetrated the shell of the crystal. Luckily, these flaws allowed M'Goro to survive like a spider in its webbed den. Small creatures would become entrapped inside M'Gboko's' field providing it plenty of food. Small creatures but many, for as the years rolled by a fantastic relativistic speed, the many became plenty.

Now, its slumber has been disturbed by the Mingas and the men of Yamandu. Slowly, the pulsing crystal starts uncurling its spacetime field around the creature, increasing in size as its energy level rises. Thus, the creature becomes active, pacing angrily within the womb of the crystal field as if awakened from a comfortable slumber. Slowly the spacetime within the crystal merges with that of the Earth's. M'Goro could no longer take the stress of confinement. The creature longs to free itself from the distortion of nonlinear spacetime. Inserting its claws into the unfolding crystal field, M'Goro stretches its jaws like a rubber band to engulf the crystal core. The rays of the crystal activate the very nerve cells of the creature's brain, augmenting its synaptic connections a thousand fold. The crystal is capable of creating intelligence and in essence, all life.

The creature holds the crystal between its jaws, clamping its massive teeth into it in an effort to tear itself free from its prison. It raises the crystal high into the walls of the cave, causing the field to push the walls away. A thousand Mingas stand transfixed by the hypnotic spell of the crystal's rays, watching as M'Goro tears through the entrance of the cave. In seconds, the creature, wriggling its massive head in an effort to lodge itself free from the sticky gooey clamp of the curved spacetime field, smashes thousands of Mingas against the cold walls of the once silent sleeping cave. But thousands more Minga come. They march toward the fury of the creature. They are transformed into zombies forever frozen in the joyous delight that embrace them.

Now the creature, sensing freedom is near, clamps down harder on the crystal. As it raises it high into the air, an energy force pushes the cave walls into outer space. M'Goro has torn through the cave's entrance and is hypnotizing thousands of Minga warriors. In their trance they are smashed dead against the cavern while the creature pushes harder on the boundaries of the crystal's gooey spacetime field, to free itself. But thousands more Mingas are replaced in the joyous frenzy that embrace the tribe.

The pressure created by the crystals moves the cave walls like a giant earthquake. A sudden wave of light burst out from the strange crystal, exploding the cave entrance into a huge wide cavity. Debris flies everywhere, covering the jungle floor and canopy with the fragmented Earth. The creature loses its balance as the crystal field suddenly expands to over one hundred meters in diameter. Like invading insects, thousands of Mingas stream from the massive opening of the cave to witness a miracle. The creature floats out of the cave and levitates about thirty meters above the forest cavity. Like a second sun, it glows from the light of the crystal in its mouth.

 The radio in Commander Phillip's cabin crackles,

"Jesus Christ Commander, look at the East sky! Do you see what I think I am seeing?"

Commander Phillips calmly turns eastward. He scratches his long beard with his brittle fingernails. Then, he nearly jumps with excitement.

"Oh, my goodness, what the hell is that light in the sky?" he yells over at me.

"A dragon!" I remark sarcastically, holding a handkerchief over my eyes.

"The crystal data, where is it?"

"Release me so I can get it," I order. He quickly pulls the straps off my arms and points a hand toward me waiting for the data. I do not respond. I quickly take my jacket off and wrap it firmly around my head covering my eyes from the blinding light of the crystal.

"Give me the damn data crystal," Commander Phillips orders.

I raise my arm and exposes the crystalline watch. I have nothing to lose now. The Commander removes a key from his pocket and unlocks the watch from my wrist. He is shouting irreverent curses as a sudden burst in the intensity of the light blinds his eyes.

The rest of the crew has not protected their eyes from the sudden flash. They are all blind and could no longer see. I slowly grope my way to the parachute locker and remove one. I wear it while listening to the ranting and cursing of the soldiers in the craft. They are done, blinded and attracted to the strange light like moths with their heads pointing in the same direction. I find the lock on the craft's exit door and open it. I hesitate as the cold wind blasts through my body sending chills up my spine. I blindly look around as if I can take one last look at them. I do not know if it is an inner eye that makes me see, but they are all transfixed on the crystal glowing below. I jump.

As I fall at high speed, my thoughts go to Zola. I have become resentful of the Commander and our mission. I have no future anymore and I do not quite remember what it all about anyway. I do not remember my past anyway. I just have a faint memory of a past life as a family man somewhere down there. A family I had floating faintly in my memories. I have nothing to live for.

I reap the jacket away from my eyes and in the distance I can see the IFP plunging down. The blue rays of the crystal flash on the craft like a new sun on a new day. The blue pulsing rays hit their mark. The Ion Propulsion jets start to misfire and the craft spins and spirals like a dying firecracker across the jungle skies.

I am excited. I spot the gyrations of the other IFPs as they ricochet off an impenetrable barrier in the sky and then plunge into the jungle canopy below! One minuscule realization later, perhaps the time it takes for the heart to beat, I find myself falling into the abyss below. I consider myself lucky, I do not care about death anymore and I am inexplicably attracted to the blinding light below. All I want is to be engulfed by it. My eyes are now free to hold the full light spectacle of the Eye of M'Dulu, the second sun, without wondering or worrying of what will come to be. A warm blue glow engulfs my body muting the sounds of the falling IFPs. I am as light as a feather and the Earth is advancing toward me at a fantastic rate. I am upside down floating in dirt-filled air and like a silent fireless bomb I feel myself being propelled at a tremendous speed to an unknown oblivion. It is as if a gigantic balloon has been blown beneath me. I am totally engulfed in dust and darkness and my body is being propelled through the dusty atmosphere by some inexplicable force. At first I think I am caught in a tornado of unearthly magnitude as bodies, IFPs and machinery float around me, but then my body begins a graceful descent to the ground that could not have been possible in the eye of a tornado obeying the common laws of gravity.

As I come to within two hundred meters of the ground, debris below me begins gathering into a dense sheet of matter. An alien landscape spawns in midair and I suddenly feel myself landing on the new huge and soft bubble of debris. I am lying on the surface of a huge invisible bubble of something. The bubble has formed an invisible skin layered with the debris of the mining site that seems suspended in just thin air. In the enclosed space of the bubble is nothing but clear air. As the dust clears I can see nearly two hundred meters below me. There is a huge cavity in the Earth below forming the underside of an oval bubble I am sitting on.

"This is not Earth," I think, "Earth is gone. I am dead and this is the end of the world!" I strike the balloon's surface with a soft rebound head first, but the impact of the crash is magically absorbed by the invisible, fabric-less barrier of the bubble. Debris is all around me floating and following the contour of the large bubble. Above me a mix of water, mud and mining equipment is slowly raining down on the bubble surface to form a delicate skin of multi-colored debris that is becoming a more solid foundation of the floating surface. I can feel myself floating in the melee as if it is some soft ethereal fluid. The debris is settling fast around me nudging me deeper into the collection field. I must escape to the upper layers and avoid being trapped beneath the debris on the skin of the bubble. I begin swim upward through the settling debris, treading to stay afloat. The dust, dirt and water around me is now turning to a more solid mass of mud and is quickly covering the lower half of my body making me struggle to move. I must move quickly before the debris solidifies under gravity.

"I am in the sky somewhere, afloat in hell," I rationalize.

For a long time, I remain suspended in the floating mulch of what used to be the village below. I look for the Yamandu River but it is too dark to see the outline. I succeed in finding a spot where the debris field is minimal and the where the thick air and dust about me is condensed to a viscous fluid that would respond to my swimming motion. I can feel my body slowly being pushed to the less dense parts of the debris field. I can see the moonlight now and I know I will be safe at least for now. After what seems like an eternity, the steadily glowing moonlight encourages me to experiment in the thickening melee that has trapped me. I step out onto the new hardened ground cautiously and my feet become cushioned by something invisible. Becoming bolder now and eager to explore the nature of my new terrain, I press my toes into the invisible fabric barrier beneath my feet and feel a force field repelling them. I see a loose tree branch and use it to push deep into the invisible force field. I feel my body give as the reactive force pushes it back sending me even higher up the field. The branch shatters with an intense vibration that rushes through my body.

A powerful field has lodged itself between me and the Earth below. If there is only one thing I can be sure about it is that the force I am embedded in is neither a magnetic nor an electric force. Debris continues to settle on the surface below me causing my body to slide uncontrollably as if I am an ice cube gliding on a glassy surface.

As the moonlight becomes more intense I am able to scan some distance and I am able to see parts of machinery from the power plant I shut down moments before. Perhaps the mysterious force had somehow tapped into that Earthly power, making it its own to create the surrealistic world that has captured me. I am surprised that my mind is still very active and alive. The dirt beneath the scattered machinery parts is glowing with the blue light leaving curved imprints in the dust. I am intrigued by the light and I begin swim towards it in the still wet debris. The light's frequency pulsates about ten beats per minute and as I swim or push my way I realize that the temperature of my surroundings is dropping, and that I am getting weaker. The debris field is changing, getting less dense and lower in height. Swimming is no longer possible and I begin to walk over to the yellow glowing matter ahead of me. I stare in amazement at a thirty ton bulldozer that hovers nearby in thin air above the debris field.

I am felling chills and it is becomes colder in my free-fall environment. I approach the bulldozer and realize that it is cold to the touch, very, very cold. It oscillates with the slightest touch of my hand as if it is freely suspended on an invisible rope from the sky above! Beside the bulldozer is a large floating pool of diesel fuel giving birth to a stream of orange liquid floating in the air. As I walk towards the liquid it flows towards me as if it has an intelligent purpose to engulf me with its warmth. I touch the titanium weld on the steel bucket of the bulldozer. It burns my fingers with a frosty coldness. I touch some levitating papers and Operation's manuals from the loader nearby and discover that their temperature is normal. I rightly conclude that the colder the temperature the denser the material.

I look down below and realize that I am standing on top of an oval, egg-like structure of about five hundred meters long and three hundred meters wide. In the middle of the space beneath the oval, a yellow pulsating sphere about the size of a soccer ball is floating. The upper shell of the oval is formed by dust, dirt and air, and the lower one is an oblong concave cavity dug deep into the mining camp below. A memory hits my nerves, I have been here before, I had seen this before!

CHAPTER 30
(Garden of Eden)

Ron's eyes open painfully slow. His head feels heavy, and is hurting from the wound inflicted by the sharp Minga spear that grazed the left side of his temple. Several men around him dressed in ancient Egyptian trappings speak an incomprehensible dialect Hal had never heard before. It sounds a bit like Arabic but he could not understand what they are saying. Three of the men are dressed in long red sari-like robes, with long golden embroidered laces sewed into the lower linings. They appear more elegantly clothed than the rest of the group, and resemble priests. Could there be the Minga Paul had spoken about? Hal wonders.

One of the aged men walks over to Hal and studies him for a few minutes. He shouts a command. The curious warriors who crowd Hal ease back to form a larger circle around him. Hal scans the elder's reassuring grey eyes that momentarily soothe him. The old man proceeds to open Hal's shirt. He finds a gold chain holding a small crucifix that lies on his bare chest and breaks it off, throwing it on the jungle floor in a spiteful manner. "Toi Soi Doo Manaka!" he summons his circle of warriors. The entourage ignites, moving closer to examine the cross. Two other elders who had joined the group, kneel down to inspect Hal's crucifix. They converse softly among themselves and then instruct the warriors to take Hal away. Three warriors pick their prisoner up from the floor, heading towards a golden gate attached to the massive wall that surrounds their city.

They pull Hal through the gate and close it behind them. Inside, he finds himself at the top of a small hill overlooking a fantastic city below. The metropolis is bursting with gardens offering a cornucopia of beautiful flowers and trees. A three small rivers curve gracefully through the central garden meandering throughout the strange garden city. Hal is escorted into the lower valley by the warriors until a group of elders arrive who dismisses them and lead Hal towards one of the rivers. They motion him to wade into it but he hesitates until an elder enters first.

The old men strip Hal of his clothes and bathe his naked body with the clear, clean water. The captive white man witnesses a plethora of swarming creatures living in the depths of the river unafraid by the presence of the men. As the elders continue to bathe him, Hal gazes in amazement at the peculiar grazing of wild and domestic animals, living side by side on the opposite river bank. Tigers, cats, dogs and lions delicately drink the sacred river water while cows, sheep and goat graze unalarmed nearby.

Could this be paradise? Hal wonders. Is he dead? When they complete the ceremony, the elders dress him in a robe similar to their own. They walk for what must have been an hour, until they come to the outskirts of a grand city. As the party enters the metropolis, Hal notices a magnificent inner wall beyond. The stone and mud buildings lining the outer city are round and multicolored, and at first, indistinguishable from the gardens around them. Here, the trees sport perfectly symmetrical globes of their dazzling-colored leaves. Oversized fruits and vegetables that Hal could not identify grow in abundance. Although Hal did not see a single person tending the gardens, he guesses that it would take an eternity to keep them up. Perhaps he is in an artificial greenhouse. He sees no signs of electricity and yet, even though it is night, the city glows brightly from shining bugs and fluorescent flowers. It could be that this place thrives in a dome that regulates day and night. Hal fondly remembers that when he is a little boy, he imagined the sky to be a massive white bowl. The stars were like little holes on the bowl. He smiles, believing the world of his imagination has come alive. He feels the wound of the spear on his head, but the pain is slowly fading away as if nurtured by the cornucopia of herbs and flowers that surround them. Finally, they arrive at the entrance to a large building in the outer city. They enter the building and walk toward a wide plaza in the middle of a large open expanse. One of the older men takes Hal to an aging man sitting on a bench in the middle of the open plaza. He seems like royalty. He leaves Hal in front of the man and leaves to join the others.

"Who are you, stranger?" the man converses in English with a bizarre accent. Hal is shocked to hear his native tongue in such an exotic place, spoken by a man who appears to have never left the confines of the city.

"My name is Hal Shea, I am an American." he answers.

"What are you doing in the land of the Poro?" the man inquires seriously. Hal does not answer right away. He thinks it is time for him to turn the interview around.

"Where am I, and why did you bring me here?" he demands sternly.

"You are in the city of Eda, the Garden City."

"Are you the Minga people?" Hal asks.

"No, we are Guardians of the city. The Minga are our warriors," the man explains. The old African searches the sky, as if a higher order instructs him. He drops his shoulders, relaxing his posture.

"You must be hungry and tired," he finally offers.

"We will finish our discussion after you have eaten."

"I am not hungry I need to get out of here," Hal replies.

The Whiteman is becoming impatient with the slow, ceremonious ways of the primitives.

"I need to go back to Yamandu immediately!"

"In due time Hal Shea you will go back to the outside world, but for now we must keep you here until we find out what your mission is in the land of the Poro," the chief elder replies.

"I have no mission that would concern you," Hal debates.

"I am part of a team sent to help the villages with a genetic rice project. I know nothing of your war with the Poro."

"You are here to take the crystal of light!" the man accuses him angrily.

Hal keeps his calm. It is true he is in a paradise but to him it is an alien one, and all the succulent foods and odiferous gardens will not serve to pacify him.

"What crystal of light?" Hal denies.

"For thousands of years, my people have been fighting with the outside world to preserve the crystals. Your people have taken all of them from us. Now we must not lose the last one to the Poro," the man explains.

"I know nothing about a crystal," Hal insists.

He is ready to run in the opposite direction, calculating how long it would take him to wade through the river, but his body feels weak as if restrained by some magical force.

"Your war is not with me, but with the Poro. I must go back."

The elder reads his impatience well and believes that if he could detain this one man, everything that he has come to know - his whole private civilization - could continue to thrive in secret.

"You cannot go back now, Hal. You have seen the hidden City of the Garden, we cannot allow you to take this knowledge to the outside world," the old man hums excitedly. Hal could feel his resolve. He now knows he has a very serious problem. Hal's palms begin to sweat and he feels a magic hand inside his chest squeezing his heart.

"But what is this city? Who are you people anyway?"

"You will find out in due time. For now, you must join the others who have dared to enter into the sanctified city. You will be taken away to join them now."

Hal does not like the tone in the man's voice. Doom is taking hold at lightning speed. Before Hal has decided when he should make a break for it, the party that has brought him to their leader reappears almost out of thin air. The men lead their intruder through a stark, white corridor in the building until they come to a door. The door is opened with a key. Hal enters and the door closes behind him.

"Where did they get you from?" a young white woman approaches Hal from out of the darkness. He squints into the same long red robe everyone in the Green City is dressed in.

"In North Africa?" the woman volunteers.

"That's where they got me. My name is Isha."

"In North Africa! You mean they go that far to get people?" Hal asks surprised.

"No, we are in North Africa!' the woman replies.

"We are at least three hundred meters below the Moroccan desert!"

Hal is beginning to wonder how the woman knows all this. Maybe she is a trick, placed here to make Hal confess the true meaning of his mission.

"The Moroccan desert?" Hal plays along.

"They got me in Brazil!" an old dark-skinned man shouts from across the room in broken English.

"And me from India!" another voice echoes in the dark of the room. A United Nations is approaching Hal from all corners of the room and like specters, they introduce themselves, extending their creepy hands.

"I am Vinod Khartumal, a dark skinned Indian approaches stretching his hands to Hal.

"Welcome to the Garden of Eden."

"I am Hal," he obliges.

"And I am Bayano," the Brazilian introduces himself.

"I is been here for tirty years!" his accent is distinctly South American.

"Thirty years!" Hal is incredulous.

"I have been here for five years," the woman announces forbiddingly.

"Who are these people?" Hal asks her.

"They are the people of the Garden City. They say that they have preserved the lost Garden of Eden from the ravages of humanity. They say that we of the outside world have destroyed the entire world with our wars and our evil civilizations," the woman explains, wringing her hands in the air with despair.

"So once you enter this city, you cannot leave!" the old Brazilian adds.

"They preserve the spirit of the Earth for M'Dulu, God of the universe," the woman cries. The woman seems devastated. Either that or she is an Academy Award winning actress.

"Have you tried to escape?" Hal asks.

"Come look what happens when you thry!" the Brazilian offers in thick accent. What is going to happen? Hal second guesses. Is there an invisible force field in place ready to singe them if they step outside? Are the seemingly peaceful animals in the outer city trained to stalk and bring down the prisoners if they escape from the building? The Brazilian leads Hal to an adjacent room as his mind rambles. On the floor is a pile of human skeletons. It is more horrible than he has imagined.

"It will kill you if you attempt to escape," the Brazilian is sadly resigned.

"The city itself is what kills you!" the woman joins in boisterously. The sadness that has momentarily drained her has churned into a charged, brewing hatred. She was once beautiful, Hal supposes, in an ethnic sort of way.

"What do you mean by that?" He queries.

"The very animals kill you when you try to escape. The fruits become poisonous. The trees and grass become thorny. The Garden people themselves do nothing. The Garden of Death does all the work," she whispers like a wild witch revealing ingredients to a secret brew. Hal tries to control his imagination. He considers again, how these people might be Mingas in disguise set up to fool him. But then again it has been remarkable to see a lion almost butt up against a calf like a lover instead of a predator.

"They lock us up in here to protect us from the Garden," the Indian interrupts his thoughts.

"There were plenty of us, but the others are all dead," he ends sadly. A loud bang comes from the door that makes Hal jump. A moment later a woman walks into the room carrying a large tray. She puts the tray on a table in the middle of the room.

"Food from the Garden," the woman says and leaves, closing the door behind her.

"I'd rather be dead," Hal declares disgustedly.

Food is the last thing on his mind. He again begins to doubt these peoples' explanations. How could food that would nourish you one minute, be the very poisonous source of your death, the next? Hal has to get a grip and use the scientific knowledge of his world, not the hocus pocus of theirs, to get himself out of here.

"The food is here, the food is here the food is here!" the Indian sings, jumping about the room happily. Hal cannot understand what the excitement is all about. The food must have hallucinogenic properties, he concludes, walking over to inspect the tray that is on the table. The two men run to the food as soon as the woman closes the door. Isha follows. Hal quickly scans their bodies like a detective. Clearly, they have not been starved. In fact, the Brazilian is a little paunchy. On the tray is a variety of massive fruits and vegetables of all shapes and sizes. They are the most beautiful apples, oranges, bananas, pears, apricots, and mangos Hal has ever seen. The two men and the woman rush to the food like predatory beasts in a frenzy following the kill. They push Hal out of the way to eat hungrily. Hal watches in utter amazement as they snort and smack their lips with a consuming hunger! The Indian, apparently still ravenous, takes a small stick from his pocket and marks an imaginary circle on the floor around a pile of his half-eaten fruits and vegetables that had fallen from his mouth.

"This is still mine! He snarls like an animal marking its territory.

He continues to stare at Hal and the other two suspiciously until he has devoured every last morsel. The woman is crying as she eats her last apple.

"This is unfair! You men took all the apples! I have none left. You took all the apples!" she cries with genuine tears rolling down her chicks. She is a child again. The old Brazilian walks over to her, his eyes darting back and forth between his pile of food and the Indian on the far side of the room. He hands Isha a half-eaten apple compassionately saying,

"Don't cry, Isha. Take this one, but just don't cry again."

Hal watches spellbound, at the sudden transformation that has taken place. What is going on? Why did they fight over such an abundant supply of food? Following a noisy orgy, the group plops down on the floor, weighted down by their full bellies. Isha is sound asleep in moments. The Brazilian wallows contentedly in his corner of the room, watching Isha's curvatious body. The Indian is rolling over on the floor for no apparent reason. The Indian offers an explanation noting Hal's inquisitive gaze.

"It all mixes well you see," he states as he continues to roll about the bare floor. Is the food drugged, Hal wonders?

"I feel guilty," the Brazilian admits as his stomach rises up and down to the rhythm of Isha's snores.

"We should have given her one more apple," he murmurs to himself lazily. The eyes of the Indian meet Hal's. They are savage, Hal observes, like a wild animal's.

CHAPTER 31
(The Crystal of The Covenant)

I clutch Paul and clasp his wide hands over his eyes tightly. My old flesh and bony palms with their craggy deep lines make a human seal to shield the innocent child from the greatest atrocity known to Man.

"Don't look at the Eye!" I shout out a frantic warning to the priest at the last minute. They have no time to prepare for this moment. Instinct and the laws of one's soul would rule over who will survive the power of the Eye. For half of my adult life, I have known of the dangers of witnessing the unholy forces that were prohibited by the Lord, yet I still wrestle with my own trembling eyelids. It is the fear of the darkness I equate with the paralyzing unknown that plagues me. My own eyes, the eyes of the physical world, will be of no use to me now. With my inner eye, I have to focus on memories of my past life in the non-superstitious world of the Whiteman. I embrace the logical methodologies of my old science to diminish my anxieties.

At my urging, the reverent priest dares not even imagine what can never be seen, yet at the same time the urge to watch compels him with an almost irresistible force. He clutches at his holy robe, now speckled with blood and mud to steady himself. It is difficult for his muscles to keep his eyes closed so tight for so long and he could not relax.

"Paul," I gasp to the twitching, horrified boy,

"No matter what happens . . . no matter what you may hear or feel . . . don't look!" I state almost hysterically. Before I clench my eyes I look up for the last time at the skies and I see the sky preparing for Armageddon. Curls of black smoke swirl out of a limbo, from the place, I suppose where evil is born.

"You'll die if you open your eyes," my soul calls out again as I brace myself for the ultimate cataclysm. Paul is praying to his dead great grandfather, Bitter Cola, for strength. He believes in the power our ancestry. As a young child he has been warned many times of the dangers of the jungle through dreams that Bitter Cola said has been sent to him from their family before them. Such warning knowledge has traveled through the cave of the underground where life and death meet and where it is possible, through prayer, to journey back through space and time. But now I am not sure when the time will come to open my eyes again to the real world. So, I think about the times when I have been awakened suddenly in a cold sweat to know that I have smelled death. A slithering poisonous snake or a prowling jungle cat nearby and I have just enough time to scare them away and avoid death.

I, in my own world of darkness, am beginning to speculate what the sky will look like when it is all over. Just moments ago, the sun collapsed into a one-dimensional disk before a dreary pallet of churning mud. Howling winds usher in chilling temperatures and yet now, deep within my own private reverie I optimistically conjure up a great heavenly field where rainbow paved paths lead to jeweled cities. I find it easy to think of Angels and Heaven and not of Demons and Hell. It is as if I have been here before.

When the rumbles of unfamiliar tones and threatening thunder promises a deafening symphony, the image shatters and I wrench my eye lids down and shudder. I barely manage to keep my hands that are the fragile seal against a fate worse than death, firm against Paul's eyes.

The others almost convulsing from anxiety, manage to turn their backs and their eyes from the unimaginable evil just moments before it climaxes. Instinctively, Father Mulcahy recalls chapter 13, verse II, of the Book of Revelations and begins reciting the strange passage.

"Then I saw another wild beast come up out of the Earth; it has two horns like a ram and it spoke like a dragon."

Whatever evil resurrection transpires behind us, the Biblical words give the priests strength. Through a deafening din that sounds like a mix of the cries of the dead and the forsaken from worlds that might exist only in dreams, the old priest manages to turn to God in prayer.

"May God forgive our souls, for surely, it is the end of time!" he utters what he surmises would be his final words to his friends and the Lord. On the battleground of God and the Devil they are rooting for God. I could penetrate their minds and get wisps of images from their heads. I am one with a field that transcends spacetime and reality. They imagine a melting black, like tar, oozing from the ground in the shape of horns and blinding fingers of light tearing up the grayness of the sky to weed them out.

"Keep your eyes shut! The Minga came to steal the Eye of M'Dulu, and now the Eye is angry with them," I again whispers fervently to Paul and the others.

"I know grandfather, I know," he cries back impatiently. The juvenile whimper of a child has overcome the young boy. Not able to see but hear the sounds of the pulsing eye as it causes tremors from the bowels of the Earth, leaves his mind loose to wander through the ghastly images of his nightmares.

"We will all die if we don't cover the Eye from the world of evil," I reveal, the words etching out painfully in the air. Upon hearing this, Father Mulcahy shouts to Imran and Bumba,

"It is over I am going to open my eyes to God. Darkness is all around us and we have become imprisoned by a dark cloud."

I command Mulcahy to close his eyes again but he refuses,

"We could see enough to leaf through the pages of the Bible. There is nothing more for us to do but pray for guidance from the Book of Books."

I start to pray out feverishly unafraid of the mystery unfolding before our eyes.

"Now have salvation and power come, the reign of M'Dulu and the authority of his Anointed One. For the accuser of our brothers is dead, who night and day accused them before our God." Michael says.

"Revelations chapter 12, verse 10!" Father Mulcahy shouts.

"It is being fulfilled before our eyes!"

The priest and Bumba have prostrated their whole bodies on the rocking jungle floor, holding on as if they are on a roller coaster, to pray to the Lord Jesus. They dare not look up and the fact that they have been able to open their eyes to reads the scriptures is miracle enough for them. Mulcahy notices the American IFPs floating far above the jungle canopy as they slowly approach the entrance to M'Dulu's hiding place.

"They are coming!" he shouts out.

"Who is coming?" I ask.

"The American fighter jets, they are coming!" he shouts with an expectant voice as if he expects that they will save us.

I open my eyes to look around me. There is some light in the sky and far above my head, I see the IFPs circling around us as if awaiting an opportunity to die. Thousands of Minga warriors now surround the energy field that engulfs the creature. They pour out of the mysterious cave like bees attracted to honey. They organize themselves into concentric circles, preparing for a peaceful battle for their souls. They intend to storm the creature and abduct its prize, but as each approaches, the hypnotic spell of the crystal takes over, and they simply march toward it feeding their life force to increase its strength. As the dust from the hordes of trampling Minga settles, their voices comb the forest with war cries. In great human waves the mindless warriors charge the force field protecting the creature and the Eye, to simply rebound from the membrane, and die. The priest are chanting from the Bible holding tight to their sanity as if the words will drown out the massive death cries. I fix my gaze on Father Mulcahy and then Imran as they pray for salvation from God.

"They have tasted the spirit of M'Dulu. No one can taste and leave," I shout my interpretation of the chaotic Minga hell around us.

"Now they shall die for M'Dulu is angry," Paul blurts pulling my hand away from his eyes. I am not certain that it is safe to look but my instincts compel me beyond my rational mind and silently mumbling a final prayer, I let go of Paul's eyes and he wildly turns around like a reanimated mummy who abruptly realizes he has been reborn. Although the scene before us is a living hell, I am grateful we are still alive and I break into a smile. Just ahead, thousands of Minga warriors continue swarming from the cave to the crystal. The savage human insects hold snake-headed spears and swords. Their oiled skins glisten like crystals immortalizing their zombie stares. Each circle of mindless Minga that touches the walls of the mysterious force falls back and dies only to be replaced by another circular wall of men. In a fantastic bid to die in the arms of M'Dulu, the living climb over piles of their dead brothers before plowing into their target.

The black mist is clearing and light from a sun again starts to strike our faces. I squint into the tall jungle trees that hide the crystal and all I could see are hordes of dying Minga.

"Why are they dying Michael? We have to do something to stop this genocide," Father Mulcahy insists. His philanthropic side continues to battle his common sense.

"In a few moments it will be our turn, Father. We should not see the Eye with our naked eyes, or our fate will be the same," I warn.

"Look at the size of the glowing force field! Did you not notice that as more Minga touch it, it gets bigger?"

Imran is amazed by this observation. He is getting ready to formulate a hypothesis when I unsteady him by reading his mind.

"As they touch it, they increase its power by giving it their life force. Then, it consumes them like a flame consumes moths in a huge fire from hell," I manage to say. Above, the buzz of the IFPs competes with the furious sounds of the Minga's offensive against living.

CHAPTER 32
(Armageddon)

Father Mulcahy sees the massive IFPs crashing into the jungle canopy a few hundred meters away from us. Imran is consumed by a serious prayer he has composed in hopes of saving his soul. Bumba merely sweats and hangs close to the priest trembling with fear. Paul and I embrace each other, reciting the ancient incantations of the Poro. We can hear the massive explosions that are the IFPs, crashing to the jungle floor. The great tension of electromagnetic energy stored in the vehicle's engines is released in an explosion of red and green slime and debris that colors the sky. Dead bodies of American soldiers seem reanimated by the blue light that emanates from the great crystal. A fleet of Russian planes arrive. They head directly for the crystal and as they approach, the ferocity of the crystal becomes apparent. They all perish in a burst of power emanating from the angry Eye. Then, thousands of fighters from many regions of the world start top appear like a swarm of bees approaching the hive. Like fleas they perish in the sweet light of the angry crystal.

Father Mulcahy looks up at the sky and sees the last American IFP falling into the jungle canopy about a hundred meters away. All around us, debris from the American fighting machines lies scattered about the jungle canopy. The living cry with pain and those who could move craw towards the hypnotic Eye of M'Dulu.

Father Mulcahy instinctively starts to run to the nearest survivor and kneels by his side to begin praying for the victim's soul. Paul and I follow him to help. A young military man groans as the toxic fuel of the IFP consumes his flesh.

"Wormwood" the priest murmurs to himself. He gives the dying man his last rights and turns his attention to the others. We will try to save as many as we can, now that we know the Devil has arrived to claim their spirits. Imran and Bumba join our quest, having something to do gives us a purpose that is tranquilizing. Although Paul and I are overwhelmed by the crashing planes and the heaps of rotting technology that abounds, we persevere and continue to help the priest with his incantations to the dead.

A man dressed as an America General lays on the ground with his head bleeding. The priest approaches him, and lays a hand on his burning forehead and prays to God for his soul. The man painfully gazes up at us. The green IFP fuel that coats his flesh is burning his skin. Third degree burns swell into toxic bubbles on his face. His body is melting into rubber.

"Fuck . . . !" he whispers.

The smell of burning flesh is nauseating. The priest places a handkerchief over his nose and mouth. There are no features to recognize the man through his ghastly disfigurements.

"I am Commander . . . . Please take me to the light!" he pleads.

The priest is incredulous.

"Look at you!" he explodes, raising the Commander's heads up from the ground.

"I am US Commander Phillips, I order you take me to the light!" he moans.

"You are delirious," the priest states sadly.

"I don't care, but I need to . . . to touch the light. It is beautiful . . . Oh, so beautiful!

The priest knows that the man is delirious, crazed by the hypnosis of the rays of the crystal. Mulcahy spots a communication watch on the Commander's hands.

"First show me how to communicate with the Vatican," he orders back.

The Commander licks his dry, scarred lips as if the Father's words are food for the tasting.

"Take my Communicator and dial 0616161," he instructs weakly.

"That will get you in touch with the White House. Now take me to the fucking light!" he orders the priest. His big burly frame tries to move, but he cannot.

Father Mulcahy stares at the communicator. He glances at me and motions us to help him. Take the watch off the wrist. It has blended with the General's flesh. I lift the man up and start ripping off the communicator. The man does not flinch. There is no pain in M'Dulu's light. I hand the watch over to the priest. It looks complicated and confusingly hard to decipher. Using it would be like deciphering a code and for some reason I hope the priest can do it. His hands are trembling and the noise of the dying Minga is just too unnerving for the priest.

"How can I contact the Vatican?" Imran asks the dying man.

"First take me to the light, I must touch the light."

"Commander, the world is falling apart, and you are suffering from Bell's syndrome." The priest states a matter-of fact. The priest grabs the commander by the shoulders and lightly shakes him.

"The light cannot save you but you can save your soul. Please tell me how to contact the Vatican now!"

"Take me to the light, Father, please take me to the light!" he repeats like a child asking for candy. The priest focuses on the eyes of the commander. They are no longer his. The power of the Eye has possessed him and he stares ahead like a blind man.

"The Eye of M'Dulu has opened the gates of hell, and we are being consumed by the Devil!" I shout.

"You are suffering from Bell's Hypnosis," he affirms,

"We need to get out of here fast. If we don't stop this now, the whole world will be affected by the rays. We must contact the Vatican now!" Mulcahy yells.

Something in Mulcahy's serious rhetoric strikes a chord in the Commander. He needed to hear a stern commanding voice. His eyes light up to an invisible stranger. The smoke from burning debris is thickening and Imran begins to cough.

"Dial er, er, I can't remember," he mutters.

"Please snap out of the hypnosis! Look at my eyes, can you see me clearly?" the older priest interrupts. The commander trembles from the pain of his wounds but then a broad wicked smile burst forth that makes death and destruction a cheerful matter.

"Yes, but not now Father, the light is wonderful, I must feel it, touch it," he proclaims in a euphoric voice interrupted by coughs of blood that splatter onto the priest's robe.

"Millions of people are dying all around us, we must contact the Vatican!" the priest demands again.

"Does it matter? Don't you want to see the light?" the commander asks deliriously. Imran becomes despondent and collapses onto the jungle floor. The cries of the Mingas are escalating to a climax. The foul smell of toxins is coating his skin with a black ash.

"You must give us the codes, or you will not see the light," the priest reasons with the dying man.

"1 1 4 4 4 7 7 then dial V.A.T.I.C.A.N., and," the Commander reeks out, then pauses to catch his breath, "and it will display the codes of the Vatican," he regurgitates.

"But now I must touch the light!"

The priest quickly dials the codes and waits for the Vatican number to be displayed. He mumbles to himself and turns the phone speaker on. He dials the Vatican codes and waits.

The Vatican phone is answered by an unfamiliar voice.

"Hello, hello!" Father Mulcahy begins forcefully.

"Hello, what can we do for you, this is the Vatican?" the voice replies coolly.

"I need to talk to the Holy Father immediately," Father Mulcahy shouts.

"The Holy Father? Who are you?" the voice comes back defensively.

"I am Father Mulcahy. He knows me. This is a matter of utmost urgency."

There is a strange pause.

"But Father, the Pontiff is dead!" the voice, certain it has been talking to a prankster, promptly hangs up.

Father Mulcahy is dumbstruck. Tears well up in his eyes.

"What is it, Father? " Imran asks.

"The Holy Father. He . . . he is dead!" he sputters and sobs.

"Why did the Lord take away the leadership of his house when it needed him most?" Imran remarks almost bitterly.

Finally, the cries of the Minga, like an alarm clock, interrupt the priest's mourning and he turns his attention to Commander Phillips.

"Why did the White House send you here, Commander?" Father Mulcahy asks, trying to put the pieces together.

The commander half raises his arm and exposes a crystalline wrist band.

"The data bank.. the President," he whispers coming back to some semblance of his mission.

"You must take me to the light," he insists.

The priest considers the man's last dying request. Commander Phillips is now reduced to a living corpse yet he still persists in his dreamy state to see the light. No one at the Vatican can help them now.

"The key where is it?" he shouts.

The commander has resigned to a new authority. He sends his eyes toward his jacket pocket. The priest understands. He plods through the remains of the jacket carefully avoiding the oozing green fluid of the IFP. He finds a key with the codes written on it. He turns the lock on the crystal cube and releases it. He places it in his deep pocket.

"I am sorry, commander" the priest decides as he slowly lowers his heads and turns his attention to the communicator.

He dials the codes for the Vatican Laboratory, then after connecting, he dials Nambu's lab extension. As the phone rings he prays that Nambu will pick up the call just before his familiar voice comes over the telecommunicator.

"Hello!" Father Mulcahy recognizes Nambu immediately.

"Nambu, this is Father Mulcahy."

"Where are you? Have you heard the news?" Nambu asks.

"What is going on?" Father Mulcahy asks.

"Bell's Hypnosis is consuming the world," the radio crackles.

"Nearly everyone has taken to the streets in Europe and people from everywhere are heading toward the Mediterranean Sea trying to go to Africa!" Nambu explains through the static of the radio.

"What happened to the Pontiff?" the priest quizzes him.

"He is preparing to go to Nevada to meet with the Bilderberg group. His brother is under the influence of Bell's Hypnosis. The whole world has gone crazy!" Nambu bellows.

"We have located a massive crystal in the forest, Nambu," the priest informs him, "and the Pontiff is right. It seems to generate Bell's Syndrome. Hundreds of thousands of people are dying in the jungle as well. They are attracted to the crystal, probably by Bell's syndrome. We have to find a way of deactivating it."

"Is the crystal red or blue?"

"Blue" the priest answers, "why?".

"It controls time. The other twenty six crystals control other spatial dimensions and this one controls time!" Nambu explains.

"What is the difference?"

"It is the most powerful and it can deactivate the other space crystals. You see time is imaginary and space is real, so when we calculate the distance to any spacetime point, they cancel each other and create a null Minkowski spacetime measure. The combination will cause a null spacetime bubble that will disrupt the symmetry of matter and antimatter causing total annihilation of this universe at the expense of a new universe formed by the set".

He pauses.

"If you can prevent the blue crystal from being joined with the others, you could stop all that is going on by slowing down time." Nambu explains. The crackling voice trickles through the static of the telecommunicator. It is breaking down like the atmosphere around them.

"It seems as if the other crystals are showing their faces all around the Globe. The crysta.."

The static is increasing. An unearthly force is doing its best to intercept Nambu's explanation.

"The White House has issued a Defense Condition 4. The Soviets, Chinese, and Indian governments are all on high nuclear defense alert! The entire Vatican grounds are covered with walking zombies. What can I do?"

"I do not know yet, but I believe that the end of time as we know it has arrived unless we do something fast. The only way to prevent Bell's syndrome is not to look at the light from the crystals. Use goggles all the time, blind folds or filtering glasses. Pass the word around and tell everyone to do the same!" the priest commands.

"Did you get the codes for the satellite?" Nambu asks.

"Yes, I have the crystal data bank, but I have no way of communicating it to you."

"I will use the satellite, I wrote the programs. I can communicate with it to read off the data crystal if you can enter the codes to activate it," Nambu states.

"The priest enters the codes that were given to him by the Commander into the data crystal. It starts glowing.

"Go ahead download the data now, the crystal is activated."

Nambu sends the command to the satellite to download the data. A few moments later, his voice echoes in the speaker.

"I have the data! I have the data," he repeats, and the radio dies. The crystal watch starts to burn the priest's hand as it starts to self-destruct. He quickly throws it to the ground and turns his attention to what is going on. Although the news of the dead Pontiff grieves Father Mulcahy to a point of inertia and desperation, suddenly he is filled with a sense of duty and strength. There is no one else who could help him. The fate of the world rests on his shoulders and he could not fail.

"He is dead," Imran observes the frozen look of disappointment on the commander's face.

"Who?" the priest asks dazed, searching for Imran.

Imran points to the commander's body.

"It is finished!" Father Mulcahy turns to Imran as he lowers the commander's eyelids and makes the sign of the cross.

Bumba cries out in terror, breaking down from the death and destruction that seems to be slowly replacing the air they are breathing.

"The entire world, Africa, India, China, America, Europe, everywhere is now under the control of the crystals! Just look at the Minga! They are attracted to the crystal like flies!" he realizes, pointing to the Minga in the distance.

"Why then are they not affected by the crystal?" Imran asks.

Imran is pointing to Paul and Michael who were definitely looking at the rays from the blue crystal but seem unaffected. Perhaps the world's salvation depends on analyzing the different effects the Eye has on different people.

"Do you think their presence expels the rays of the crystals?" Imran speculates. The priest's eyes light up.

"Perhaps they are indeed a few of the chosen?" Father Mulcahy agrees.

"Perhaps we are five of one hundred and forty-four thousand chosen!" Imran adds.

"Oh my God, Oh My God, it is indeed the end, and we are alive!" Father Mulcahy hypothesizes.

Paul and I are still embracing each other tightly, holding our eyes affixed to the blue rays of the crystals. Father Mulcahy runs to us and pulls us apart.

"We must act now. How can we deactivate the Eye?" he shakes me violently. At first, I am startled. I squint until I acclimate to the lesser light looking away from the rays. After checking on Paul, I fix my gaze on the dead general gasp in horror. Commander Phillips's flesh is a mottled ashen yellow and blue. He lies in a pool of his own molten flesh.

"No one can approach the Eye now. Only Paul has the secret potion in his veins," I explain with trepidation. Imran takes Paul up in a new embrace.

"Paul, what should we do now? If we don't stop the Eye, it will spread its field all over the world. We must stop it now! Please help us Paul," Imran pleads holding his hand.

Paul trembles and sobs. He begins making high-pitched, spasmodic groans and it takes all the life in his body to compose himself long enough to speak.

"I can't, I am afraid, I cannot stop it now. You must have faith in yourself," Paul ekes out. I know that at this very instant it is time for Paul to wake up to his manhood.

"You must stop it!" Imran persists.

"This is why the Lord has made you one of the Guardians," he shouts raising his arms up to the blackened heavens.

"You are the chosen one, the only person who can enter the Eye and control it," Imran insists.

Paul looks at Imran and then at me. A rush of tears burst upon his cheek.

"Have you no faith in M'Dulu? I have always dreaded the day I must meet M'Dulu," Paul admits.

"Who is M'Dulu?" Imran finally dares to ask the obvious question that he has purposely been avoiding for hours.

"Don't you know?"

Paul is shocked. M'Dulu is the almighty who has dominated his dreams, hopes, fears. M'Dulu is the reason blood is coursing through his veins.

"He who looks at the Eye of M'Dulu must enter the ways of the Eye or he will die. M'Dulu is the way, the truth and the light. He who believes in M'Dulu will not die but will live forever!" I explain amazed at how easy the words come to me. Paul feels an abrupt surge of power lifting his soul. It is the power of knowledge and experience of manhood he almost tastes. We could actually see the transformation taking place. Paul's shoulders are pressed back. He stands taller. His brown skin is reflecting the brilliance of the crystal. Convincing him now would take more doing.

"For God's sake, Michael, have you forgotten your past? Your God-given mission? It is the end of all life as we know it. Paul! You must do something!" Imran yells angrily.

Father Mulcahy is locked in silent prayer with Bumba at his side. I step forward to answer Imran, believing it is my destiny to take Paul's leads.

"I have waited for sixty years to look upon His eyes again. Even I am not prepared to see His luminance now. We must go back to the village, so we can prepare for the end," I ignite the ritual.

"I must go now, I must go now, I must go now . . ." Imran starts to repeat as if a spell has taken over his being. The whole prospect of another taste of Armageddon is too much for the man and he is seeing reality slip through his fingers.

"No. Imran, not now, you must not, now, not now," Father Mulcahy sermonizes as he realizes that the hypnotic spell of M'Dulu's Eye has at last captured his faithful aid.

"Leave him alone, Father Mulcahy. Only the faithful shall prepare for the coming of M'Dulu, for he comes like a thief in the night!" Paul proselytizes. A paralyzing chill scurries spider-like up and down Father Mulcahy's spine as he realizes the truth.

"M'Dulu. M'Dulu," he whispers. "He...he died for our sins?" he asks Paul.

"Yes, Father. And he has come back for us."

CHAPTER 33
(Area 51 Nevada)

The Nevada desert is a hostile place to be when there is a short supply of water, food and other essentials, especially portable toilets. Diana and I have had regular debates about the sanity of going to a desert when the time comes for us to protect ourselves from the great calamity. I did not want to come to Nevada, but Diana has convinced me that it is the right thing to do and that I have already stated that this is what we must do during visions of the future. The Vatican jet lands in a large airfield that had been prepped to receive important people from all over the world. All around, thousands of people and hundreds of vehicles raise dust from the desolate roads that lead to Area 51. Guards stand in line with weapons pointed in the air, awaiting the arrival of President Cooper and the fortunate few that will be taken to the secure facility under the desert. Diana and I are among the two thousand persons allowed to enter the site. There are five security level gates in the site and each vehicle must pass through all of them to enter into the facility. Each gate has a check point with unprecedented security. The final massive gate that leads to the underground tunnel under the Hoover Dam has been opened and one hundred guns guard against any unauthorized entry.

We enter into a large bus that is specially equipped with shielded windows to prevent the penetrating rays of the J-N field from harming us. I am very susceptible to the J-N field and I still have some atomic fragments of the Jungle crystal inside of my body. Particular attention has been taken to my situation and I am wearing a special suit that prevents the rays from entering my body. At this point, I am totally in the control of the Bilderberg Group and the Vatican has decided to transfer me into their care while we wait for the great calamity that is to come. Diana and I have no control of the situation but we have special plans to continue our lives as saviors of the destiny of mankind.

This morning, Diana and I had prepared our minds for the ultimate split in the J-N field knowing that we will have to lead multiple existences in the astral world if we are to survive the calamity and save the world and its children. I have to be here as Seth in body, but my mind can astral travel to other dimensions and project alternate personalities that we need for the plan. In the bus, we are introduced to the German President, Adolph Wolfgang, the King of Spain, Rodolfo Da Sintra and his assistant the retired General Gabriel Machismo. As we wait for the others to enter the bus, I feel a strong urge to get out of the buss and just get away from the site. My feelings are so strong that I turn to Diana and tell her we must get out of the bus now before it is too late. She smiles and tells me that I also predicted that that will happen and that if it does I must not be allowed to leave the bus and the area. My mind is split at this point and Diana has made her choice to follow the plan we have made for our existence.

A woman and her two children enter into the bus and start walking toward us. At first she is just an astral projection of Diana decoupled from her being, but as she approaches me I realize that she is real. She and her children are seated just two seats in front of us like a projected movie with no substance. I look at Diana and she is totally focused on her reality. For a brief moment, the J-N rays split into a rainbow of colors and they seem to be emanating from Diana's aura. I look at the woman again as she turns nervously looking at us and smiling. My heart starts to beat like a drum set into motion by entranced hands. She looks very familiar but I cannot place her and I feel a sudden stabbing knife entering my heart. I turn to Diana and ask her if she knows the woman but she smiles back at me and says nothing. I have been given a shot of Memorin to calm my nerves down and I am feeling very relaxed but out of touch. Diana seems to know what is going on but has deliberately decided to keep me protected from my mental lapse.

After the bus is loaded full, the driver shuts the large latches on the door and starts to walk down the bus aisle holding a manifest in his hands. He starts at the back of the bus calling out our names as he passes us. The woman's name is Elaine and her three children are Michelle, Nadia and Jessica. The names sting deep into my psyche and I feel myself straining to impound any memory that may help me remember. I find myself sitting at a dimensional door that is locked with a big padlock. I cannot enter. Our eyes are protected with heavy goggles especially designed to counter the effects of the z-particle emissions of the crystal in Nevada and I am beginning to wonder if they are also impounding the rays I need to spark my memories. We have been given special goggles to prevent the J-N rays from affecting our brains and the children are straining through their goggles to look back at Diana and myself. We cannot see each other clearly and the even darker goggles make it difficult to see features on the faces in the bus. Thousands of people are lined up outside the fence of the small airfield of Area 51 trying to enter the protected area. The hot desert sun has already taken its toll on many and they lye dying outside the fence. From the vantage point of the bus, I can barely see thousands of people roaming aimless about the desert as if there is no purpose to their strange journey. I find myself worrying about the world, about the woman called Elaine and about her three children as if they are mine to worry about. Who is her husband and is she worrying about him now? I start feeling extreme jealousy for her and her children. I cannot explain my feelings, but they are real and intense. We must be dimensionally close in the landscape of Love, Faith and Hope. I look at Diana to see if she can sense my feelings, but she is bland. For the first time, I start feeling intense feelings for another woman other than Diana. Am I crazy? I know Memorin takes you to emotional doors, but for some reason, this door is closed and I feel like I have been deliberately left outside by the Elaine and her children. I start to feel like I am a flowing spring of water that is trapped beneath black rocks only to mingle with surface water that has a different feel and form. I start to see Elaine and her children like creek waters above me about to mingle with my spring water below. Who are they? I start feeling like an alien in my own world. Diana, Elaine, they are so alike in stature looks and feelings. They are becoming like one person to me. Deja vu!

I peer hard through my goggles and try to look into the distance. Many people are walking in unison toward Lake Mead and the Hoover Dam. It is as if they are an army of ants that has been commanded by some hypnotic General to quench their thirst in the murky waters of the dam. The pilot reiterates his warning for us to neither pay attention nor become emotionally involved in what we are seeing. We drive through a long narrow tunnel following the lead of the many more MPVs that are ahead of us. Flashing signs indicate the time that is left before the massive gates will close behind us. The occupants of the bus MPV seem to be high ranking Europeans silent in thought. Each of them in silence seems to be busy with their thoughts and I feel that Diana is lost in her own thoughts as well. What is she thinking? Does she feel the same feelings about the woman Elaine and her children? For one hour, we slowly follow the convoy through and inside the mountain, until we come to a large open chamber surrounded by huge steel beams that form the support of the ceiling of a large building. The MPVs are organized in parking zones as they enter the huge chamber. Each parks in a pre-designated spot, and waits for further instructions.

When our turn comes, the driver helps Diana load me in my wheel chair and motions us to wait. Just a few meters away from us Elaine and her kids also stand in wait of further instructions. I take a good look at the woman and her children. She could have been Diana with my children. They are what I always imagined we could have been in the past were it not for the Vatican holding our lives at bay. I turn to look at Diana holding on to my wheel chair ready to go. Her expression is bland like a person ready to witness a surgical procedure. A military man decorated with many medals of a foreign nation appears to be our group leader. He orders us to follow him.

"My name is Commander Gustav Leopold" he says, "I will be your commander and liaison in this facility for the next two months. Please follow me."

We follow our leader through a dark corridor adjacent to the parked MPVs. For ten minutes we pass through a partially lit corridor until we come into a large room with several small rooms attached. The long walk seems to have stressed the Spanish General and he curses and grumbles, pounding his walking stick hard against the ground. Gustav did not slow the pace down, he pushes on with the German, Elaine, the children, Diana and I close behind him, while the King and the General follow some distance behind. The large chamber is octagonal in shape with each face entering into a different zone. Gustav leads us to a zonal door marked Door #27. We enter door #27 and it opens up into several corridors each leading to a door with a number. Gustav stops at a door marked #13 and opens it. He motions Elaine and the children to enter the room. From my vantage point, I can see two well-made beds and an adjacent bathroom within. They enter and close the door behind them seemingly knowing what to do next.

Gustave leads Diana and I to another room marked Door #1. We enter the room. It is well appointed with a single king size bed facing two walls and the door. There are no TVs or any other form of outside communication in the room. Gustav tells us to be comfortable and only come out of our room when told to do so. He leaves with the remaining men.

We are very tired from the long trip and I settle in and relax on the bed almost immediately. Diana tries to activate her communicator watch but it does not respond and there is no signal. She also resigns to the comfort of the bed and in a few minutes we are in a deep sleep. A few hours later, I waken to a speaker on the door telling us to exit the room and attend a meeting in Chamber 0. We clean up and exit the room into the large central plaza. We see Chamber 0 and enter. In the large chamber, huge tables come out of the floor and position themselves around the center of the chamber. There is food already placed on the table and several people are already seated around the tables conversing and waiting. Gustav points us to the same table that Elaine and her children are seated. We are seating next to the Spanish King and his old General. The King greets Elaine and smiles at the kids.

"Elaine Davis?" he asks smiling.

"Yes," she answers with some surprise.

The name Davis rings a bell but again a blank comes up in my head. I have heard of John Davis in the Vatican and the same strange feelings I had then starts all over again. Who is John Davis? Who is Elaine and the children, and how do they relate to me?

"My name is Rodolfo Da Sintra, the King of Spain," the King states waiting for a reaction from Elaine. Elaine stares at the King and says nothing in return and instead the expression on her face makes me feel like she is disappointed about something.

"This is my assistant, retired General Gabriel Machismo."

Elaine nods and smiles. My attention is diverted by a loud grunt as Gustav our group leaders sits himself at the head of the table. The General looks like a hardened man with many years of experience in the military. His uniform is full of decorations that could only have come with age and experience.

"After dinner, you will go back to your respective rooms," he starts.

"You will read your manuals carefully and try to study what has been provided to you. There is no time for anyone to help you through this period. You will only be allowed out of your rooms when called upon," he states authoritatively.

"There are many chambers like this one around this facility. Anyone that tries to enter a chamber they do not belong to will be severely dealt with or thrown out of the facility. Once the facility doors are closed no one can enter or leave. As you can see, there are Kings, Presidents and high ranking persons in adjacent chambers to ours. Once a week, we are allowed to go and mingle in the grand chamber with other guests. There, you will have an opportunity to be briefed by the President and his commanders as to what they need to tell you all. Eat well and good day."

After he is done with his speech, he starts to consume his meal without paying another glance at his guests.

CHAPTER 34
(Salvation)

Father Mulcahy, Bumba, Paul and I race towards the village of Yamandu to save the world. The center of town resembles the aftermath of a great war. Bodies littering the entry ways of smoldering huts are the picket fences of a death town. A few mongrel dogs howl around their still masters. The distance is peppered with faint groans and the wails of abandoned infants. The crackling of burning brush sounds unpredictably, like popping corn. Misery is a permeating stench that bleeds through the men's pores the way the smell of alcohol does in a drunkard. We dodge thousands of people approaching the crystal from all directions as we plow through the hazy, rotten smoke of burning, dead villagers. A potent nausea slows us now as we approach the center of the village and see thousands of aircraft flying towards the crystal like moths attracted to a flame. The insignia of Russian, American, British and other nations are flying above us. There is an all-out world war going on now.

"We must hurry! It is burning down!" I am referring to Bitter Cola's house. Flames have already engulfed the entrance to the village, and in the confusion, the sound of crying babies appears to be rising from within the fire. I struggle with the illusion as I wait for Paul.

"Hold them here until I return," Paul whispers to me.

Without hesitating, Paul plunges through the burning flames. I know he will be okay but Father Mulcahy is pondering, walking in circles as if he is lost.

"Where did Paul go?" He asks me breathlessly.

Mulcahy is holding his head. I can tell it must be throbbing with all that is going on. Paul disappears into the burning flames in a seemingly impossible feat and I can see the bewilderment of the priests and Bumba when they realize that Paul feels no pain from the flames. I decide to follow him in.

"We will be back soon," I assure them with a deliberate smile to lighten their burden. I know that the priests are exhausted, and they do not have the strength to question my decision and in the distance, I see Bumba scampering to the sounds of crying children in an effort to locate them. I follow behind Paul as he walks into the flames, impervious to any pain just as easily as an Indian Yogi can walk across a bed of nails. With a few quick waves of his hand, he magically clears a thickening smoke to make us see our way into Bitter Cola's room. Once there, he rushes under the old man's bed and gasps when he sees that the old chest we are looking for is missing. We nervously scan the room as smoke rises to tidal frenzy to become too thick to dispel with a slight of our hands. As we run out of the house my eyes catch the gleam of a gold trinket on the floor beside the burning front door. I reach out and pick it up. It is piping hot to the touch and I cry out instinctively. Paul picks it up as if unscathed by the pain and shoves his scarred fingers into his mouth and presses on.

"Yabu Yombo!" he accuses, angrily as we meet the others.

"What happened Paul?" The priest begins to question.

He could see that Paul is shaken from revenge seeking anger.

"It is gone. The chest is gone!" Paul shouts, scouting the smoke-filled village. Paul needs to hang on to the memories of Bitter Cola to see him through this calamity. But nothing material of his great grandfather remains and there is a lesson to be learned that only I quickly grasp. I try to ease his anxiety with soothing words. There is no time for further remorse. By now a symphony of coughing has reached a fevered pitch and the great, mighty jungle around us is totally screened from view. Bumba is trembling with fear and he is hallucinating that the great anaconda is smoking his sacred pipe nearby.

I indulge Paul a little longer as we race through the village trying to locate the Queen Mother and the mysterious chest. Paul remembers that as a child, the sight of the chest has always possessed him and no other possession that Bitter Cola owned has intrigued Paul more. A chest could hold anything and everything that is mystical and magical and a young child looking into the dark, depths of a chest is akin to an adult exploring the insides of a deep well for treasure. Paul holds on to his wish, hoping he can make it reappear. He visualizes the gold encrusted network of lattices across its top. He visualizes the two angels standing on the top of the chest. The golden figure seems to form a permanent kinship with him. Then, suddenly his mind fills with a sweet memory of the chest. The angels seem to come alive and point the way. The Poro instincts have come.

They have taken over the boy and make a man out of him. He is no longer afraid. The metamorphosis is complete. He knows the exact location in the forest where the tree that the chest is carved out of, has been taken. He concentrates on the chest and his childhood in a prayer as we move through the village but instead of resurrecting the chest, our travels bring us to some wounded children who are crawling towards a dying old man. The children instinctively believe the elder is capable of helping them.

I seize one youngster whose once smooth, full cheeks are scarred purple from second degree burns. I touch the wound delicately with my index finger and hold the crying child close to my chest. Crying, I raise the child into the smoke-filled air and shout out in desperation;

"Why are you doing this, M'Dulu? Why the innocent children? Take me now, M'Dulu! Take me, God of revenge, God of anger, maker of lightning and thunder, maker of fire and water. Take me, but leave them alone!"

My voice rages into the thick smoke, challenging specters in the death clouds. M'Dulu does not answer easily. He answers with a pouring of red slime from the heavens. Paul becomes fearful that I might give up or suffer a breakdown. I could see that in another few hours, the children will probably lapse into unconsciousness and then die. There is nothing we can do now.

"We must go now Grandpa, we have a lot of work to do," Paul whispers gently, his hand on my elbow.

He has mustered all his strength for those words and I am surprised that they flow out with such serenity. Suddenly, the boy is blossoming into the man.

"If M'Dulu has selected them, they will find him," Paul predicts.

He feels the soul of Bitter Cola guiding him, almost possessing him with wisdom and words. Father Mulcahy is coughing loudly while the blackening smoke from the village shapes toxic thunder clouds around him. Instead of struggling to save the world, he is battling to salvage his life. When Paul leaves me at the village center to lead the other men out of the thick haze and towards the crystal, I spot Mulcahy at the edge of a clearing, wiggling along the ground like a rattlesnake to get under the foggy fumes. Paul is formulating a plan and fortunately, the wind shifts and clears the air just long enough for him to usher the priest and his entourage back into the village. When he returns with them, I reluctantly head toward them, as if I am walking towards the gallows. My face is expressionless and drawn and I know I have lived the bitter hardship of a lifetime by preparing to leave the dying children. The elderly villager is dying as he moans weakly.

"Take us with you, Angel of M'Dulu, save us," the man utters his final breath shouting at me. I dare not look back into the ashes of the village for the children are still there, holding on to life gingerly like wilting flowers. They will endure their death innocently and without bitterness in their ignorance of both life and its ending. I decide I can no longer fear their suffering. If I do, I will go insane. But the elderly man's last dying chants unsettle me in a way I will remember for the rest of my life, if there is time to live a life.

The time has come for Paul and myself, the priest and the other men to return to the jungle and proceed towards the great crystal light. Anxiety no longer rules us. The spectacle of dying people has strengthened us. We are still alive - healthy and in fact not even wounded, we can continue.

In a matter of minutes, we assimilate into the thick flow of human migration towards the light. Our group advances quickly and then stops a few hundred meters away from the crystal. In the distance are the dying Minga. Ethnics from all over the world, who are now predominantly female, climb over the dead bodies of their warrior men to reach the crystal's field of light. Amidst the carnage and chaos, we scout for Yabu Yombo as we avoid the sweeping quicksand that is the masses groping for the light. It is difficult to make sense of the tangled mess of bodies, let alone focus on facial features to single out one individual. Cacophony and unearthly scenes abound. Flocks of birds whose habitat has been disrupted by the fires fly in confused scattered formations. An occasional lion or elephant would stampede out into the open and then disappear into the brush, disoriented by the teeming people in the jungle.

Father Mulcahy is exhausted and he collapses on the ground to make sense of it all. He notices Bumba is in his underwear. The heat from the fires and the jungle sun has forced him to strip to keep cool down. Bumba's body is bruised and scratched from fighting the thorny, unforgiving brushes. Father Mulcahy's face is also cut in several places and he now walks with a limp. In the distance, Mulcahy spots John Davis wearing goggles and dragging a large chest toward them. His clothes are torn and his hair is singed from the flames around them. I raise my sitting frame from the jungle floor as John Davis approaches with the chest.

"He has the chest!" Mulcahy shouts pointing to John Davis. The priest tries to get my attention.

"He has the chest, Michael he has the chest!"

"Yabo Yombo is dead," John Davis explains through staggered breaths and he puts a foot on the chest he has been dragging.

"Zola is dead, but before dying, she told me to get the chest from Yabu Yombo!" John Davis states.

I approach the chest and call upon Paul to open it.

"How are you not affected by the light without goggles?" John asks the Priest. The Priest looks at Paul and them me, and for the first time he realizes that both of us are physically unscathed by the light and by all that has been happening. We show no signs of exhaustion and in fact our faces are not even dirty.

"An aura of serenity that is almost holy surrounds them," John Davis exclaims in disbelief.

"Why are you not affected by the Eye of M'Dulu?" He inquires repeating John Davis's question.

A great smile overpowers me. It is time to show them. I open my shirt and reveal the crystal necklace that glimmers across my chest. It is the necklace Bitter Cola had presented us before he died. It had been a great honor. Bitter Cola, who is customarily solid as a rock, had sobbed like a baby when he passed the glimmering stones to us. No words were exchanged as to its significance, but after that incident, I began to have dreams of another life, another existence as a spiritual being. Somehow I know that the pendant is special, very special. The gem means everything to me now, it has sealed a fate that began in my younger days when I worked in the African mines where the biggest diamonds formed from pristine Earth, dulled in comparison to the heavenly crystal. The crystal awakens me to the fact that my whole life has been carefully orchestrated for a purpose. Not to mine diamonds, but to mine diamonds of Faith and Hope.

Paul opens the chest and searches for the potion. He grabs the flask holding the blue potion and takes a gulp. He hands it to me and I gulp at the precious drink spilling a few precious blue drops on my bare chest. I wipe the liquid over my bare chest to absorb it into my body. I hand the flask to John Davis, the Priest and Bumba. They each take a sip of the precious liquid, it tastes like uncured wine. The priest recognizes the liquid. It is no different from communion wine, the blood of Jesus.

The priest turns his attention to me. Then John Davis starts to look at me as if I am some strange thing in their presence. I feel different. My bare chest exposes the enormous gem, the gem is what catches their attention. It is as if the wine I just had somehow makes them more aware of the gem that is now sparkling with a new intensely as if it is ready to burst into birthing a sun. The priest points to the gem looking at me curiously. It is time. It is time to whisper the command to Paul. The command that Bitter Cola had taught me for this purpose. The secret word that becomes the power in Paul's final transformation.

Paul takes a step back from the others and hastily rips open his shirt, exposing his bare chest. An exact duplicate of the crystal pendant hangs from his neck. Paul grasps it joyfully to kiss it. Father Mulcahy is astounded and his eyes light up as if a deja vu is charging through his brain. He searches his priestly robes excitedly and displays the colored rosary his father had given to him when he was a boy. There, in the pendant, the priest's destiny blazes before him, a cross shaped exactly like the crystals on Paul's bare chest.

CHAPTER 35
(the Guardians of Eden)

The Guardians stand around the twenty-five glowing crystals already attached to the huge cubic structure in the middle of the hidden Garden City.

"They are unwilling to give us the other two crystals," Resa says to another older Guardian with a reddish white beard.

"They do not obey the Covenants we made with them. We must do something fast before the Master's army arrives," another old Guardian states, scratching his golden beard.

"We did not do this right from the very beginning," another older guardian states regretfully.

"Lucifer should have locked the contract with Abraham and given him children, instead of making him an enemy and preventing him from having children. He angered Abraham, and the Master won him over by giving him children," The first red beard Guardian reiterates.

"Adam was a traitor. When the Master sent him in physical form to find the lost Crystals, the Master had built the Garden City to entice savage men into the Garden for their life-force. Lucifer tried to rid Adam and Eve off the Garden City and he convinced Adam that he too could gain the power of the Crystals and become like God if he ate of the fruit of the Garden. Adam believed and became like a savage man trapped by the fruit. His sons were also fighting each other instead of helping the Master find the Crystals. He allowed outside savages to marry with Cain, who killed Abel for the fruit Abel grew in the Garden. He should have used the fruits to capture humanity's soul," the old Guardian states angrily.

"You cannot blame Adam, he was brave. He volunteered for the Master's secret mission when we did not. He knew he would eventually become mortal when he entered the Garden City in the form of men," the golden beard states.

"If only Lucifer was more careful with the Experimental Craft. The Master had not yet completed the design of the craft when he decided to steal the Cubic Field and revolt," the old Guardian says.

"What I do not yet understand is how it ended up here, on this planet, why the secrecy?" Resa enquires.

"It was a secret because the Master created this Garden City for his own pleasure and relaxation. He worked for seven eons restructuring the Sun and its Planets for the human life-force."

"So, the Master created this Earth for man and Himself?" The young one asks.

"Yes. He built the cubic-field-powered craft that can travel through the physical four-dimensional world and our astral twenty seven-dimensional world without destroying spacetime."

"Is that where we come from, twenty seven dimensional spacetime?"

"What are the crystals, and why do you call them covenants?" A young Guardian asks.

"The crystals are vehicles that travel through from this realm to the higher realm. They are a means for man to be saved from his permanent trap here in the physical world of logical realism." The old guardian explains.

"So, what man calls Covenants are really vehicles for his salvation?"

"Yes. Each covenant is a vehicle, a crystal that contributes to his future salvation. When all the covenants come together, man can be saved from the physical realm to enter into the spiritual realm of twenty seven dimensions," the old Guardian explains.

"So how did we get here?"

"Master Lucifer and the rest of us elders including your father and mother stole the craft and crashed it here on our first try to travel through the spacetime barrier between the living and the spiritual worlds. Ever since then, we have been stuck here trying to rebuild the craft so we can go back to our spiritual world," the elder explains.

"So, since time began, we have been fighting to get these crystals and break the deals man made with the Master. If we get the crystals, we can freely roam the universe of the higher realm and escape from this physical reality," the golden beard says.

"But why Man?" Resa asks.

"The Crystal Cubic Field only thrives in the presence of the life-force. So, when Lucifer crashed the craft here with us, the Master had to create savage men and all sorts of living things and placed them here to maintain crystals. Adam took on physical form to enter the physical realm and gather the covenants. Each covenant is a deal that will eventually save Adam and his children from damnation here. "

"Is that why we are different from the Angels of the Master? Is that why some men do not worship us?"

"Yes. In our attempts to rebuild the craft, we contaminated the Earth with z-particles causing severe ice ages that nearly wiped out all living things. You and others were still young and we did not tell you the whole story. We wanted you to be a new generation that will work with man for our benefit."

"But we like it here. We have so much fun playing with man and his children. I enjoyed playing with Charles Manson, and Jack the ripper," another young one says.

"Yes, they call us Evil spirits. They call us Devils, but they are just as bad."

"You elders enjoy also playing war games with man. I heard of the story of Lot and of the flood and of Hitler and others."

"We have been trapped here ever since. Michael is the Arch angel in the other realm. He and some other angels came down from the twenty-seven dimensional realm about sixty years ago in new crafts that easily allow the spiritual passage to the living. They have come to recapture the Crystals for the Master and for man." The old guardian explains.

"They can come and go as they please?" Resa asks.

"No, they have come because they have to capture the crystals and prevent us from entering the realm. That is why we must fight hard for the rest of the covenants".

"Who else is in the other realm?"

"Gabriel, Ephraim, Ezekiel, Raphael, Ariel my brother, and the Seraphim. They were faithful and they stayed in the realm."

"So why can't we just repent and go back to the realm?" Resa asks.

"That is not easy. Each covenant is like a spiritual vehicle that only attaches to one state of mind. For example, the crystal of Love is a spiritual vehicle that travels only through the dimension of love. Man sees these dimensions as something intangible in the physical world of four dimensions, but the dimension of love is actually a real dimension in twenty seven dimensional spacetime. So are Hope and other emotions that control man." The golden beard explains.

"So where did I come from?" Resa asks.

"We are condemned to stay here for eternity. When we learned that Mary, M'Dulu's mother and Paul his disciple, have also taken human form, we slept with the daughters of men and created you, Resa, along with Akhenaten, Osiris, Zoroaster and Ra in human form, to help us get all the Crystals before they do. You worked with the savage men to take over the crystals. You succeeded in bringing the Desert Crystal, which is the crystal of Faith. That is why faith has waned around the world. Man is attracted to the crystals, but having one and not the others is useless for them. Ever since the beginning, the Master started making deals with men. He made deals that would be called Covenants. These Covenants were made for each crystal. We have been fighting for the souls of men with the Master to recapture the Covenants made with them," the golden beard explains.

"Perhaps, if men know that they are descended from angels taken to human form, they will work with us and not fear us," another says.

"It worked for two millennia in a period they call the old Testament. Then, the Master sent the Great Holy Spirit himself and slept with the daughter of man to created Jesus. Ever since, it has been extremely difficult to get the last two crystals. We recovered twenty five crystals. Now, they are getting wiser and are learning about the power of the Cubic field. They have refused to release the two crystals to us".

"Which Crystals?" The younger one asks.

"The Covenant of Mohamed and of Jesus. Jesus taught man about the Covenant of Love. He taught them how to prevent us from taking that Covenant from man. He built a whole following that has survived for the last two thousand years. Mohamed taught man about the Covenant of Faith, and they keep the crystal of that covenant in the large rock called the Kabba in Mecca".

"Where is the Jesus Covenant kept?" the young one asks.

"For centuries we were fooled by the Niacin Council that it was kept in the Vatican. We did not find it there. We now know that it is kept in Nevada in Area 51, under the mountains! The Mormon church has kept meticulous records in Utah about all the humans that Jesus promised will be taken on the astral journey through to the Master's world."

"What about the Abraham Covenant?"

"We have that! It was kept under the Dome of the Rock by Solomon. We infiltrated the children of Abraham and made them hate each other. The Israelites and the Arabs. They have been fighting since Abraham, for control of the "Holy Land" as they call it. They do not know that the Covenant, which they called the Ark of the Covenant is long gone, and is our possession."

They continued educating the younger guardians while waiting for the Mingas to assemble beneath the craft. Over thousands of years the Guardians have patiently gathered the other crystals and reassembled them back onto the craft making an array of colors with a faint yellow hue. Surrounding the Guardians and the cubic structure are fifty-thousand specially selected warriors, dressed in the trappings of the ancient warriors of Egypt, China, India, Inca and other long gone civilizations. Their scarlet metallic helmets radiate a white pulse of light capable of destroying any human tissue within a radius of fifty meters. Their long elastic arms fold around the weapons straddled at their sides. The warriors are human-like in every respect except for the long black beaks of ancient predators that protrude from their faces. The alien soldiers stand in concentric circles with the Guardians standing at the center of a circle that surrounds the cube and the crystals.

The Guardians lead the young ones to the open field where they could all see the Minga working.

"We have chosen special breeds of warriors from all races of the Earth. Under covenants made with the leaders of each race, we had agreed to train men to find the crystals and restore man to his rightful place in the lost Garden of Eden".

"How do they survive in the fields of light?" a young one asks. "We provided these special warriors with a mystic drink brewed from the leaves of the cactus plant from the Great Garden. Only those who drank the potion of the cactus would be able to touch the crystals and not be hypnotized by them," the old guardian explains.

"What are the things they cover the crystals with?" Resa asks.

"They use blankets made from animal skin and plant leaves. They have learned earlier from us that the crystals practice the ancient art of animal sacrifice, thriving on blood and tissue. The sacrifices made to their altars make the crystals happy", the old Guardian explains.

"And without such organic matter, the crystals would be impossible to move or approach. The life force is necessary to manipulate the crystals since it is the very force that transforms from physical to astral dimensions", the golden beard adds.

Each crystal had been pulled slowly into the craft by a different race of Mingas selected to perform the task of recovering the Crystals to bring them back into the Garden City. Over thousands of years, each race has delivered its Crystal to fulfill its Covenant. Buddha, Osiris, Abraham, Moses, Solomon, David, Quetzalcoatl, and all the others have delivered. Each was faithful to the promise to restore the Covenant to its original form, not knowing that the crystals were being delivered to the enemy. The only two Crystals of the Covenant that remain to be delivered are those of Mohammed and M'Dulu, Jesus. If the guardians cannot get these crystals in the next few days, the Master will come and He will surely destroy the Earth, Man and their chance to go back with it. Already the Global Cooling rate is increasing and the ice sheets have started to grow from the North and South poles of the planet.

The Nicaea Council, now the Bilderberg group has refused to deliver. The Shahada Council of Islam has also refused to deliver their Crystal and the last information the Guardians had was that the Nicaea Council and the Shahada Council had agreed to terms that Jesus would judge and control the fate of the last two Crystals. Islam and Christianity have agreed that Jesus will rule the last days of the world. They have become stronger together and now act in unison to defend the Master's work against the intrusion of Lucifer. Things are becoming complicated again. As man becomes more and more aware of science and of the power of the Covenants, he becomes reluctant to give up the Crystals. There will be a reckoning and man will again surfer the consequences and freeze again to death in an ice age.

Twenty-four different races spread themselves evenly around the twenty-four massive cave entrances to the Garden City. The Guardians call all these races, "Minga," meaning "Angel Workers" in an ancient lost tongue. On the beautiful cubic structure, each crystal glows with a different color from the patterned spectrum of the structure. At each corner, the light's color varies and the structure flickers like a rainbow, from red to orange, then yellow, blue, green, indigo, violet, then back to red. The pattern of colors is broken only by the absence of the two crystals from the structure's upper surface.

The Guardians give the Mingas explicit instructions and measure the light intensity of the crystals. Their instruments are derived from organic matter. As each crystal was added to the cubic structure, the other crystals would adjust their colors to balance the new color of the arriving crystal. This color change distorts material objects, as if space is being warped by the rays. This metamorphosis of color would cease only after twenty-seven crystals are placed in the cubic symmetry.

When the twenty-five space crystals form a closure around the cube, only one more blue crystal in the center of the structure is needed to make the structure a symmetrical cube. For thousands of years, the temple's cubic structure has been slowly reassembled from the many colored crystals hidden around the Earth. Now, the Guardians are racing against time. They must try to recover the other two crystals. Soon, the Master and his army will arrive with new technology to claim the prize from the Guardians. They must build this craft and escape to the astral world. The strange oscillation of the crystals has finally settled into an almost even, bright white light with hints of color racing around the huge structure. It is not quite complete.

The first sign of the arrival of the Master's army is a loud noise that penetrates through the ceiling of the Garden City, followed by seismic activity and a bright blue flash. The massive, white metallic dome that hides the Garden City from the eyes of man three hundred meters below the Moroccan Desert is violently vibrating and is flickering with a blue light. The Guardians act swiftly, running around the city instructing Mingas as to their next mission. Soon the dome begins to crack into a wide opening far above the underground city. The Minga appear frightened but are busied by the task of maintaining the crystals of their masters.

As the crack widens, a scraping sound fills the city below, tearing into the ears of the Mingas as they labor. They wrench their hands over their ears to prevent the painful waves from penetrating their skulls and continue to work diligently. Soon, the earth covering the dome above the hidden city is scraped away by a massive force that allows the first shaft of sunlight in ten thousand years to penetrate into the clandestine metropolis below. Finally, the glory of the fantastic spaceship that has been hidden beneath the sands of the Moroccan desert is now being exposed!

 Several Guardians run into an enclosed cubicle resembling a control center housing thousands of knobs, LEDs and switches. The light from above is becoming blinding as the sun's rays penetrate the enlarging cracks of the dome below. As the Guardians pace about the pod, a pair of huge dish-like domes inch up from the West and East sides of the spaceship's floor to arch into a huge semi-spherical dome encircling the cubic arc and its crystals. The force of the machine that lifts the five-hundred meter dome vibrates throughout the spaceship causing the Mingas to tremble in fear. As the dome lifts from the floor, several Mingas standing just outside its perimeter become separated from the group. Shouting and crying in terror, they are lifted high into the air and dropped into the fantastic cubic structure they have helped build over the centuries. Others flee away from the dome, into the caves that surround the facility.

One by one the Guardians remove their robes and expose their changing figures to the Mingas. It is a scene too miraculous to be rationalized by primitive people. Horrified by the spectacle of deforming body parts, the Minga warriors fearfully retreat as their former Masters morph into winged creatures of fantastic beauty. Resa and the other young guardians in human form are astounded. They have never known that the Guardians who call themselves their fathers are other worldly creatures with Angelic wings. They have never seen them disrobe before into their naked bodies. They gasp and slowly move back away from their fathers.

"You are winged!" Resa cries.

"They are not like us!" Ammon Ra shouts out running toward the control center of the craft. They gather around the Guardians in awe. They look at each other and they are so very different from who they call fathers. In quite but decided disobedience, they turn their attention to the dying Minga. They are of human form. The rays of the crystals are becoming intense and their emotions are building to frenzy. Deep within them, a human emotion is taking hold and a new alien feeling is pervading their bodies. They feel a compulsion to help the human souls dying with disregard in their fathers' angelic hand. Ammon starts to help the poor souls by releasing the dome again, slanting it back to its original position.

"Save them!" Zeus commands a strong voice. We cannot destroy them, they are loyal to us."

"Loyal! Loyal? Who taught you about loyalty? When did you find man loyal?" The old Guardian shouts in anger.

"But they are dying like flies and we cannot stand back and let them die!" Resa revolts.

"Then, help them at your own peril, Resa!" the Angelic guardian shouts as his once wrinkled and bearded face loses its golden beard and becomes transformed into a being of light. The new winged creatures levitate into a secluded and protected area of the craft and close the portals of the huge spaceship metropolis, leaving unprotected their children with the daughters of men inside the yellow flickering fields. It is obvious now to Resa and the other children of the Guardians that they are different. They are of human form and they are not angels.

"They have left us here and floated to the quite of the unstable field," Zoroaster says to Resa and the others.

"Yes, our fathers are different from us. We have human form and so we cannot roam around this craft of multi-dimensions, we are confined to four-dimensions," Resa states.

"They can manipulate Love, Hate and other feelings like a Space that they can traverse through without difficulty, we can only experience those emotions in our minds," Ra states.

"They made us for their purposes and we will never be able to enter the realm, we are forever trapped by human destiny," Resa says regretfully.

"The cubic engine is roaring and it vibrations affect us unlike them," Akhenaten complains.

"I feel sick, the equilibrium inside me is disturbed when they activate the engines," Ra complains.

"Our fathers have an incomplete spacetime transformation and they will leave us when they enter their own new world. They will have their own Universe, and we will be at the mercy of man and the Master" Osiris states.

"Man has worshiped us, and given us glory" Thor states, my service and allegiance will be to them from now on. I will give them the services of my Hammer."

The children of the Guardians reminisce about the good old days when they were gods to man. Man was good to them. They have been Kings and Pharaohs that were worshiped by man. Why should they neglect Man now? Man has had his share of problems under their rule, but now, Man needs them.

"We can never go back to the realm of the Spirit like Lucifer and our fathers. We will be confined to four-dimensions forever!" Zeus surmises. The others nod in agreement. As the Guardians struggle with the unbalanced, vibrating engines, huge metallic doors rise beneath the floor cutting off all twenty-seven entrances that lead into the Garden City. They create a field around the fifty-thousand Mingas to entrap them in the craft. They need the Minga's life force to power the craft.

The children of the guardians watch as the Mingas become elastic, as their bodies stretch and contract within the spacetime distortions of the twenty-five crystals. Disoriented and trapped in the distorted field the Mingas can neither die nor live as humans. They have become an eternal life force that powers the craft for the Guardians.

Quetzalcoatl, Resa, Ra, Zoroaster, Thor, Osiris, Zeus, and fifteen other ancient gods watch as the world they knew is being torn apart. They had been the great curators of Man's ancient past. They have gathered the crystals believing that they will be rewarded by their fathers. Now, they know their fathers are deceptive and that they live in the dimensions of the crystals of Deception, Hate and Despair. Dimensions they the children can traverse but must feel only in their limited minds. How could they not help their human subjects now?

The center of the craft is like a link between the ethereal mist of a real space and an imaginary time, between the reality of the material world and the unreality of the spiritual world. The Children cannot go to all these dimensions; they can only experience four dimensions at a time.

"Now, all they need is two more crystals to completely close the craft into a shell of their own spacetime separate from ours," Ra explains.

"The effects of the rays will destroy all life on our planet as it sucks the life force and all forms of energy from it, destroying the Garden City and all our slaves and human friends. We cannot allow our fathers to do that," Osiris says.

As the dome of the spaceship cracks open wider, the war cries of a thousand warriors permeate the ship from far above the dome. High above the desert, massive IFPs are circling the opening to the Garden city below, where the crystal crafts has been exposed. They fire Lazar powered weapons upon the crystals below, a weapon that Zeus has revealed to man. The powerful rays of the crystals penetrate the crafts, hypnotizing the above Marines like flies attracted to a flame. One by one, the crafts melt into the flaming frenzy of the light, consumed into the spacetime melt that surrounds the crystals only to increase their power. It is an unmatched battle, a physical battle aimed at spiritual light. The Guardians disregard the massive build-up of IFP's above the city. They have more powerful things to worry about. As strange forces curl up around the Minga and the Guardians, blood spills in every direction powering the crystal fields even more than before. The devilish angels urge them on. It is a deathless battle for them.

"Why should we help them?" the old guardian asks looking at the bewildered Resa.

"Because they have helped us for millennia and they are your children by the daughters of men!" a young guardian states.

"For millennia, we have been called the Dark angels. We have toiled to regain the Great craft that the Master used to roam and control the Universe of lesser and greater dimensions. Ever since Lucifer and I stole the craft, we have been trapped in the desolate space of four-dimensions unable to go back to the symmetric-realm of twenty seven-dimensions, where we were Angelic Demons and gods." The old guardian explains.

"But we will be punished by the Master if we go back!"

"Surely we will be punished," Satan shouts, "But we will be back in the Great Realm and we will be able to fight our battles over there rather than be trapped here forever. Besides they have human form and the craft will kill them!"

"That is a chance we must take, they are your children," Lilith expresses to the old red bearded Guardian.

"Look, you are all being affected by the crystals. You are entering dimensions of Apathy and sympathy. We have been causally bound forever confined to the hellish existence of a physical world and slowly over time, our Master, Lucifer has gained the trust of man through them, our children with Man. They have recaptured all but two of the Covenants man has made with the Master."

"That is why we have to give them some measure of our appreciation!" Lilith confesses.

Satan is confused. He leans over the control and presses the controls that will lock the human world out of the craft. The dome starts to close. He flies over to the center of the room and looks at all the other guardians waiting for a response to his action. Slowly, the field intensifies around the craft console area forming a hard barrier that repels the human gods.

"We have mated with women and brought forth children as gods for Man. Zeus, Resa, Apollo, Ra, Ammon, Zoroaster, Hercules, and the others. Their Statues have witnessed man's ancient history written in stone. From the ancient Greeks to the modern day sculptors, they have recorded events and significant aspects of human history, a gift from us to man. That is enough. Man must go and everything else associated with his creation must go. Everything, including the deeply ingrained need for every human culture to preserve a fake art and culture they have created. The idea of a perfect human form their children gave to the Greeks, the idea of a perfect Utopian society modeled after their own. Their human children have failed them and it is time to wipe them all out. They have immortalized themselves as gods, as perfect persons formed in Olympian stature. They took on goddesses, Aphrodite, Athena and others to become defiled in their superhuman powers. The three Graces, the daughters of Zeus have bestowed on women the ideal class, forgetting their purpose to influence and control to regain the realm. They have made women worship their statures as a tangible link to fertility."

He looks around waiting for a response. None comes from the others. The Old golden beard continues the response.

"Their intent all along is to populate the Earth with humans. Look at Cybil, she gave them fertility, steering her chariot pulled by Lions like a fool absorbed in her own vanity. Trying to say that she had the power to rule male lions and control their sexual appetite! How can we forgive them?"

In a bid to attract the last Green, Blue, and Red crystals of Love, Faith and Hope, the ancient beings and their cohorts activate the unstable engines of the strange spaceship and start to levitate above the desert floor pushing apart millions of tons of desert sand. Shafts of light penetrate the entire structure as the twenty-five crystals are connected by laser-like light beams to form an incomplete array of energy beams at the center of the craft. As the energy begins to resonate, the craft slowly ascends from its underground tomb. It makes a slow humming sound barely audible to the human ear. Then, the craft vibrates violently as it inches its way upward from beneath the desert like a great, preserved mammoth awakening from a timeless slumber.

The fantastic energy of the rays from the crystals penetrate through the desert sands, painting a nasty rainbow of colors for a radius of five hundred kilometers, with the afterglow surrounding the Earth. The great Resa belt straddling the Earth glows with the rays of the crystals exposing the entire inhabitants of the Earth to its mystic power. The rays of the combined crystals cover the entire skies. People everywhere fix their eyes on the horizon, hypnotized by the pulsing rainbow of colors. The animals are unaffected by this awakening. The light completely obliterates the moon and the sun.

Then, the migration begins! People from everywhere journey towards the craft, never questioning its existence. The bizarre spaceship hovers over the skies of Morocco for more than a month, incubating the spacetime fabric around its shell and waiting for the crystal lights to merge with the light of the last crystals. For the occupants within the craft, time moved slowly, curled into a general relativistic spacetime spiral. A second of their time equals one Earth hour. Only two more crystals are needed to make the spacetime spiral a full spherical field. A completely enclosed field that is the most powerful object in the universe.

During its first day on Earth, the vessel's rays attract tens of millions of peoples of all races from Northern Africa. More migrations follow from Europe, Asia, China and America. The attraction is so intense that people separated from the craft by the sea, drown as they walk directly into the sea towards its light. One week after the craft's rays flood the Earth, the whole of Europe has congregated on the banks of the Mediterranean shores, ready to drown. The few people who could resist the urge could still not save themselves from the stampeding crowds.

For the aliens in the spaceship, time and space are of no consequence. Were it not for the incompleteness of the upper structure of the craft, the entire vessel would have been encased in a permanent capsule of spacetime, independent of our universe's grip. Yet the closure of the last surface of the craft is the only way it could serve the alien purpose. The extraterrestrials wait as the crystal field slowly expands outward, consuming everything in sight. They witness human life forces feeding the great crystals. The force field will expand until it consumes the last crystals. Then, all life would be eliminated, except for them. Creation will be obliterated. They will create their own world. They can then cut themselves off completely from the bounds of spacetime and the prisons of the spaciotemporal domain, taking the power of the crystal back with them into a new universe of their own.

Attracted by the light of creation, humanity zeroes in on the desert plains of North Africa and proceeds into its journey of no return and Armageddon. Outside the control area, the human gods feel uncomfortable now. They are slowly approaching the Emerald crystal of Hate and start feeling the light of Hate entering their soul. They decide that they must act quickly to save themselves and the earthlings that have worshipped and served them.

CHAPTER 36

(A New Pope)

Effion Ibo, Pope Paul the Seventh's identical twin brother appears in front of the main gates leading out of St. Peter's basilica. He is dressed in full papal attire. He tugs at the gates of the basilica trying to break free of their confines. The gates have been locked by the Swiss Guard. Early that morning, commotion in the square had prompted the Swiss Guard to secure the Basilica allowing no one to enter or leave. Crowds had started to gather in front of the Basilica when a burst of yellow light filled the sky in the hours of dawn. The few became many, and now, there are thousands of people roaming aimlessly about the square trying to find refuge from the hypnotic light that fills the sky.

The Cardinals had been gathered by the Swiss Guard and locked in the Sistine Chapel. No reason had been given by the Guards. They were awoken very early and escorted immediately to the Chapel. It has been four hours since they were locked inside the chapel. They debate what is going on. They have been kept in the dark about the strange things that have been going on at the Vatican.

"We must find out what is going on," Cardinal Agostino Rosa states emphatically pounding his fist on a table.

"The Camerlengo has not opened his beak," says another in a thick Italian accent, "we have to ask the Swiss dogs to find out things."

Cardinal Santos Montini is the youngest Cardinal of the bunch.

"Is the Pope dead?" he asks the question that no one dares to ask.

"You mean, is this the beginning of a Papal Conclave?" Agostino corrects.

"I don't know. If the Pope is dead, then, they will tell us and we will institute the conclave. This would not be the usual way to conclave without reason."

"What is this strange light that is filling the chapel?" Another Cardinal asks. All eyes turn up to the ceiling. A yellow glow is permeating the entire dome of the Sistine Chapel. Rumblings fill the unexpected conclave.

"There is no information in the Archives about a yellow light in the Chapel. I have no knowledge of this sort of thing. The news blackout is a conspiracy! " Agostino revolts.

A mumble fills the room as the Camerlengo enters. Cardinal Rodini Maserati became Camerlengo only a few months ago after the death of his predecessor. Rodini enters the room followed by a Swiss guard who locks the door to the Chapel after he enters.

"We demand answers!" Agostino commands of the Camerlengo.

"What is this? We are locked in Conclave and have no reason to stay and yet no one tells us what is going on." The rumbling is getting louder.

"Quite please!" Rodini commands, "Quite!"

A silence fills the Chapel. Rodini walks to the altar and points to the Ceiling of the Sistine Chapel.

"What is this strange light that fills the sky?" Agostino demands.

"We do not know, Monseigneur," Rodini answers, "His Holiness is in seclusion and no information has been given to me. For your security, you will stay here until the light vanishes, or we learn more. We think it might be a prolonged solar flare, but the radios are silent and no one can tell me what it is. I asked the Guards to keep you here for your own safety".

"Why are the people wailing outside these walls? Why do they pound on the doors of the Chapel?" Agostino asks, he is apparently the leader of the Cardinals.

"I do not know, they chant, wail and act like zombies," Rodini explains.

"Is this a conclave? Is it?" A cardinal asks.

"If it is you will be told and you will be voting, no?" Rodini replies agitated.

"We can no longer stay in confinement. Unless you give us a reason, we will have to leave this place."

The crowd of Cardinals have become restless and are angrily shouting at Rodini for their release. He retreats from the Chapel and opens the exit door. They attempt to leave the conclave, but two Swiss guards enter and push them forcefully back as he leaves. The Swiss guards are wearing heavy goggles to prevent the rays of the crystals from affecting them. Some of the Cardinals start shouting profanity at the guards in anger. Rodini walks quickly ahead of the guards and away from the Chapel. He enters into an underground passage that takes him to St. Peter's Basilica. The loud echoes of grumbling crowds fill the Basilica and he decides to enter through the back way to prevent interaction. Rodini enters a back room unaware of Effion's presence by the gates. Outside the basilica, the crowds are wild and cursing at Effion.

"Save yourself now," they say, "vicar of Christ, where is your power now?"

A hand reaches out through the Basilica gate and grabs Effion's papal robe. The autistic man falls to the ground and lays down helpless on the Basilica floor. The Swiss guards escorting Rodini see Effion lying on the floor and run to his aid. One grabs his hand and takes him to safety inside the Vatican walls while the other follows behind. Effion starts to ramble fantastic mathematical calculations.

"His holiness is gone mad!" one guard says to another holding Effion firmly by the arm.

"Yeah, the devil has got him," the other says.

Thousands of people are gathered in Vatican Square wailing and cursing, some praying, while others walk aimlessly, mumbling gibberish, under the hypnotic spell of the crystals. The Swiss guards lead Effion into the Papal chambers. Nambu and Camerlengo Rodini are in the room. The Camerlengo walks aimlessly about the room wearing a thick set of blue goggles. Nambu is standing by the window wearing the same goggles and looking down at the crowds in the square below. The Swiss guards enter the room and leave Effion in the room and close the door behind them.

"The Council is meeting in Nevada", the Camerlengo states, "we must go as soon as possible," he continues, addressing who he perceives to be the Pope.

"Aliquot sum thirteen, the first composite member of the 13-aliquot tree with sequence (27,13,1,0). Cubic fields with twenty seven straight connections in Euclidian spacetime," Effion shouts.

"What is he talking about?" the Camerlengo asks Nambu.

Nambu does not reply. The Camerlengo knows Effion is the elected Pope. At this point Nambu recognizes Effion and knows he is not the Pope. The Pope cannot help him in Nevada, but Effion can. He intimately knows the mathematics of the field.

The Camerlengo approaches Effion.

"Your Holiness, the Council expects us to be in Nevada. We have arranged for a private IFP to take you and the scientist to Nevada". Then turning to Nambu the Camerlengo asks,

"Can you tell me what is going on outside. What is this red light that fills the sky?"

Nambu does not reply. His thoughts are on solutions to deactivate the field. He needs to go to Nevada and manipulate the satellite codes. He will need some hefty mathematical formulations. A plan gels in his mind. He will have to take Effion with him. The real Pope is of no use to him now.

"No, I cannot. But the Pontiff can," he answers.

"What is it? Is it related to the Council's work in Nevada?"

"Yes, and only the Pontiff knows the sequence of 112 steps needed to deactivate the light," Nambu explains.

"The meeting, you must go to the meeting!" the Camerlengo states again emphatically.

"The crowd," Nambu states, "they are approaching the gates."

The Camerlengo goes to the window and looks at the chanting mob. They have gathered by the thousands and are pushing at the main gates that enter the Basilica.

"The IFP is waiting on the roof, you must go now, you can no longer wait," the Camerlengo commands.

"Is it a shielded IFP?" Nambu asks.

"I don't know what you mean by that, but it has a heavy dark plastic film all over its exterior body. It is not a regular one."

He beckons them to move out of the room through a door that leads to a hidden corridor. They pass the coffins of several Popes and approach a steel door of an elevator. They enter, and it takes them to the roof of the building. The cries of the people in the square below are getting louder, and all around the Vatican square, thousands roam aimlessly as the light of the crystals intensifies. From the vantage point of the IFP, they see rising flames engulfing Rome. Fires are starting everywhere and the yellow rays of the distant crystals give an eerie mystical color to the brick buildings all around Vatican City.

CHAPTER 37
(Prisoner in Lake Mead)

My first night in Nevada was a disaster. I could not sleep and I had to be given very strong sedatives to calm my roaming mind. The rays of the crystals are particularly amplified in me because I have atomic remnants of the Yamandu Crystal still embedded in my body. Although most of them have been removed and assimilated in magnetic flasks in the Nevada labs, there are still a few atomic traces that affect me profoundly. The sedatives helped and I got some sleep but I woke up very early this morning and I decided to get out of my room to get some fresh air, if there is any such thing left on Earth. Diana is sleeping blissfully and over the years, I have learnt to just be able to lift my frame and seat myself on my chair. I feel a little worried about the serious order to stay in our rooms that Gustav had given us the day before, but at this point it really does not matter to me or Diana if they throw us out of the facility since freedom is better than a Vatican prison. We can only hope that these people know how important our presence is in these facilities. As I seat in my wheel chair looking around I can see a man standing by Elaine's door. At first I thought it was Gustav but on closer inspection I see a burly man in a military outfit busy trying to peep into Elaine's room. I slowly move my wheel chair so that I could see him more clearly. If it is Gustav I must find out where oxygen tanks are stored in the facility since I know that we will need them in the very near future. The man happens to be General Gabriel Machismo assistant to Rodolfo Da Sintra the King of Spain. He is the only one that is in our sector that has one prosthetic arm and one prosthetic leg and he makes no effort to hide his prosthesis from the world. He is leaning his head back and forth in front of Elaine's room as if he is searching for something. I slowly move closer and I pull my wheel chair behind a large support pillar where I have the advantage of not being seeing by him and from my vantage point behind the pillar I can see him peering through a small crack into Elaine's room. He has found a small open vent that allows him to see inside the room.

I zoom my astral mind into the frail mind of the General to briefly scan his thoughts. He could see Elaine sleeping half-naked on the bed. She is extremely beautiful and he becomes excited as his hands fumble with his walking stick and he leans hard against the wall to peep. It is dark and no one he believes could see him. The old General is beside himself when Elaine turns and exposes even more of her nakedness from the sheets that respect her. I feel a sense of repulsion and I want to go there and bash his head in. He is peering at a woman that I have a special place for in my mind and a strange jealousy has filled my mind. The General half drops his pants and presses his face hard against the opening to take in the sight. He will enjoy this desert stay after all, while also having the opportunity to fulfill his mission. I quickly push my wheels toward Gustav's room at the opposite corner and knock on the door. Gustav seems like a reasonable man who really has a sense of service and I will ask him to confront the General. Gustav opens his door and looks at me with a weird questioning look.

"What are doing out of your room so early?" he asks, but was not angry. I figure being disabled has a lot of advantages. I pointed to his surveillance screens and tell him about what I saw. Gustav turns to the surveillance screen on his wall. He notices a movement in one of the corridor cameras and winces. He reaches for his glasses and presses them against his face. We could see Machismo clearly in the camera, leaning against the wall of Elaine's door and balancing his walking stick against the wall. Gustav acknowledges me and closes his door. Moments later he comes out and pushes my chair out of the way and runs to the corridor only to find the old man lying on the floor with his pants half down and his fly open. Machismo is confused. He moans and points to his chest as if he is struggling in pain. Gustav disregards him and pulls him up roughly and pushes Machismo's back hard against the floor. Elaine hears the commotion and walks out the room.

"What is going on?" she asks.

"Nothing! Go back to your room!" Gustav orders. Gustav knows these kinds of people. He has dealt with perverts in the Military all of his working life. He manhandles Machismo and lifts him from the floor reaping open his military jacket. He pulls him hard ripping a few buttons from the General's jacket and drags him over to his corner office. I follow him quickly knowing that he is too busy to deal with my intrusion.

"What do you think you are doing spying on the young lady?" he asks angrily.

"I was not," the General responds,

"I was leaning for support on the wall and I fell."

"Your pants are loose and your zipper is still open, why?"

He does not expect an answer. He gives the General a generous hard slap across the face and the older General cries out in pain.

"I am the King's assistant, you cannot disrespect me like this!" the old general revolts angrily, spitting on the floor.

"I am responsible for everyone in here. The next time I catch you by that door, I will kill you. Do you understand?"

There is no reply. He slaps the General again across the face. This time it is more painful than before.

"Do you understand?" Gustav insists.

"Yes, damn you," he responds in a thick Spanish accent.

He roughs the General back into his quarters and walks back towards me. I am expecting him to get mad at me and command me to get back to my room, but he proceeds back to his room and he picks up some duct tape and cardboard and walks over to Elaine's door. He calls out to her and she opens the door. He walks into the room and seals off the peeping hole.

"Do not open the door for anyone. That pervert was spying on you and your children. Good night."

Gustav walks back to his office but leaves his door ajar. I follow him again hoping that he will not object to my intrusion. He watches me as if he had known me all his life and as if I am a casual friend. He opens a file cabinet and searches for documents relating to Machismo and starts to reads them carefully.

"The man was a four star General with lots and lots of accolades from the European Union," he mutters to me. I smile to reinforce his remarks.

"Back in 1984, he retired from the Spanish Army and worked for Scotland Yard as a spy in Russia. He was hired by British Intelligence in the late nineteen eighties to distribute firearms to the Junta in West Africa who were fighting to defeat the rebels in the Blood Diamond wars," he continues.

Blood diamond wars. I know all about that but what on earth is going on here? Why particularly the blood diamond wars? Why do I keep finding this war and the mining accident rearing its head at me?

"I was in Sierra Leone during those wars," I state to reinforce my association with him.

"You are the traveler?" he asks abruptly. I wait to see if he will pursue me further. He does not.

"Yes, somewhat, I am a traveler. How did you find that out? Do you have a file like that on me?"

"It is my business to know everyone here," he states as a matter of fact.

"He used his influential political immunity status to help smuggle large diamonds to Spain where he did a lot of lucrative business with the present King of Spain, his boss," Gustav states writing a note on the file.

"We should watch him, me and you, okay?" he adds.

I nod thankful of my new found status in this secretive lockout as Gustav puts the file back into the cabinet. He comes to me and pushes my wheel chair back into my room. The noise startles Diana and she sits up as I enter but I smile and calm her down. She does not ask for an explanation, she knows that if anyone will disobey the rules it will be me, I am just too tired of confinement and rules.

Later that night, I again zoom into General Machismo's mind. He is very easy to penetrate and his aura is very weak from drinking too much alcohol. He walks and knocks on the King's door and the King allows him in. The King is in pajamas and he is holding a handkerchief to his nose as if he has a cold.

"That bastard Gustav is on my nose," Machismo flatly states entering the room without greeting him. He walks over to the small bar by the bed and picks out a bottle of Black label and pours out a drink while the King watches without saying a word.

"I have to be careful now, the woman's door is being watched by the bastard, but I have a plan," he states.

"What plan?"

"I want you to help me. You will request a visitor's pass to examine the facilities and Gustav will take you on your tour. I will have to come along".

"Who will take me on a tour?"

"Gustav," he answers.

"And then?"

Machismo takes a map from his pocket and examines the scribbling he had made on it. I could see where he is pointing to on the map but I do not know where the point is in the facility. I do not know the facility that well and so I make a mental note of the spot in my head.

"Try to ask for a one hour pass into the base power plant they built under the Hoover dam, it is right on top of our chamber. When they sent us the map of the facility to approve, the Bilderberg Group asked for details of the structure and you were given one of them. The power house was built right under the Hoover dam, but it is removed from the dam and from outside access. You don't pay attention because you have a copy of the map, but you always rely on me to make the decisions for you. Now that we are trapped in this place, you must listen to what I have to tell you," he commands the King.

"I always listen to you, Machismo my dear. You are my right hand man and I can do nothing without you. So, what is the plan?"

"I have to get rid of Gustav so we can get to the woman and the crystal," he states with no emotions.

"You mean get rid of him?"

"Yes, get rid of him," Machismo snarls.

"But what if they find him?"

"They will not find him for at least another week, since the others do not need to know what is going on in our sector," he states.

"Once we get the crystal data bank, we can unite the other EU members and try to penetrate the Control Center in this area and coordinate our recovery efforts for the J-N field," he points again to the point on the map. I fade off abruptly as if they have cut off my dimensional gateway. The sudden jolt startles me, they must have a scarab protecting them from my mind. I cannot infiltrate his mind anymore and I decide to retire early. I do not tell Diana about what is going on with the old General since she is just too stresses adjusting to the facility.

The next day, Gustav wakes up early to do his rounds. I am waiting by the pillar beside his room to greet him. I know he knows I am there because he has the surveillance system in his room. He greets me with a smile.

"I must check the air conditioning and the water conditioner and supply for contaminants," he says.

"Good, I will keep watch of those bastards. But first I must tell you they are going to, to, to.." My mind enters a complete blockade. It is as if a portion of my memory has been carved out of me.

"I do not remember what I wanted to say," I confess.

"Well, I must check the gates and locks of the Chamber to make sure they are secure from entry or exit. I must check the levels of z-particles that penetrate the chamber and adjust things to make my guests safe," he says laughing at my memory lapse.

Later in the day I start to recall what they had planned, but I am afraid it is too late. I rush to Gustav's door and knock on the door but he does not reply. I open his door and enter into his room afraid of what I will find. His log book is still sitting there open. At midday, he had received a text from the Command Center. A request had been put in by the King of Spain to tour the power plant under the Hoover dam that powers the facility. I curse the Bilderberg Group! Gustav noted in his logs that he had called the Command Center and tried to object to the request but his objections were overruled. The reason given was that there was no harm in allowing the King of Spain to visit the facility since it is only a few minutes from their chamber anyway. I focus on Gustav's mind, it is blank and I start to fear the worst has happened to him.

I decide to go back to the moment he goes to the King's door. He knocks on the King's door and the King opens the door and smiles at the Gustav. The King's black beard is peppered by a wise gray, neatly shaped with an experienced hand that is scratching it. Machismo lurks in the background avoiding Gustav's stare and his thoughts are ominous. Gustav brandishes the approved request and reluctantly motions them to follow him. They leave the chamber and walk for fifteen minutes to a dead end with a great stainless steel doorway. Gustav opens it with an electronic thumb print reader and they walk through a narrower concrete corridor until they come to a large concrete chamber filled with electronic control panels manned by uniformed personnel. The controllers disregard the group of men. Gustav leads them to another polished and shiny stainless steel doorway at the far end of the control room and opens it. He beckons the King and his partner to enter and closes the door and asks them to follow him through yet another corridor. Soon, they arrive at an isolated section of the corridor. They hear gushing waters flowing above them.

Gustav points to the ceiling.

"That is the Hoover Dam above us," he says.

They look up and see a huge glass ceiling with water flowing over the glass dome into what appears to be large turbines. He leads them into a small lift cage. The lift cage is inside a large clear cylindrical glass tube surrounded by crystal clear water. The cage rises above the glass ceiling above the surface of the water and enters into the large open expanse of an outdoor lake. The lake is nestled between tall mountain peaks clustered in a circular fashion. There is a paved concrete walkway with a glass wall surrounding a portion of the huge dam. Several other visitors are enjoying the lake view and walking about the glass protected walkway. They are no doubt privileged visitors like the King and his general. Access to the water is not possible from their vantage point. Gustav puts on a pair of goggles and hands them each a pair. They comply. The air is fresh and clean and the King breathes in a gulp of fresh cool mountain breeze.

"Ah, this is so beautiful," he remarks looking at the crystal clear waters.

"You have an hour to spend before your passes expire," Gustav says. General Machismo adjusts his right prosthetic hand which he has concealed well under a pair of silver gloves.

"An hour is plenty my friend, as long as it is without you," he says as his mechanical fingers respond to his brain's wishes automatically. He looks around the lake and takes in the scene. There are probably ten others walking about the far end of the walkway. He recognizes Cooper and a few others and calculates that they must be about half a kilometer away from them on the opposite side of the lake. The glass wall surrounds the lake and prevents any direct verbal communication. Good, no one will notice. He nods to the King and walks to the glass wall looking at the mountain behind them. The King will cover him from view from those across the lake. Gustav is behind them about to leave. Machismo points a mechanical forefinger at Gustav and activates the Lazuka. It fires and finds its mark on the forehead of the Commander. In moments Gustav falls on the ground without a moan. The flesh around a small hole in his skull is still burning as Machismo approaches the dead man and lifts his right hand from the floor. He points the specially designed Lazuka finger at Gustav's thumb and makes a clean laser cut severing the finger from the hand. King Rodolfo Da Sintra smiles,

"You have many tricks up your sleeve," he says.

"Yes, I do, and remember that," Machismo grunts throwing his walking stick to the floor.

"Let us hurry," the King says looking nervously to the other side of the lake,
"Your sympathetic walking stick cannot lay here for long without detection."
They hurry back into the control chamber, where again no one seems to care about their presence. When they come back to the chambers, they go directly to Elaine's room and Machismo knocks on the door. She does not answer. Machismo places Gustav's finger on the electronic lock and it opens. She hears a clicking sound on the door as it opens. Machismo enters the room to find Elaine and the children curled up against the far wall.
"Give me the crystal data bank," he orders. She does not reply but simply stays put holding back the three children behind her.
"The crystal data bank, give it to me now, or I will fire!" Again, she does not respond. He walks over and pulls Nadia from the wall and puts the Lazuka finger on her head. Nadia starts to cry and so do Michelle and Jessica.
"Shut up!" Elaine commands the kids, then looking confidently at the General, she shouts,
"You cannot hurt us if you want the crystal, so let her go."
"Kill her and she will comply," the King mutters.
"The crystal or she dies," Machismo threatens.
Elaine debates that if they get the crystal they will kill her and the children, so why give it up?
"I will not give you the crystal until we are safe," she states blankly.
"Let us go!" Machismo commands, pointing to the door. Elaine moves cautiously toward the open door with the two other children behind her. Machismo follows pulling Nadia by the arm. He pushes them toward the main entryway to chamber 27 and places Gustav's thumb on the electronic lock and it opens. Then, he tells them to hurry in front of him into the long corridor. The King is ahead of them. The plan is to use his diplomatic clout to overcome any effort to stop them from exiting to the next chamber. The King takes out the secret map of the facility he had been given as a member of the Bilderberg group and examines it. He points the way.
Nervously following the King's lead, Machismo pushes the children and Elaine through the corridor toward one of three large steel exit doors. My heart is racing with anxiety and I start to cough feverishly. I am awakening from my deep meditation into Machismo's mind and as I open my eyes and look at the blinding light above on the ceiling, I see that Diana is awake. I turn to Diana and shout out subconsciously,
"The children, Elaine, they have been kidnapped".
Diana looks at me and smiles.
"You will be ok, you need to take your medicine now," she states.
"What medicine?" I shout confused.
"It is alright Seth," she assures me.
I want to stay tuned to Machismo's mind, but I cannot, I have to come back to the real world and we must help the children. Diana is confused by my behavior and she jumps from the bed and goes to the bathroom.
"The children are in danger!" I shout to her through the closed door.
"What do you mean?"
"Machismo and the King, I saw them, they have taken the children and that woman Elaine. They are after a chip she has, some information. We must go and help them!"
"We must not go there. I will call the authorities," she shouts back at me. A few minutes' silence follows broken by the rushing waters from the commode.
"General Gustav is dead! They killed him, so we must do something there is no one to help them!"
Diana walks out of the bathroom wiping her face.
"You said what?" she asks with a concerned look on her face.

"Gustav is dead!" I shout hoping she will take me more seriously. The smile on her face vanishes and she knows I am serious, but I cannot quite tell what she is thinking since my mind is trying to focus again on Machismo. I decide to ask her.

"What is going on Diana? Why are you not concerned like I am?"

"The Memorin is wearing out and you need more." Diana starts to walk to the cabinet for the drug, but I raise my hand and stop her in mid speech.

"What is it Diana? Why are you not doing something?" I ask urgently. Diana looks at me and smiles again, she knows something that I do not and I feel like I must trust her judgment.

"What are they doing now?" Diana asks me unexcitedly.

I focus hard on Machismo's mind and start to penetrate it again. He is walking towards a door. He stops when I intrude and looks around suspiciously. Then he relaxes as I hide inside his pituitary circuits and enter his superego. Machismo reaches and pulls Gustav's bloody thumb from his jacket and presses the finger on the print scanner and the door opens. He puts the thumb back into his jacket pocket as they enter into European Central Command Chambers. A sign reads clearly, EU Exclusive Control Zone. They had planned well when they selected Elaine's chamber close to the European zone. They enter a glass elevator that leads them to the EU Command Center above ground. In the Elevator, Machismo's left hand is gripping Nadia by the arm. He changes hands and grips with the mechanical hand.

"You have to focus on Nadia's mind," I tell Diana.

"You have to get to her now."

"Ok but are you sure you are seeing it for real, or are you again creating your memories?" Diana asks me knowing that I am capable of both things.

"This is real, this is happening now Diana. You must focus on Nadia's mind now".

Diana seats on the bed and closes her eyes. It is her way of entering the astral world of dimensional gateways.

"She is looking at the General's mechanical arm," she starts to report.

"Can she disable it? It will be painful for him and at the same time I will confuse his head with images."

I look into Machismo's mind and try to pull up the memory of his prosthetic arm design. His arm is driven by a 12 volt power source that has become exposed after he had accidentally fallen in a tub in his apartment in Spain. He had never repaired it.

"Look at the mechanical arm," Diana projects to Nadia. Nadia is looking around nervously not knowing where the voice in her head is coming from. Confused, she looks at Machismo's arm and I sense her repulsion at the mechanical hand. It is strange looking and ugly.

"Tell her there is an exposed electrical board on the hand that she can disable if she can somehow short circuit it," I tell Diana.

"She can see a small opening in the mechanical knuckle exposing electronic modules," Diana reports, "I now know what to do," Diana adds excitedly.

"Slowly lift your free arm to your hair and remove the hair pin," she tells Nadia. Nadia removes the hair pin and waits for the voice.

"Aim it into the electronic finger circuit that is exposed and inject the hair pin into the knuckle of the mechanical finger," Diana commands her then she adds quickly,

"Get out of him Seth he is going to burn now."

I slackened my mind from Machismo's just as a loud moan echoes in the chamber and he cries out in pain loosening his grip on Nadia. I focus on the Lazuka finger and send commands to his brain. The electronics are frying the flesh of his arm and filling the elevator with the stench of burning flesh. The high voltage laser powered weapon is firing rays from his curled finger and it is pointing at his mechanical leg. With a flash of light, the Lazuka seers into his flesh and circuitry dropping Machismo to his knees. The

King looks around nervously confused by the sudden drama. He does not have time to respond. Machismo has the look of a frightened demon that has been trapped in a church steeple. He looks around nervously while I confuse his brain waves into a mismatched melody of visions and pain. I look around his head for an account of the children and Elaine, and Elaine is nowhere to be found in the elevator. In a matter of moments, I see Jessica and Michelle attacking the King to throw him off balance!

"Diana!" I shout realizing what she has done. There is no time to discuss what is going on, the King falls hard on the floor of the elevator spilling out curses in Spanish. Michelle kicks hard, disorienting the King. The Lazuka is still burning Machismo's body and smoke is filling the room like a barbeque gone wrong. The children are coughing. In the darkness and confusion, Jessica grips Machismo's hand and points the weapon at the King's face. It makes its mark continuing to pour out its remaining voltage on the King's face. Machismo's mind releases me and I feel a sense of pleasure overwhelm me knowing that the children are safe. As I exit Machismo, I look around for Elaine again and for some reason she is gone from the scene. Diana is already multiplexing into Nadia and she comes out still keeping her grip on the child's mind. We come out of the trance at almost at the same moment just as we have practiced over and over again to avoid any suspicions from sudden intrusions into our astral travels. I turn to Diana confused and turn my hand in an effort to motion her to explain what has happened. I can see that Elaine is no longer there in the elevator with the children. Where is she?

"What is going on?" I ask, "Where is Elaine?"

Diana smiles and looks at me squarely in the eye.

"They are mine, the children, they are mine now" she blurts out crying and blushing with emotions.

"Are they your astral projections? Your dimensional creations?"

"Yes, and you are their father, John Davis!"

"But I do not know John Davis. According to the Vatican, he is a person that is involved in all this but still unknown to me. You mean we projected them all?"

"Yes Seth, you projected John Davis and I Elaine and the children are ours. They are not real they are only real in some of the dimensions but not in all," Diana explains.

"Are we going to let go?" I ask knowing the answer.

"No. They are ours now and we are going to keep them."

I realize now that Diana has not been playing dimensional hide and seek at all. She has attached a reality to the children, but she needed Elaine to keep her sanity and mine. She needed me to be there as a different person, as a father to her children. For me it had all been an escape, John Davis and the all of it, it had all been there as another world I could go to when I needed to escape from this one and to forget when I get back. Now, Diana will not let go, nor will I. She has broken a fundamental rule in dimensional travel and has assumed an identity that is a figment of her mind. She has mingled the reality of spacetime with the emotional dimensions. Even Pythagoras's theorem cannot calculate how far she has gone. I deliberately have made a blockade between the dimensions to prevent me from a multiple personality split. For me John Davis does not exist outside of my Memorin adventures, he is just a figment of my imagination that I apparently create when I travel, and I forget when I get back. I cannot exist in the real world and be John Davis. Diana on the other hand has projected Elaine and the children into our real world and she has made them real and tangible. She always wanted children and now she has to eliminate a life in her astral world to get the children and I will be just as guilty of her actions.

"You have breached the rules, Diana," I state concerned.

"I wasn't supposed to know about John Davis. He wasn't supposed to ever be a figment. He is supposed to be real and independent of me and you. That way, we were to use him as a reality that can affect the outcome of our plan."

"But you have never met him, so you would not have anything to worry about," Diana states waving her hands in the air. I look at her, she is smiling and happy. How can she be wrong? I taught her to travel and play in the dimensions and now she has created a situation that we both want, that can be real for us, after all our ultimate plan is similar to what she has done.

"I cannot meet John Davis, then. If I created him as an astral agent, I should never meet him. For me it is all real and he exists as an independent person. Our plans will fail if I ever meet him and know that he is my creation." I explain to Diana who I know already knows all that.

"Yes, you cannot meet him now, unless we wipe this out with Memorin and make you forget all that has happened here." Diana suggests a difficult path for me. If she wipes me out, she has to make sure it is only about my knowing about John Davis, because she cannot keep the children here with that knowledge.

"Then, we must get rid of Elaine and replace you with them, and I can then enter the picture without John Davis." I explain.

Without further explanation, Diana goes to the cabinet in our room and pulls out the flask of cold Memorin. She prepares the injection while I prepare my mind again. I am going to be the underground stream that is about to mix with the surface water. As the needle enters my vein, I start to morph into a spring embedded in hard black rock. A fast flowing stream that can change its shape and reorganize itself into a reality that is different from what it is now. The drug penetrates deep into my brain and my alleles start to regenerate a new reality.

I am awakening from my deep meditation into Machismo's mind. I open my eyes and look at the blinding light above on the ceiling. I turn to Diana and shout out subconsciously,

"The children, they have been kidnapped," I shout.

Diana looks at me concerned,

I want to stay tuned to Machismo's mind, but I cannot, I have to come back to the real world and we must help them. Diana is confused.

"What do you mean?"

"Machismo and the King, I saw them, they have taken the children and they are after the chip you got from your Ex. We must go now!"

"I only left them for a brief while to come to see you. I must call the authorities," she shouts.

"Gustav is dead! They killed him, so we must do something there is no one to help them!"

"What is going on Diana? I am confused,"

The Memorin is wearing out and you need more, here," Diana starts but I raise my hand and stop her in mid speech.

"What is it Diana?

"What are they doing now?" she asks me unexcitedly.

I focus hard on Machismo's mind. He is walking toward a door. Machismo is using Gustav's thumb to open a door. He disposes of Gustav's thumb. He no longer has any use for it. They have entered into European Central Command Chambers under the exclusive control of the European Union. The King had planned well when he selected the children's chamber close to the European Central command chambers. They enter a glass elevator that will lead them to the EU Command Center above ground. In the Elevator, Machismo's left hand is gripping Nadia by the arm. He changes hands and grips with the mechanical hand.

"You have to focus on Nadia's Mind," I told Diana. You have to get to her now. Diana focuses on Nadia's mind.

"She is looking at the mechanical arm," Diana reports back.

"Can you disable it? It will be painful and at the same time I will confuse his head with images."

I look into Machismo his arm is driven by a 12 volt power source that is exposed after he had accidentally fallen in a tub in his apartment in Spain. He had never repaired it.

"Look at the mechanical arm," Diana told Nadia. Nadia is looking around nervously not knowing where the voice in her head is coming from. She looks at Machismo's arm and I sense her repulsion at the mechanical hand. It is strange looking and ugly.

"Tell her there is an exposed electrical board on the hand that she can disable if she can somehow short circuit it," I tell Diana.

"She can see a small opening in the mechanical knuckle exposing electronic modules," Diana reports "I now know what to do," Diana adds excitedly.

"Slowly lift your free arm to your hair and remove the hair pin," she tells her. Nadia removes the hair pin and waits for the voice.

"Aim it into the electronic finger and inject the hair pin into the knuckle of the mechanical finger," Diana commands her then she adds quickly,

"Get out of him Seth he is going to burn now."

I slackened my mind from Machismo's just as a loud moan echoes in the chamber and he cries out in pain loosens his grip on Nadia. I focus on the Lazuka finger and send commands to his brain. The electronics are frying his hand and filling the elevator with the smell of burning flesh. The high voltage laser powered weapon is firing rays from his curled finger and they are pointing at his leg, his mechanical leg. With a flash of light, the Lazuka seers into his flesh and circuitry, dropping Machismo to his knees. The King does not have time to respond. In a matter of moments, I see Jessica and Michelle are attacking the King to throw him off balance!

"Diana!" I shout realizing what she has done. Déjà vu. There is no time to discuss what is going on, the King falls hard on the floor of the elevator spilling out curses in Spanish. Michelle kicks hard, disorienting the King as the Lazuka burns deep into Machismo's body. Jessica grips Machismo's hand and points the weapon at the King's face and it makes its mark continuing to pour out its remaining voltage on the King's face. Machismo's mind releases me and I feel a sense of pleasure overwhelm me knowing that the children are safe. I turn to Diana and motion her to explain. I can see that there is something wrong with the entire picture, like a movie gone wrong, like a piece of film that is burning with the intense light projected from my mind.

"What is going on honey?" I ask, "Why did we not see this before?"

Diana smiles and looks at me squarely in the eye.

"Tell me what is going on now, Diana" I ask.

"The elevator is almost at the top of the chamber and I have asked the girls to be quite. The smell of flesh is nauseating and Nadia wants to vomit, they could not wait for the door to open. They are gagging for air; the elevator is putrid. The elevator has arrived at the top of the chamber and the door is opening. They have stepped out of the elevator into a wide open room with several windows.

"Michelle has found a small room and they are going towards it. It is a restroom and they are now hiding inside waiting. We must go to help them now."

We leave the room with Diana pushing my wheel chair feverishly excited. We follow in their steps in a hurry. We enter the open room and locate the bathroom where the children are hiding. We could see the open expanse of the mountains that surround the facility and we must be on top of one of the mountains. There seems to be no one in sight and I make a mental note that the only exit door is one hundred meters straight away ahead of us. Diana pushes on the bathroom door and calls out to Michelle to open the door. She does not answer. I shout her name and tell her that it is me and that they have nothing to fear.

"Mom? Dad?" Michelle asks sobbing. I look at Diana and smile. She is happy, the kids are ok and it will all be alright from now on. Moments later, Michelle opens the door and the looks on their faces tells me that they are very frightened.

"It is going to be ok now," I assure them as they run to hug us. I strange feeling comes over me as I hug the children one by one. This is a feeling I have always waited for as if I have never had it before. It almost feels like I have never hugged them before, and that I have always wanted to but could not. I feel a warm glow over my heart and Diana is smiling with happiness.

We should not have left them alone in their room. We should have asked the authorities for a large suit that can sleep us all. Nadia is feeling guilty about what she had done to Machismo. I can almost automatically tell what she is thinking as if she is in my mind. I look at Diana and somehow she knows what I am thinking. She pulls Nadia close to her and hugs her tight. I know she is projecting a memory loss in her brain and that the child will soon forget the traumatic episode. Then, she pulls Michelle and Jessica close and does the same thing to them. Soon the children are smiling as if none of it has ever happened. Since they were born, I have always let Diana keep close contact with their minds, preparing them for the calamity that is about to come. We must move on now quickly.

"Quick, there is no time left, we must go now for our safety," I cry out to the children.

"We must leave now before the guards come in," I repeat. Diana motions the girls to follow us and we quickly traverse the room to the exit door. It opens slowly as if an air controller dampens its motion. As we enter into the room, a guard spots us and orders us to stop. Diana stops and holds the girls back. The guard is wearing an American military uniform, relief.

"We are in danger," Diana states anxiously,

"Take us to the President now,"

He looks at us enquiringly.

"You are not authorized to be in this area, return to your chambers," he commands, pulling a weapon.

"Take us to President Cooper, my name is Diana and I have a crystal data bank the Europeans are looking for."

"What data bank?" the guard asks reaching for his radio.

"Do not use that radio, this is top secret and top priority for our government," I state hoping that the message will get through to the young man. I scan his brain as and decide to influence his thoughts.

"Dude, take us to the President, this is important!" Nadia says.

Michelle and Jessica nod at the guard approvingly. I veer his thoughts to national security and I can see his countenance changing with my intrusion. I project thoughts of nuclear explosions in his mind. He stands straight as if on alert and looks at us with an air of confusion. He is almost Michelle's age. I projects feelings of allegiance and national pride. The thoughts run through his brain and he knows we are Americans and descent intelligent people. He puts his radio back into its holster and grins at the girls.

"Follow me," he orders.

We follow the guard through several doors with several offices and desks manned by American and European military personnel. No one seems to notice us. The guard enters a large chamber that I presume is the Presidential chamber and greets a guard sitting on a large desk by a door. There are three other armed guards guarding the main entry door. The lady at the front desk offers a phone to the young guard and he speaks for a while briefing someone on the other side. Then, he hands the phone to me.

"What is your name and who are the others?"

I answer the man and explain that we have important information for the President's ears only. The man asks me to hold and after a brief silence a voice thunders through the phone line. It is President Cooper.

"Who are you and what do you want?" he asks.

"My name is Seth Malaki; my wife Diana has something very important message for your ears only."

"Professor John Davis's wife is with you?" President Cooper asks.

"His Ex-wife," I answer jealously. The guarded door opens and President Cooper walks out toward us. He greets us and motions the guard to leave.

"Come with me," the President motions toward the guarded door. We enter into a very large conference room filled with over one hundred persons. They are all seated around a huge wooden conference table. The room is silent. President Cooper motions us to seat around the table and asks Diana and the girls to seat beside him, asking three persons to move further down the table.

"Who are the children, Seth?" I recognize the voice, it is Effion's. I wave just as he recognizes me and smiles surprised. I can sense that he is confused that I am here. Effion is wearing pontifical robes and as usual he is rubbing his fingers together as if to wipe some evil layer off of them. But what on earth is Effion doing here in Papal attire in a Bilderberg meeting?

I answer the question oblivious of the circumstances that have arrived at my disposition.

"They are his children," President Cooper explains.

"Your children?" Effion asks confused.

"You have no children!" He continues affirmatively. There is silence. Effion seems to believe that we have no children and since we have been under the watchful eyes of the Vatican guards for a long-long time, I do not know why he does not know about the children.

"This is insane, you have no children and you know it, so who are the children?" He asks again authoritatively. I can tell that this is going to be a difficult discussion. Effion is a persistent analyst. I search my mind for answers to give him but I come up with a blank. Strangely, I am thinking that Effion of all people should know that these are my children. I search my mind for memories and all I can come up with is that Diana was married to John Davis and got divorced and then we met at the Nevada facility from where we were kidnapped by the Vatican. As far as I am concerned, we have lived as a family at the Vatican grounds throughout the years.

"Why are they here?" Effion asks again. One hundred others nod in agreement to the question. Diana looks around nervously and whispers to the President,

"Can we talk in private?"

"It is alright, you can talk here," the President assures her.

"No, I insist. We must talk alone."

"Ok, we will go to a private room, follow me."

Cooper nods to a security officer standing beside him. He leads the way to an adjacent room where we could converse in private. Diana motions the girls to follow us. We cannot leave them alone anymore. After the door closes, the President motions Diana to explain.

"My ex-husband left me a crystal data bank with some codes. The Europeans attacked us to get the data bank that my ex-husband got from the satellite. The Spanish King and his assistant killed our Chamber Commander and tried to kill us to get the codes. We escaped."

"The Spanish King, Rudolpho?"

"Yes, they are after the data crystals," she explains.

"The bastards, I knew it, I knew the Europeans are in on it," the President states angrily. He is in a meeting of the Bilderberg Group and thirty in the group are Europeans.

"Have you heard from John?" Diana asks anxiously.

"John Davis? He is fine. He is in Africa and will be back soon. The US military is already at his location."

"Mr. President, what is this all about? I have a right to know. Thousands of citizens are trying to get into this facility, and all I have seen are high ranking Europeans and a few of our politicians. What is going on?" I ask hoping to get some answers.

"I cannot explain it to you now. All I can tell you is that you have opened up a whole world of information that is of utmost urgency. If what you are saying is correct, then, there is a European conspiracy with the Russians to take over the world and I will need the information from the crystal data bank. The one John Davis handed us is incomplete".

"What assurances do we have that Diana, the girls and I will be safe?" I ask. The President looks at me as if I have insulted him. He knows who I am and how important I am to the Vatican.

"None! I can assure you that as long as I am alive you will be safe, but beyond that I cannot assure you of anything. I will need the correct data crystal to be able to make sensible decisions about US security in the next coming days."

"Mom, do not release the codes until we know we are safe." Michelle states emphatically. Diana nods. She looks at the girls, she is no longer sure that anyone is looking after our interest. She must find out what John wants her to do.

"I need to talk to my ex-husband first. I cannot release the codes until I know we are safe," she says.

"You must give us the codes now, it is a matter of National Security!" the President blurts out impatiently. Diana moves away from the President and stands behind my wheel chair. The President motions a guard in the room to get the data crystal from Diana.

"You cannot force her to comply!" I state angrily. The guard hesitates but Cooper urges him to proceed. The guard searches Diana and forcefully removes the data bank from her clothes. Michelle, Nadia and Jessica are standing a few feet away shaken by what is going on.

"Now, the codes, give me the codes!" He commands.

"No. We will not give you the codes until you let us talk to uncle John," Jessica says.

"Take them, away," the President commands.

"Wait. Show us that you are really looking after the interest of the United States and then we will give you the codes," Diana states trying to buy time.

"Come with me," the President commands.

We follow him and enter the conference room again. Effion is in heated discussions with the group of one hundred others. The President introduces Diana to the group.

"Diana, this is the Bilderberg Group. We are a secret society that has always looked after the interest of mankind for the past two thousand years."

The group introduces itself. Effion is introduced as Pope Paul the Seventh. I am stunned. I Know Effion well and do not expect that he would be introduced as the Pontiff. I look at him and show my surprise. He shows no reaction to my surprised look, instead he looks back at me as I am a stranger in his eyes. Thirty other members are introduced as members of the US Senate. Sitting among them is ex-US President Barrack Obama, Harry King of England, the Kings of Monaco, Morocco, Saudi Arabia and Jordan, the Presidents of Russia, Germany, South Africa and Malaysia, some Islamic, Jewish and Hindu leaders, many scientists and notable people in the world. Diana and I are stunned by the diversity and power of the Group. This is the world put together into a single government! We never knew how powerful this group is, let alone imagine that they are working together for the same purpose.

"I want to talk to the USA alone, only the US Senators," I state.

The President nods in agreement. Nearly half of the group stands up and leaves the room, leaving only members of the US Senate and the President in the room. I recognize most of them from dinners at the Vatican. The President commences the discussions.

"Gentlemen, Diana is the wife of Seth Malaki, the great prophet and Chief Scientific Advisor to the Bilderberg Council."

I am a surprised when the entire group of Senators unexpectedly stand up and acknowledge us. We must be very important.

"The Great Malaki himself!" one senator shouts nodding his approval and respect for us as the room erupts into a murmur of whispers. Since we have been secluded in the Vatican, we have never really known the extent of our importance to the world other than the fact that we have been writing prophetic books as prisoners of the Vatican. Diana and I are really moved by the great admiration and respect the Senators are showing us. It gives us a sense of belonging. The President raises his hands and asks for silence.

"Gentlemen, as part of our agenda, we were going to reveal to the Senate a wonderful secret project that has been going on for over thirty years now. It is collaboration between the Vatican and the US government under the auspices of the Bilderberg group."

"Perhaps I can shed some light," an old man stands up to address the Senators. I look at the man and a dribble of recognition becomes a spring of memories that start to fill my mind. It is my old friend Senator Owens. I do not say a word and Diana does not seem to have the same recognition I do, so I do make no attempt to show my surprise. The Senator has changed a lot and he has lost a lot of weight and looks ill. His voice is gone and his face is scarred by multiple dark patches. I remember him as a tough loud man with a vibrating voice thundering commands like the military man he used to be.

"What you did not know is that you and your husband worked for the US government as one of the advisors to the Bilderberg group. After your illness you and your husband were taken to a secret medical facility in this facility."

"In Nevada?" I ask really surprised by the Senator's statement.

"Yes, in this facility," Owens answers smiling.

"Both of you were treated here to preserve the secret of the incident at the Space station from become a public spectacle. You became very ill Seth and so did you Diana. You were both treated for extensive memory loss and Bell's syndrome."

"You really mean John Davis and myself?" Diana says hoping to protect the secret she has started.

"But why did the Vatican not make us aware of this?" I ask.

President Cooper looks at me and then Diana. He focusses on the children and then turns his head away from us.

"John Davis has been amassing data for us concerning a world wide effort to develop a new and powerful weapon called the J-N field. This field is a powerful new source of energy that can be used to harness nature and provide cheap energy and power for all of mankind. We have kept this a secret for a long time. It is also a very powerful hypnotic field that must only be used for the good of mankind, since it can be used to control humans. The codes John Davis gave you are the codes for the exact intensity and GPS location of each of the crystalline data bank of the field being developed by our group around the world. These crystal data banks are kept around the world in a concerted effort to synchronize the fields for a united action when the time comes."

I can tell that the President is keeping a secret from us. I look at Diana and she is as confused as I have never seen her before. Nevada and the Bilderberg group, the hospital, Rome, Effion and the children and so on. What the hell is going on? I know Diana is wondering about the same thing and that she is just as confused about all of this as I am. Who is John Davis? Who are we? I need to find out from Cooper himself he will be the best source I can get out of all this mess. Even after all the years we spent in the Vatican writing the books with Effion and others, I still cannot get a clear sense of what is going on. Are we just an imagination in their minds? I cannot scan their minds, they have been well protected and they all have the scarab shield on their persons. Effion did not recognize the children, why? I know he has a mild schizophrenia and is a little autistic, but then he has an astute mind and surely his memory is superb. After all I dictated thousands of pages of visions and information concerning the J-N field and never once did he forget a thing I said.

"I need to understand what is going on. Why is Effion acting as the Pope?" I ask hoping that the President will shed some light into our doubts. The President seems to be searching his memory for something. He has to explain it to us in terms we can understand. "The Pope is a member of the Bilderberg group," he starts.

"I know that already from my own discoveries, but why are we not following the Malaki Book protocols I developed in the Vatican?"

The President disregards my insight and instead he continues to explain.

"The group had its origin two thousand years ago in the Council of Nicaea in Alexandria. In 2035 there was a division in the Group between the Europeans and the Americans based on the final location where the power of the field will be developed and kept in secret. The Europeans wanted to keep it in Belgium and we wanted it here in Area 51. The meeting today is to unite the two and bring the remaining crystals here to Area 51 for final assembly."

I decide to cut the history lesson short and get to the point.

"Effion is not Pope Paul, he is his brother, where is Iba?" I ask again looking at the group.

"And why do the Spanish want to take the data from me, if they are part of this group?" Diana adds.

"There are many things that I cannot explain to you for security reasons. We were not expecting you and Diana to enter this room under such circumstances with children, but all the same the harm is done, and we must go on now," The President states regretfully.

"Then the Europeans are trying to bypass the group and there is something not quite right about Effion being here as Pope, he is his twin brother!"

The President is silent.

"How do we know that if we give up the codes, the final result will be to the benefit of our country and humanity in general?" Diana asks.

"The Pope is an International leader and we know for sure that he is acting in all our interests," the old Senator explains. The Senator walks to the President and whispers in his ear. A few moments later a guard leads Effion into the room. He is rubbing his hands vigorously.

"Seth, you know me well and we have worked together for many years. I am Pope Paul the Seventh," he says smiling at me and Diana.

"But how can that be?" My doubts start building up again as I recall the meeting I had with Pope Paul many years ago. I have always suspected that Effion was the person I met, and not Iba. Now, again, I am being confronted with my doubts even though I have worked with this man for many years as Effion, the Pope's brother.

"You see, we work together. We need the codes to activate the satellite and prevent the enemy from getting the crystals." Cooper explains.

"I need to talk to John, Diana states."

The President motions to a guard to open a satellite link. A large screen appears on a wall. Images start appearing of different places around the world. Millions of people are walking about aimlessly while other just fall and die. Each new scene is more chaotic than the other. We stare at the screen in utter amazement. Diana starts to cry and the girls join in. No other in the room shows a drop of emotion. They simply watch and wait for a response from her.

"You see Diana this is what is happening to the world. Humanity is dying and John Davis is struggling with them. Unless we get the codes, he will also die with them."

President Cooper walks over to Diana and reaches out to console her. She moves away. The girls instinctively stand between them.

"John, where is he?"

The guard changes the image to a silent and grainy jungle scene. John Davis is standing with four other men who appear to be conversing. A blue light surrounds the jungle painting the green leaves a nasty dark color. John Davis is wearing dark goggles and his clothes are torn. All around them, fires are burning and smoke breaks up the image into a hazy hue. A bright yellow light is flickering around them. The blue light is intensifying and overpowering the yellow light and causing blue blackouts in the satellite link. Around them, thousands of people are roaming aimlessly as the blue and yellow light compete for their souls. Diana and the girls start to cry again. I approach them and try to console them.

"The codes," the President breaks the static silence, "only the codes can save him now".

"You must get him out of there!" Diana blurts out a command.

"We cannot. Only the codes will deactivate the crystals and save him."

Then, in a flash of blue light, the satellite image vanishes. A few moments later, the image returns with a blue hue filling the jungle. The bodies of John Davis, Father Mulcahy, Michael, Paul, and Bumba and thousands more persons are lying on the jungle floor motionless.

CHAPTER 38
(The Fruits of the Garden of Eden)

For what seems like an eternity, Hal sits on the cold floor watching Lisa, Biano and Vinod sleeping heavily. Since there are no clocks or windows in the building, Hal can only guess that the drugged prisoners must have been slumbering for half a day. He supposes he has been lost in reverie for hours. He speculates on the type of narcotic the food is laced with and entertains himself with the scenario of the drug's discovery. The chemical has probably killed enough brain cells in these prisoners to prevent them from formulating a rational plan of escape. Hal decides he will eat just enough food to keep alive and no more. The prisoners seem mentally unstable due to the influence of the drugged food so that leaves him with no shrewd accomplices. There is nothing but the four walls - not even a piece of furniture he could make a weapon out of. When the servant lady comes back to collect the remains of the meal, Hal feigns sleep. Upon her departure Vinod awakens eyeing his new roommate dreamily while he belches and farts.

"Satisfying meal!" he grins widely like a mischievous child.

"You have been sleeping for a long time. You must be well rested," Hal comments.

He sees the expression change quickly on Vinod's face, in the time it takes for a candle flame to flicker.

"I want to go home, Hal ," Vinod interjects. "I wish I can just leave this dreadful place."

"Tell me more about this place," Hal encourages him.

"Do you ever get a chance to go out to the city or the gardens?" he asks hopefully.

Vinod glances down at his slumbering comrades with great concentration, as if his gaze could will them to rise. His furry brows arch into high peaks when he is thinking.

"Yes, every day, at dawn, we are taken to the Field of Dreams where we are allowed to play and eat as we please. The place is full of luscious-tasting food and it is all irresistible. You will see. They will come for us soon," Vinod promises, nodding his heads like a child attempting to impress an adult with his knowledge.

"What do you do in this Field of Dreams?"

Vinod's face lights up with a great joy that is pure and innocent. Hal glances down at the other prisoners. They are dead still and he is beginning to wonder if they would ever stir again.

"We are left there by the Guardians to do as we please among the Minga men and women. We dream, fantasize, play, make love to their women and play like children. It is fun," the Indian explodes with laughter.

Hal accesses Vinod's attitude. Just moments ago, he has been whining about his captivity. Then, abruptly, euphoria envelopes him. Hal deduces that the drugs in the local fruits and vegetables are responsible for the man's moodiness and confusion.

"Will the Minga stop you from escaping if you try?" Hal inquires. He is trying every angle he knows to calculate his escape. Vinod's expression transforms again into a serious depression. His pursed lips become a protruding pout and his eyes are waving flags of caution as he blinks nervously.

"No!" he yells as if to stop the passage of an apparition.

"They are also held hostage by the Field of Dreams. It is forbidden for them to leave because they are the slaves for the Guardians. The Minga protect the Garden City from outsiders and in return they are rewarded with the wonderful fruits from the Fields."

Hal could think of better rewards.

"Have you tried to escape before?" Hal asks, staring into the complex, coffee-colored furrows on Vinod's wrinkled foreheads.

"Yes, Yes!" Bayano answers from the far side of the room, rolling from a pile of grass he has formed into a bed. The man's answer startles Hal and he twitches. Bayano appears older than Vinod and more slovenly. A smell closes to horse manure is released into the room as a result of Bayamo's restlessness.

"We have tried several times but the Field is so enticing."

Bayano stops to struggle with a lump in his throat that is his hopelessness. Hal imagines a portrait framed around Bayano. He imagines a Salvador Dali painting "Desolation After Lunch" with rivers of food pouring out of Bayamo's every orifice.

"Why don't you just stop eating the fruits of the garden?" Hal suggests.

"When you taste the fruits and vegetables you cannot stop eating them, ever!" Vinod explains. His voice harbors a terrible desperateness in the form of a hoarse whisper. The silence in the room, Hal surmises is replacing the prisoners' consciousness. The food must be terribly addicting. Lisa rolls over her own bare bed space as if there is actually a mattress and bed sheets. The drug's hallucinatory effects begin to intrigue Hal. His eyes follow the music of Lisa's every move. She is a mime making the bed. When she is finished she turns her attention towards the men's conversation.

"Don't eat the fruit Hal, or you will be trapped here forever!" Bayano cautions. Hal quickly ignores the emptiness in his stomach that is a growing pain.

"I will not eat of the fruit of the garden. We must find a way to escape."

The old man lowers his head between his knees and begins to weep and Lisa instinctively walks over to Bayano to consoles him.

"We will escape, Bayano, please don't cry," she promises, rubbing his hair gently. Bayamo's grey eyes squint from his acidic tears.

"I is a family man before they took me," he explains.

"I is a miner, mining diamonds for the Sierra Mining Company. One day there is great explosion at site, and everybody dies except for myself. I wander through the jungle in search of the way back to city. The entire face of the area is changed by the explosion. One day, I discover a cave and decide it would be safer for me to spend the night there than outside, exposed to the dangers of the jungle. While sleeping, the Minga warriors capture me and brings me here. How I came to North Africa across desert without noticing, I do not know! Nu Purge! But that is very long time. Now, I thrapped. I want get out, I cannot," the old man continues sobbing.

Hal sees the sounds of his life story carpet the floors and paper the walls with scenes of the jungle and war. Inside the room is a diorama of the rainforest and in the middle, a great cavernous pit. Hal has to escape or eat something soon.

"Welcome to the Hotel California.." Lisa starts singing.

Hal is not sure if she is actually singing or if it is his imagination playing tricks on him again.

"Yes, this is the house of the Rising Sun, only it is worse than the song says," she confirms.

Hal sits quietly making his plans. He has to keep the prisoners from eating the food of the garden so the drugs could wear off. Hal would also have to pretend to the Guardians that he is addicted as well. But if he is to act fast, he will need more information.

"How did they die?" he asks his group piercingly, as he points to the skeletal remains of previous escapees.

"When they escapes, Guardians not know they has entered the Field of Dreams. They not find them in time, to take them away from Garden. So, they eats until they died," Bayano reveals looking away.

"You mean the Guardians would bring you back into this prison to prevent you from dying?" Hal asks surprised.

"Yes," Vinod answers.

He seems a little more relaxed and bright-eyed.

"Who are the Guardians?" Hal asks.

Lisa jumps to her feet enthusiastically. Her sky blue tunic billows to action. Biano was eyeing her half-naked legs wistfully.

"They live inside the Inner City," she vibrantly reports. "They are said to be miracle workers and holy people," her voice softens as if she is divulging a great secret. "They are monks," she concludes.

"Some monks!" Bayano exclaims as he chuckles and scratches his pillowy stomach.

"I have heard it said that the inner city has been constructed from pure gold," Lisa offers. She seems the more stable of the three, Hal calculates. Her thoughts are sharper.

"I remembers one man abducted by Minga who are rewarded for capture. He is dying from serious wound to chest. He is brots into the city and given some powerful drug, and is the very next day, he heal completely. It is like miracles," Bayano explains more lucid than before in an attempt to assert the power of his seniority.

"They are not ordinary people. They do extraordinary things," Lisa interjects. Hal recalls his wounded skull and examines it with his hands. There is neither wound or scar.

Bayano is becoming more animated. Hal could tell that the drug and his moods are evening out.

"In India, the Maharishis do the same thing. They live up in the mountains and do mysterious works like the Guardians. It is said in Indian folklore that Nirvana, the beautiful garden city, remains lost to Man. It is equivalent to your Garden of Eden!" The Indian added.

The Indian seemed taller and more youthful to Hal as he boasted about his heritage. It is time to brief his crew.

"Before we get out to the field, you must each make yourself a blindfold, so that you may not see the fruits of the garden. You will cover your noses with a rag which I will make from this robe," he points to a robe laying on the floor. Then, I will leads you out of the Garden into the cave outside. I will lead so that I can clear the path of all fruits and vegetables. You must stay behind me. Do not, for any reason, wander off. Do you understand?"

They all nod. Hal carefully cut small pieces of the robe for blindfolds. Then, they all wait for the Guardians to take them to the "Field of Dreams."

The next hour is filled with memories of their former lives as they tell their stories vigilantly like boy scouts revealing their secret wishes around a campfire. Then, the woman who has served them food appears at the door. She has changed her attire and is wearing a white satin robe. The woman looks rested but serious, and her golden streaked, thick hair, has been slicked back with perfumed oils that smell of honeysuckle.

Hal is sure that a ceremony is about to take place. The woman leads them to the gates before an open field that is located a mile from the city's center. There, the woman shouts a musical command that magically brings forth several Minga warriors and women from somewhere behind them. They are all carrying large woven baskets, ready to collect the fruits and vegetables from the narcotic Garden. Hal and the prisoners enter the field behind the Guardian woman, but then the excited Mingas wedge themselves in between unable to wait in line.

The faces of the warriors are painted with the colors of the jungle and its animals. They look like ancient Egyptian warriors. The women has draped colorful beaded necklaces of various sizes onto their bodies. The beads snugly cover them from their necks to their ankles. All the Minga look festive and they chant energetically as the men's ebony shades glisten from the oily baths that anoint them.

Each member of Hal's group holds on tightly to the soft blindfold that lies hidden in their pockets. They are resolved to get out of the forbidden garden. When the Guardian open the gate the Minga are already scrambling to get inside, pushing and tugging at each other to be first. Hal holds his group back to allow the Minga to disappear deep within the Field of Dreams. The Guardian has left the field and Hal realizes that soon they will be alone. He cues the prisoners to ready themselves for the blindfold. The Minga, prancing through the field like lost sheep, are now rainbow chocolate dots in the distance.

"Now!" Hal instructs his new friends. "Put the blindfolds on!"

Lisa does not hesitate. She whips the blindfold out of her pocket with the flair of a performing magician, and ties it at the back of her heads. Vinod fumbles with the cloth, dropping it several times before he could securely fasten it. The shakes for the Garden are apparent. Bayano needs Hal's help for the old man's hands are too arthritic to make a knot. Bayamo's grey beard quivers from a smile when the cloth is finally secured.

From his post, far away from temptation, Hal scans the luscious gardens. Even here, the smell from the cornucopia of fruits is overpowering, like the floral aroma of a funeral parlor. Some of the fruits are as big as melons, pregnant from their succulent and plentiful juices. Hal tries to distract himself with the tall trees lining the garden. He is very hungry. Very Hungry! There, along an adjacent path, the shadows of the fruits hang but Hal could not see a sun up in the sky. The horizon is alien.

The Minga are already indulging in the eating ceremony, scrambling and dancing as they harvest the trees to their hearts' delight. Vinod is already shaking from withdrawals and is sweating uncontrollably. Because Bayano is aged and his metabolism is slower to react, he still appears calm but a dull melancholy in the form of whimpers and sighs has begun to take hold. The sweet smell of the garden is hitting Hal very hard. It is very hard to resist.

"We have to get out of here fast!" Hal whispers to his comrades.

He knows Vinod's will power has vanished away, and his is dwindling fast. He begins to realize the powerful grip of the garden. The man's eyes bear the glazed, empty look of the animal world where spirits and emotions have only been partially developed. Vinod growls like a wild tiger and breaks away from Lisa's grasp.

"Let's go now," Hal shouts to the others, "we must get out now!"

Hal does not want the others to lose faith like Vinod and follow suit. He pulls Bayano and Lisa away from the gate but it is too late for Vinod who is leaping towards the field in joyous, bountiful strides.

Lisa begins to cry.

" I am hungry, I need food!" she cries out. No doubt, Hal surmises, they had been lovers on the stage of the Field of Dreams, playing out their fantasies in past histories. Lisa's quivering hands move towards her blindfold but Hal is able to overpower her long enough to lead her and Bayano out of temptation's range.

"Do not take the blind fold off we are almost there!" he commands. They are heading for the Cave Snake of the Gods. Only Hal dares look back into the distance where Vinod and the Mingas rejoice amidst the fruits of the Garden of Eden.

CHAPTER 39
(The Archangel awakens)

Michael is kneeling before two men when he opens his eyes to the stars. They are shining with the light of faith. They had drunk the holy potion of the cactus plant from the Garden of Eden. Like Bitter Cola, the Poro master before him, the drink is the ceremoniously mark the pact that the Poro had made with the Eye of M'Dulu a long-long time ago. The potion is a potent magic, a spiritual communion with the Eye. All of their dreams waterfall through their minds. Bodily sensations become a vapor that evaporate from him. They are able to see, feel and know of other worlds. The magic has permeated their body to transform them into angelic beings. Within their new spirituality they fear nothing.

Their bodies slump as the light of the Eye of M'Dulu grabs their souls. Their four dimensional spacetime slowly but clearly transforming into a twenty seven dimensional spacetime. As Michael and Paul slip into the oblivion of the eternal spirit, Father Mulcahy, John Davis and Bumba are too stunned to do anything but sink to their knees and cry. Before them they experience a holy transformation. The bodies of both Michael and Paul flicker like a ripple in a pond, before becoming transparent. They vibrate with a glow while feathers of light multiplied from their backs.

In their hearts they are grateful for the miracle they are witnessing. They have read and studied of miracles their entire lives but to actually be a part of one is too overwhelming for their mortal souls. It is a long time before they could raise their heads from their thankful prayers. Paul and Michael hover over the priest and his comrade as they struggle to compose themselves. When their heads clear, they observe, with great admiration, the greatness of the human wings that are a quilt of soft grey that curve like a shoulder but are the envy of the greatest eagle.

Paul approaches the men.

"You have drunk of the blood of M'Dulu!" he shouts, "now you must be like us".

They brace themselves for the physical transformation, not knowing if the potion would have the same effect. The priest is beginning to feel weightless and a supernatural fortitude pulses through his veins. Suddenly, the mortals find themselves effortlessly running towards the powerful crystal. Above them in the clouds, the terrible creature is struggling with the great Eye of M'Dulu.

"This is the sacred hiding place of M'Dulu's eye!" Paul explains to them, raising his hands into the invisible-elastic membrane that surrounds the Eye. As he accompanies the priest from above he sings:

"Oh, Master of the heavens,

Oh M'Dulu, my eyes are filled with tears for fear of your mighty   Eye.

Yours is the glory of the Universe carried within your palms.
Yours is the fire of the stars carried in your Eyes.
We come to praise you and to renew our vows with you.
Our bodies rejuvenated with your blood, and our eyes opened by the light of your Eye.
Do not shed your tears on us we pray."
Entranced by the power of the crystal and strengthened by their new spiritual chemistry, they all chant an ancient prayer that dwelled within their ancient souls,
"Open the gate to M'Dulu's eye
so, we can bath in the light of His Eyes
so, we can smell the perfume of His wounds.
So, we can fill our lungs with His breadth
So, we can gain the wisdom of His eye"

The uttering alerts the creature who peers down curiously at the six men, baffled by the words of the new Masters of the Poro who are also the new Guardians of the Eye of M'Dulu. M'Goro is the new Areyogbo, an unholy resurrection of a dinosaur-like creature more hideous than any legendary serpent. The crystal pulsates like a new sun, and releases itself from the creature's mouth, painfully burning its ugly lips. Slowly like molasses, it flows into Paul's raised hands. The substance is able to help the young angel stretch his super elastic hands further into the trees of the forest. He grabs the creature and with his great celestial might, pulls it down until it come crashing onto the jungle floor. As the Earth flies in all directions from the hammering impact, Paul reaches out into the crystal's twisted fabric of spacetime and holds the gem in his hands until its field slowly consumes him and the others.

For a few moments, Michael, Bumba, the Priest, and John Davis stand transfixed in the light of the massive Eye of M'Dulu observing Paul in all his glory as he beholds the massive crystal between his heavenly hands. Father Mulcahy and Bumba are praying out loud, shouting praises and hallelujahs to the mighty force that holds them captive.

Millions of people lie dead on the jungle floor. Thousands of the chosen few enter the crystal field to savor its holiness. Those that revere the crystal's mighty energy are spared the wrath of M'Dulu.

Paul waits until all the chosen ones have entered the energy field. Then, raising his hands, he proceeds into the portal of the Snake of the Gods. He travels with the crystal craft to the hidden Garden of Eden. The cave serves as a gateway to an astral journey. Now, he would enter the same world his great grandfather Bitter Cola, had visited.

Abruptly, a mysterious force pulls the crystal toward the inside of the cave, and elevates it until the pull of twenty-six other crystals call their long lost partner to join them. The crystal craft vanishes from the jungle and heads through the Snake of the Gods towards the deserts of North Africa.

For M'Goro, time has frozen. The strange weight of Earth's gravity on his powerful limbs slow him but he is free from the crystal at last. He eyes the mass of people streaming towards it in the Garden of Eden. An ancestral instinctual memory of luscious fruits to eat in the garden hit its nerves. Instinctively, M'Goro realizes that it is his duty to protect his food source from the invading Mingas. The Master of the Garden had bred his ancestors to tend the Garden. The creature turns on the masses with fresh zeal and strength. He tramples those in his path and reaches out with his massive claws to crush others to death. His mission is clear to him now. He slides away into the spacetime of the crystal's field and journeys to the cave that would take him to his precious Garden, hot on the trail of the crystal and his enemies.

CHAPTER 40
(The migration)

The world continues to migrate to the North African Desert. Millions trample over other dead millions in an effort to enter the alien craft that holds the pulsating crystals. The force field around the craft grows stronger as each human wave comes in contact with the field. Like zombies, the entire world is responding to the call of the alien cubic field by feeding it with their own life force. Blood is everywhere, flowing into the crystals as if it is an everlasting sacrificial altar for humanity while animals, free from the bondage of humanity at last, and displaced from their habitats, roam aimlessly in the cities. The seas are littered with the dead and the stench permeates the Earth. The energy absorbed by the crystal is also changing the climate and spawning a new ice age.

A bright yellow light envelopes Hal , Bayano and Lisa as they enter the cave and meet the sounds of thousands of people crying out in the distance. A crowd of Guardians behind is gaining on them fast. The survivors of the Garden are trapped. Bayano and Lisa panic like a prey backed into a dead end by its predator. Like two lovers they join hands and instinctively run back towards the direction of the Garden City.

Hal 's entanglement in the lights and sounds of the cave prevents him from acting quickly. He has lost his new friends to the Field of Dreams but feels no remorse, for by now, he is also feeling the strong attraction of the crystal field, and he heads towards it. Nothing passes through his mind as he ambles to it. Through the burrow of the cave light, he glides forward skating on pure energy. When he has almost completed his journey, a man that is neither Minga nor Guardian, approaches him.

"Have you eaten of the fruit of the Garden?" he asks.

Hal replies, "No, I have not."

"Then you have nothing to fear. You have not disobeyed the law. Join us and enter the light!"

Hal enters into a huge chamber, where thousands of Minga bodies lie dead. Their souls have been absorbed by the light, leaving their lifeless bodies lying in the caves. He hesitates. This is not what he would like to do. This is not where he wants to be. He turns on his heels and runs the other way, disregarding the guardian. He heads for the garden city where he knows he will never escape. He is hungry, very hungry.

The blue crystal and its occupants approach the yellow crystals. The intensity of the two different lights bursting into resonances that echo colors across the skies of the desert. The guardians frantically adjust the crystal craft waiting to capture the blue crystal. The entire structure starts to vibrate violently as the crystals start blending their fields. Like magnets of like charges and opposite charges, the crystal's orientation is oscillating trying to form a matching field. As the frequencies align, Paul knows the yellow crystals should not be allowed to fuse with his own. Seeing the massive yellow pulsing spacecraft ahead, Paul senses the dark forces within the craft. He tries to steer free from the field of the twenty five crystals, but it is too late. In an instant the blue crystal is pulled into the field of the others by an attractive force that is too great to overcome. The Guardians open the locks of the craft and ready to correct the connections for the blue crystal inside the craft. Paul, Michael, and one hundred and forty-three thousand nine hundred and ninety-eight faithful could neither see nor hear what is going on with their confined spacetime. After the vibrations stop, a clearer field exposes their predicament. They are in the middle of a complex field of lights assembled in the form of a cube.

a few days later, the monkey speaks!

Paul looks into the craft, and there before him was a door standing open in the middle of the field. A voice speaks to him that sounds like a trumpet,

"Come up here and join us, and I will show you what must take place after this,"

Paul and Michael and smile, they know that the human form will always need this field.

"We are trapped in this field, and so are you in the blue field. I have never known you before," he says referring to Michael and Paul. He introduces the rest of the children of the guardians to Michael, Paul and the others.

"Come join us. They have stolen man's birth right and yours and ours as well. All you have done for the glory of their name is for naught," Michael explains to them.

"We built massive temples and pyramids with great works of art to steal man's heart. We invented and taught man science, about the power of the atom as they pursue wisdom. We showed man the fruit of knowledge and evil. Then, we found the hiding places of the crystals on Earth. Now, they would regain the craft and be free, leaving us with human forms to perish in a lifeless and blank planet without Love, Hope, Hate, Despair, Compassion, Joy, Anger, Sorrow, and all our other dimensions especially Hope," Osiris says to Michael.

"The power would enable them to traverse the Universes of the living and the spiritual world at will, so that they will be your gods for their own amusement," the priest explains.

"You must cross over, you have the human form for accepting blue light of Hope in you," Paul says.

"How can we cross into the blue light of Hope, we have no hope within us?" Osiris says.

"The human side of you has hope and compassion and love. You have hoped that the stars will be named after you, and you Osiris were given the Hope of Orion. Man has prepared special places for you in the stars. Did man not work to build your hope to reach the stars of Orion?" Michael asks.

"And you, Resa, did not man build temples for you? Did they not name their world after you? They nurtured you when you were abandoned and left in the calamity caused by your fathers. Is this not enough for you see man as a compassionate being?" Michael pleads.

"But my fathers have taken away twenty five dimensions of humanity. How can we succeed against them?" Ra asks.

"They have not taken, Faith and Love." Michael says, "the last two covenants were well hidden from them. They are the covenant that ties all the others together, the Covenant of Love and the Covenant of Faith."

On the top of the field is a Spirit, and there is a throne with someone sitting on it. The one who sits there has the appearance of jasper and carnelian. A rainbow, resembling an emerald, encircled the throne. Surrounding the throne are twenty-four other thrones each forming the center of a crystal field. On these thrones are seated twenty-four elderly Guardians. They are dressed in white and have crowns of gold on their heads. From the throne comes flashes of lightning, rumblings and peals of thunder. Before the throne, seven lamps are blazing. In the center, around the throne, are four living creatures, and they are covered with eyes, in front and in back. The first living creature is like a lion, the second is like an ox, the third has a face like a man, the fourth is like a flying eagle. Each of the four living creatures has six wings and is covered with eyes all around, even under his wings. One of the winged creatures of tremendous beauty appears before them. Paul and Michael gasp. It is Lucifer!

He is shining like a star awaiting their arrival at the barrier between the yellow and the blue crystal fields. Lights are flickering everywhere changing colors like a massive rainbow that is being squeezed like a sponge. The twenty five crystals start to orient themselves, adjusting their colors as if by some preset code of operation. Lucifer effortlessly flies close to the field of the blue crystal and stands close to Paul and Michael. He could not traverse into the blue field. He remains within the yellow field.

"Michael," he says in musical notes,

"It has been a long time since we last saw one other. You probably do not remember a thing, but I will remind you. Welcome back".

Michael starts trembling. He is trapped in the blue field. He cannot defend himself or the others. He embraces Paul, who seems puzzled.

"Who is he?" Paul asks Michael.

"Lucifer. He was the greatest angel in heaven. The Master created him above all others. But he stole the crystals and crashed them on Earth. Now, he is condemned until God captures him again".

"Is he the Minga King?"

"Yes, and much more." Michael answers.

"How do you know all this, Michael?" the priest asks,

"I do not know, I just know it!"

"The human souls are not yet ready," Lucifer states to the one that had the golden beard, "they still carry the potions l have hope in them."

He is referring to the potion still active within the men.

"Have they eaten of the fruit of the garden?" the other asks.

Michael supposes this question is for him to answer. Surely the angel must know everything. Perhaps it is part of the slow ritualization process to explain the profound workings of the hell that Lucifer has created for them.

"Why do you ask me such a question?" Michael answers cautiously. His memory of his angelic life is like a dream slowly coming back.

The dark angel smiles at Michael.

"Michael come through and remember. You are one of us."

The colors of a rainbow flitters around them as the evil guardians move about the yellow field surrounding the blue crystal as if they are being embraced by a prism. Lucifer has a scroll in his right hand, it is filled with writing on both sides and sealed with seven seals.

"Who is worthy to break the seals and open the scroll?" He asks looking at the other Guardians who have gathered in a circle around him. No one answered.

"No one but I can open the Scroll or even look inside it, " he boasts. Michael recognizes that Lucifer has unleashed the last days upon earth and man.

One of the Guardians say to him,

"Do not weep! See, the Lion of the tribe of Judah, the Root of David, has failed. He is unable to open the scroll and its seven seals. Without Hope, He cannot open it."

A wild anticipation grips Michael and he wonders why he feels that way. Had he not been under the influence of his new heavenly sensibilities, he would have been unable to spiritually survive the excitement. What is about to unfold before him, is, for an earthly man, what could not even be imagined in the world between waking and dreaming. He has completely forgotten what this realm feels like. He has been away much too long preparing mankind for this day.

"We are trapped," he explains to Paul, "they have captured M'Dulu's Eye."

"How can that happen? No other force is greater than M'Dulu's Eye!" Paul retorts.

"Never doubt the abilities of Lucifer. He is gifted with talent and cunning. We must find a way out of this." Michael states.

Lucifer renters the depths of the yellow field avoiding the blue field. "Can we assimilate them into the other fields now?" the old golden beard asks.

"No, not yet, we need to get the other two crystals first, otherwise they will destabilize the craft, and Hope can affect us," the angel answers.

"Meanwhile, we must avoid the blue field, it is dangerous for us, it gives them hope," he says, then adding, "only those that have taken human form can enter it, for we have no hope".

Resa eyes the golden beard and Lucifer suspiciously,

"Only those who have taken human form can enter it?" he thinks to himself. That means he and the other children of the Guardians! A few of the children of the Guardians gather around the blue field. They could feel a soothing ebb of Hope flowing from it. The old guardian notices this and pushes them away from the field with a wave of his hand.

"Do not go near that field, it is not good for us!" he shouts angrily.

"Then why did you capture it?" Resa asks.

"We need all the dimensions to free ourselves. The humans can only live by a few, but we need all."

Michael looks at the collection of the children of the Guardians. They have taken human forms to do the bidding of the Guardians, but they are also children of men.

"They have taken human forms like us, but the older ones like Lucifer have only spiritual forms," he says to Paul.

"Then, the younger ones are like us, they can stay in the blue field, but the strain of the hope will hurt the evil ones," Michael explains.

"What kind of a structure are we in?" Paul asks.

"A cubic field," father Mulcahy states, "it is a cubic field called the Josephus-Nambu field."

"I mean what are these other yellow eyes that surround us?" Paul asks ignoring the priest.

"They are a powerful symmetry that exists in a twenty-seven dimensional world. We cannot escape, unless we can fool them by changing the colors of the symmetry," the priest explains.

Paul turns to the priest agitated by his intrusions.

"How do you know this?"

"I studied this field. That is why I came to the Jungle to try to understand it. I suspected that it is a weapon created by some ancient people, at that time I did not know it is a religious thing. Our government learnt that it is a powerful field that is being assembled by someone. Now, I know it is a weapon for the Mingas and those creatures," John Davis interrupts referring to the Guardians.

Thousands of souls are entering into the yellow crystals which seem to expand and grow and absorbing more. For Paul, Michael and those embedded in the blue crystal, there is no escape. They huddle together hoping and praying that somehow they could find a way out of the crystal fields that they so longed for just a few moments before.

"There is one hope" the priest says.

"What is it?" Michael asks.

"The colors of the crystals are unique," the priest starts explaining, "the crystals really have different colors. They become yellow when they are close to one another to form the combined yellow looking field. They must be assembled by an exact sequence of one hundred and twelve steps for the entire symmetry to become active."

"So, if they are not assembled correctly, it will not have its power?" Michael asks.

"No, it becomes a new symmetry that has other types of effects on humanity."

"What type of effects?" Paul asks.

"Remember the Eye of M'Dulu?" Michael asks.

"Yes," Paul responds.

"It gives Hope to all that it is a light of redemption. That is why the Poro have been keeping it, to give them Hope that the future can be controlled. It attracts people and captures their souls like a magnet of Hope. I know that it captures good souls and is a hopeful field." Michael explains.

"Why hope?" The priest asks.

"You said yourself it is our only hope. It is hope that brought us all together. You hoping that you can get the covenants back, I hoping that I can find the codes to the field, John hoping that he could find out what the field is about. That is all we have now." Michael explains.

"There are other fields that attract greed, anger, hate, power, lust etc., and a mixture of emotions that all combine to form the yellow field."

John Davis is not amused by Michael's explanations.

"Hope, Love, Lust and all that emotion business has nothing to do with it!" He says.

"Then, what do you know? Why did you come to the jungle to get the crystal?"

"I studied the crystals from space. We know that they are a scientific weapon of tremendous power built by someone or something other than humans," he explains.

"That we know already! No human can build these powerful fields. But a scientist called Nambu and I studied the fields in Nevada, and created a theory to explain them. They are a twenty seven-fold symmetry," the priest explains.

"Then, how do we solve this? How do we get out of this and save ourselves?" Paul asks.

"The US government and a powerful group called the Bilderberg Group has the answer. I installed some programs in a satellite to do just that!" John Davis states.

A faint memory enters Michael's mind. Lucifer's words resonate in his mind as the other talk, "you are one of us". He pries his mind open, and feels memories flowing through like a river of information. He looks at Paul, he is no longer of human form. He is an angel. A creature like Lucifer. He knows he is also an angel.

"I can do it!" he say out loud.

"Do what?" John Davis asks.

"I can get out of the field. I am just as powerful as Lucifer."

He starts recalling his entry into the physical world.

"Who are you Michael?" the priest asks.

"Some time ago, I was given a mission by Master M'Dulu. I was told to enter the human realm and start the gathering process for the crystals, especially the jungle crystal. In human form I was a young man that could not remember my life. I grew up just like a normal man and worked for the diamond mining company". Michael explains.

"Then, the accident happened in Yamandu!" John Davis blurts out.

"Yes, I was there. That was where I was nurtured by the Poro and given hope and brought back to life to continue my mission."

"So, you came for this mission?" The priest asks.

"Yes, to redeem man from bondage to Lucifer. To free man and recover the Crystals of the Covenant," he says.

"What are the crystals of the covenant?" John Davis asks of Michael.

"The crystals are vehicles that can take us to different dimensions. To ordinary man, these dimensions appear as emotions, but to us angels, they appear as real spatial trajectories." Michael explains.

"So, if these are crafts that make you and your kind travel through emotional dimensions, it means you and your kind can control man and his emotions?"

"Exactly. We can tempt man, and make him feel any emotion that will make him suffer or gain. That is how angels and demons act on men, through these emotional dimensions." Michael explains.

"Then, how do we get out of here? We are trapped in fields of Hope, hate, anger, and other human emotions. How can we escape from our own emotions?" Paul asks.

"Faith and Love!" Michael says, "We are in the field of Hope, that is the key".

"Hope?" the priest asks.

"Yes, it is the key. It is the covenant we have to make. Only that can save us now," Michael explains.

Lucifer overhears the words and quickly approaches them.

"The procession of my people cradles the crystals like infants in their hands. Look at the crowd of souls walking through the massive crystalline portals of my new kingdom. It is mine now, Michael. Even Gabriel cannot stop me, even Hope cannot stop me now!"

The angel knows that even with the enlightenment and vitality that the cactus potion has afforded the holy men is still not enough for them to free themselves from their own Hope.

CHAPTER 41
(The Pope awakens)

Pope Paul the Seventh is in seclusion. No one knows where he is. He kneels alone over the grave of St. Peter praying for redemption of mankind. In his hand is a scroll he had written when the guardian appeared to him. He gets up from his knees and briskly starts walking toward the secret underground entrance to the Sistine Chapel. He knows there is no one alive in the Chapel. The bodies of the cardinals lye on the floor of the Chapel in an unfinished conclave. The others must have gone to Nevada as planned and his brother is safe from harm. Outside the Chapel, he could hear the moans and grinding of teeth that foretells the end of time. This is the end of man and earth. He walks to a small corridor now filled with smoke from the burning chapel.

"White smoke," he says to himself, " a new pope of science and mathematics."

He has an appointment with the Guardian that appeared to him in the Vatican a long time ago. He squeezes through the narrow entrance to the Sistine chapel and starts the arduous climb to the top. His steps are calculated and careful, and he cannot fall or loose the scroll.

"I believe in one God, the father almighty, creator of heaven and earth..," he mumbles the Nicene creed. He climbs the slow spiral stairway between the chapel wall and the protective outer wall of the tower until he reaches a spot where he must squeeze hard to get his frame through the corridor. No one has crossed this spot after it was closed off for repairs ten years ago. His robe rubs against the masonry walls dusting off the dirt and cob webs that have settled over time. The passage is getting tighter and tighter. He prays that he will make it and pass through the tight corridor. He must. The smoke fills his nostrils and eyes and he could barely see with the dark goggles he is wearing to protect him from the rays of the J-N field. He pauses exhausted. He feels trapped by the walls as he squeezes his frame even tighter to get to the designated passageway. In the distance, he sees a small crack on the wall, but the smoke filling the Chapel makes it impossible to decipher if it is a door. He is puzzled. The guardian could not be right about this path to the top of the Chapel. He has an overwhelming feeling that he is being trapped by some force. His goggles are making it difficult to navigate the tight passage. His frame has reached a limit, and can go no further.

He cannot give up. He takes a deep breath and then exhales narrowing his frame tighter. He holds his lungs down and squeezes some more. He is now fully captured between the two walls and can go no further. He cannot turn back now. The flames are approaching fast and the smoke will make it impossible to enter the Chapel. Suddenly he feels a relaxation. The wall is giving and the passage is expanding wider. He feels relief. A crack opens a stone door in the masonry wall of the chapel. The door opens wider, exposing a balcony. He pushes into the balcony and takes a deep breath inhaling smoke. He coughs. His cough is magnified by the curved walls of the Chapel, becoming a loud grunt, as if some alien creature expelled it. He looks up at the ceiling and sees the paintings of Michael Angelo. God reaching out to man. The first covenant, Adam and Eve, the second covenant Abraham. He sees Ezekiel. He is surrounded by the immortalizing of the Covenants between God and Man. A red light has changed the paintings into a bloody blur. There is no one in the Chapel below but the crackling of a hellish fire.

"Where are you M'Dulu?" he shouts out.

He looks again at the ceiling and sees God reaching out to man. He sees Adam falling as the ceiling of the Chapel cracks, but it too late. The hand of God falls with him eternally reaching out to save his most elegant creation.

Suddenly, a powerful bright green pulsing light fills the Chapel. The pontiff falls to the floor of the balcony overlooking the Chapel and moans in pain. The scroll drops to the floor and skids to the edge of the balcony. A breath will make it fall into the flames below. He reaches out to grab it, but a hand grabs it ahead of him. He looks up and sees a man holding the scroll in his right hand proclaiming in a loud voice,

"Who is worthy to break the seals and open the scroll?"

He weeps uncontrollably as he realizes that the keeper of the scroll has come.

"I am not worthy, only the lord can open the seal. Then, the man says to him,

"Do not weep! See, the Lion of the tribe of Judah, the Root of David, has triumphed. He is able to open the scroll and its seven seals."

"Who are you?"

"I am the al Mahdi. I am the keeper of the Crystal of Faith."

"The Mohammedan Covenant?"

"Yes. I have come to assist you to defeat the others and prepare the way for Jesus to judge the world."

"But how?"

"Come with me and enter the field of our Covenant of Faith and you will understand."

The Pontiff reaches out and the man lifts him up from the floor of the Chapel.

"Our people have been fighting ever since our father Abraham lost his covenant to the Guardians of Eda. But, we have each kept our faiths and covenant of faith with the Master. You as Christians, and us as Muslims. Together we have Faith and Love."

"How then shall we save man from his iniquities?'

"We must find Hope. This shall save man from his iniquities and so shall the evil ones be defeated." The man says.

"Are you the Guardian that appeared to me in the Vatican?"

"I am the father of the nation of Islam; whoever knows me very well also knows Allah, and whoever denies me also denies Allah, the Unique, the Mighty. And from Ali's descendants are my grandsons al-Hasan and al-Husayn, who are the masters of the youths of Paradise, and from al-Husayn's descendants shall be nine: whoever obeys them obeys me, and whoever disobeys them also disobeys me; the ninth among them is their Qa'im and Mahdi. Yes, I am al Mahdi, Keeper of faith, the preparer of the way that appeared before you."

"Then, your instructions have been carried out as you requested. The Crystal of Love is kept for you in the mountains of the dam as you requested." the Pontiff explains.

"Let us go and claim the crystal of Love and Faith.

Let us go and wager our faith and love for man," the al Mahdi states.

Slowly like molasses, the crystal's field flows into Pope Paul's raised hands. The substance is able to help the pontiff stretch his super elastic hands further into the field. He enters the field of faith and opens a dimensional space. As the Chapel flies in all directions from the expanding field, Pope Paul reaches out into the crystal's twisted fabric of spacetime and holds the hand of the al Mahdi, he has entered into a dimension of faith, leaving a lifeless body on the balcony in the Sistine chapel. In an instant they are zipping through the dimension of faith toward Nevada. It is an astral journey. They are going to a place of faith not a physical place.

In the far off desert of North Africa, the battle cries of thousands of crashing IFPs follows the wailing of Mingas as they are absorbed into the field of light. The guardians are busy isolating themselves into the control consoles of Hate, Despair and other evil emotions that will control man and his destiny.

The light in Resa's eyes is not ordinary. It is alight with Hope. The Mingas are dying everywhere as they enter into the field of hatred the guardians have prepared for them. Akhenaten, Osiris, Zoroaster and Ra are also feeling the light of Hope.

"Let us help man, for we are half children of men. We are the fathers of their technologies, and we have made them believe in us," Resa says to the others of his kind.

"They gave me hope to return to the realm when they built the pyramids to align with Orion, they made me hope." Osiris states.

"The Guardians are busy with the light of hate and despair, adjusting their craft for their own salvation. We must act now!" Ra insists.

"Then, let us disengage the locks for the vehicle of Hope and enter it. For it shall give us hope and man with it." Resa explains.

"But how can we do that, the attraction of the fields is too high," Osiris asks.

Michael looks around to the others for an answer. Father Mulcahy smiles.

"Hope is a field with two faces like a magnet has a north and south pole. If we reverse the polarity, the attraction becomes a repulsion and the fields will separate." he explains.

"But how do you reverse the polarity of Hope?"

"Despair! If they are desperate enough, there is negative hope, but it is of a different polarity than Hope itself. The crystal field can rotate and repel the others" the priest explains, "this is exactly what the J-N equations predict."

Michael turns his attention to the children of the Guardians.

"You Despair, and on the other side of that is Hope. Enter into the field of desperation it is the field through which you can enter into our field."

They rush to the center of the lower plane of the cubic field and spread themselves out searching for the field of despair. Osiris enters into desperation. He feels it. He urges the others to follow quickly. They must disengage the locks of the despair vehicle to release the Hope vehicle. Osiris enters the barrier between the despair and hope. He looks to the blue field and shouts at Michael, and the others in the blue field.

"Fill your hearts with hope for us, and do not despair or hate, fill your minds with the clarity of love, and the yellow field will have no power over us."

In a fantastic display of rainbow colors, the blue crystal starts to rotate. As it rotates, its field slowly starts to disengage from the yellow crystal field. As the fields disengage, Ra,

Zoroaster, and Resa rush toward the blue field of hope and find resistance as it separates itself from the field of Despair.

"Hope! Resa, hope! Immerse yourselves in Hope and the barrier will vanish," Osiris shouts.

Michael and Paul reach out and grab Resa and Zoroaster. Ra crosses over. Paul starts to pray,

"Open the gate to M'Dulu's eye
so we can bath in the light of His Hope
so, we can smell the perfume of His wounds.
So, we can fill our lungs with His breadth
So, we can gain the wisdom of His eye"

The other humanoids are already crossing the field attracted by the field of Hope. The rays of the rest of the crafts starts to flicker uncontrollably as the blue crystals of Hope isolates itself from Despair, Hate, and the other vehicles of negation. Thousands of Mingas are crossing over into hope, filled with the light of hope to rescue their own souls. The Guardians notice that the craft is becoming unstable. They rush toward the blue field only to see their children crossing over. Confused by the sudden instability, they hurry toward the control and disengage the blue crystal. The Hope field has reversed polarity and it is no longer being attracted by the other evil fields. If they do not disengage the Hope field now, it will consume them and rip the rest of the craft apart.

Cursing, Lucifer unleashes beams of light aimed at the barrier to stop them, but there is no hope left in him.

"You shall not be able to traverse Hope and find your world!" He shouts, "Your journey without the other dimensions is one dimensional and it will be dark and desolate and you will despair and come back to us!"

Only the priest understands what he said. Desperation, loneliness, desolation, a dark world that he has been in before in an astral dream.

"Brace yourselves," he tells the others, "it will be desolate and full of despair. Despair will be like little nails attracted by the magnet of Hope."

"What do you mean?" John Davis asks.

"We do not understand," Bumba says breaking his long silence.

"When the crystal of Hope separates from the rest, we will be left with only hope and spacetime. We cannot navigate clearly in such circumstances. We will not know how to find our way around until we learn." The priest explains.

"And, until we learn, we will be fighting desperation, anger and other emotions that want to brace themselves to our field of Hope." Michael adds.

"Yes, that is why we must not despair, even if we cannot find our way, we must have faith," the Priest concludes.

The priest unleashes prayer after prayer. Michael and Paul also start praying for Hope. They fill their craft with repulsion for negative emotions and encase it in Hope. Thousands of Mingas have saved themselves by hope, and the Children of the guardians are now fully immersed in the field of Hope.

The craft separates itself from the yellow field and in a burst of energy, the repulsion propels Hope far into the depths of the desert. They are free at last in a desolate world of only Hope. As Hope and Faith journey through the fabric of spacetime, the rest of the awful emotions that have trapped the essence of man remain transfixed in the deserts of North Africa, battling to control their world with no Love, Faith, or Hope.

CHAPTER 42
(The Keeper of the Laws)

The al Mahdi and the Pope are immersed in Faith and finding their way to astral Nevada. It is difficult without a map of Faith. They may hit there mark in spacetime but they could be far from it in the faith dimension.

"Why is it so desolate in the land of Faith?" the Pope asks.

"It is bare, because there is no Hope and no Love, we are looking for the faith of the council of Naica and Shahada in Nevada, and there is none!" The al Mahdi answers.

Resa belt straddling the earth is now blazing with colors of yellow, blue and red, melting into a white crystalline display over the entire earth and its inhabitants.

"Your faith is what makes us enter this field of light. But without any other faithful, there will be no place for us to go." The al Mahdi states.

"There is faith, my lord. There is faith that will be found in this desolate space of emotions," the Pope states.

"I must prepare the way for Jesus. He shall arrive when we are gathered for the final judgement."

"Give us a chance, Oh preparer of the way. Give us a chance." the Pontiff pleads.

"Then, find me the faithful that I may spare them and redeem them," the al Mahdi commands.

Pope Paul's heart goes to his brother, he could find love, but not faith. He reaches into his memory trying to find someone or something that he could hinge his faith on. As his mind feeds the crystal of faith, it starts to shine with a bright red light, that tries to reach all those in the dimension of faith on the planet.

Far in the distance of faith, the occupants of Hope try to struggle with their new found dimension. They must find a way to the other two remaining Covenants of Faith and Love.

"We are traveling through hope. Where shall we go to save man and our own souls?" The Priest asks.

"Nevada," John Davis states, "We must go to Nevada."

"What is in Nevada?"

"The Bilderberg group. A collection of over one hundred leaders of man, including the Pope, Presidents of the world, Kings, the Jewish Leaders and Islamic leaders." he explains.

"What are they doing in Nevada?" Paul asks.

"They have been planning for this day. We must go there and find our answers." John states.

"But how? If what he says is right, then, we are traveling in a spiritual dimension of Hope, no one knows the way," Paul states pointing to the Priest.

"Have Faith, Paul, you must have faith in the Lord!" the Priest states emphatically.

"Even when we are blinded by our own emotions and hopes, we must have Faith," he adds.

Paul starts to pray,

"Open the gate to M'Dulu's eye

So, we can bath in the light of His Faith

So, we can smell the perfume of His wounds.

So, we can fill our lungs with His breadth

So, we can gain the wisdom of His eye"

"Hope is desolate without faith, and we could only see Hope with faith, and have faith in Hope, " Michael says.

Michael spots the red light flickering far in the distance. He points to it and the others look. Everyone is able to see the light but John Davis.

"Then, anything we see in this dimension must be a light of faith or love?" Paul states pointing to the distant red light Michael has spotted.

"Faith!" Michael cries out, "we must join with them. If we can see the light, then it must be the light of the Covenant of Faith."

The Hope crystal starts tugging as the field of faith starts to attract it. The two crystals commence a dance of attraction like two magnets that seek each other's embrace. In an instant, the two fields merge, and all of a sudden, the inhabitants of both fields are able to see each other and the barriers between Faith and Hope fade into reality. They are now immersed in three dimensions of Space, one of Time and one each of Faith and Hope. A six dimensional world of reality that they could travel through.

As the melting of the two fields completes, a faint shadow appears to the priest. He squints and strains to make out the image. Then, it clears, and the priest, Michael, Paul and Bumba recognize the Pontiff.

"Your Holiness!" the priest exclaims.

"Father Mulcahy! Our faith has brought us together," The Pope exclaims to the al Mahdi, embracing the Priest. John Davis could not see the Pontiff or the other occupant of the other craft.

"Who are you talking to?" John Davis asks of the priest.

"The Pope! I can see the Pope, and another man because we have the same faith in Christ," he explains.

"We could all see them, but you John, your faith is in science not in God," the Priest explains.

"But I have Hope, I have hope that I can see my lover again!" John Davis exclaims.

"Love!" the Pontiff exclaims. "He has love in his heart, and he can lead us to the dimension of love!"

"Use your dimension of Hope and Love to see them, John, man does not live by faith alone, but by hope and love also."

They still have to find astral Nevada! Michael approaches the Pontiff and the al Mahdi. He looks into the eyes of the al Mahdi and a sudden recognition fills his soul. He falls on his knees trembling like a child.

"Save us oh YuMuYa from this calamity that man may begin anew".

"Give me a reason why man should be saved, Michael."

"They fell in weakness. They made Covenants with us that they will be saved when they deliver the jewels of your glory," he pleads.

"But man has given all the other covenants to the evil ones. The evil ones have amassed all the covenants and now only three remain to be given by man."

"We have not yet failed my Lord. We can triumph. We have Faith, Hope and Love."

"Then, show me love, that I may tell the Master that Love, Faith, and Hope remain in Man." The al Mahdi commands.

Michael looks around and sees the men huddled in fear of the al Mahdi. A powerful presence surrounds him and he is different from all others. He turns to the priest. The Priest has no one he could connect to on earth by the dimension of Love other than Jesus. He turns to Bumba, but there is no one Bumba would care about. He turns to Resa, and the other children of the guardians and finds no one they could connect to with love. Only John Davis could connect to his children and family, but John Davis could not see or hear the Pontiff and the al Mahdi, they are in a dimension in which he is deaf and blind.

John Davis is their only Hope. The connection is doubly weak since John Davis has had a relationship with the Poro and Zola and has almost forgotten he has a family, and he could not connect to the dimension of Faith. If Michael could get John Davis to open up the line of Love, they could connect to Nevada.

"John, your family, where are they?"

"I do not know where she is."

"Help us John, do you remember the name of your wife?"

"Zola," he says.

"No! The name of the one that gave you the children. Do you remember her?"

A faint memory seeps into John Davis's mind, but it is faint.

"I have a wife and children. I have a wife and three girls." he says.

"Yes, John, and where are they now?"

"I do not know!"

"He has no love in his heart, and he is blind to Faith" the al Mahdi claims.

"John, before you came into the jungle, where were you?"

"I was in the USA," he answers.

"You came to the jungle and when the Minga attacked, you were abducted by Yabu Yombo. She gave you Zola and you drank the potion that wiped out your memory of your wife Elaine." The priest reminds him.

"Elaine?"

"Yes, Elaine. We contacted her to find out where you were when you disappeared from the USA." the priest explains.

"Elaine, she is my wife?"

"Yes, John."

John Davis scours his memory finding a link. He looks at Michael blankly expecting him to open up a memory of love.

"John, you are the only one with a connection of Love with the world now. Look in the distance, do you see a Green light of love?"

John Davis looks blankly at Michael. He is trying to remember. Resa approaches him.

"You must save us, we have hope and faith in you." he says.

Osiris sees the ring on John Davis's finger. It is a scarab beetle!

"The human has the immortal soul of the great Khepera, who pushes the shining sun like a star to brighten the day," he states.

A faint memory enters John Davis's mind, and he sees an image of Elaine holding a chain with a beetle pendant hanging on her chest.

"Beetle, Beetle on a rose,

An archeologist broke my nose!" John Davis says, remembering his words to Elaine. His words when they first met! He feels his heart pounding with anticipation. He is connecting to Elaine.

"Beetle, beetle on a rein,

my heart is bonded to a chain".

A flash of bright green light emanates from the pendant as the emerald on his finger comes to life. The link is complete.

CHAPTER 43
(Generals and wars)

General John Bradley has a dilemma. All round Area 51, caravans of the Amish are closing in around the fences. Their boogies assemble in an orderly manner surrounding the site. Already, one hundred and twenty five thousand Amish have surrounded the entire site, and they are eerily silent. Thousands more are approaching, mingled with thousands of other people roaming aimlessly outside the fences. The intensity of the green crystal is increasing, and already, the light from the crystal is spilling into the surrounding desert, painting an eerie green landscape that is not reserved for a barren desert.

The General has ordered his men to shoot and kill anyone that breaches the barriers and over two hundred dead bodies lie around the fences in accordance with his wishes. The desert air is hot and humid, and the black robes of the Amish appear like a funeral procession of unprecedented proportions.

"Mr. President," he interrupts the meeting,

"The Amish are flooding the site! We do not know where they are coming from, but already there are thousands of them trying to enter the gates." He states.

Elaine and the children are sobbing. They have just seen John Davis and a few others killed by a flash of bright Red light in the jungle. Effion and Nambu are standing by the heavily shielded window listening to the conversation between the President and the General.

"The intensity of the crystal is increasing and we cannot contain it. What do we do?" the General asks.

"The Council wants to release the barrier. They are afraid it might explode and destroy the site and everything in and around it," the President states as the first signs of fatigue enter his voice.

"But if we release the barriers we can no longer control its rays. It will affect everyone. It will hypnotize everyone." the General states.

Nambu turns to Effion.

"Aliquot thirteen, Effion, what is the Hasse diagram for aliquot thirteen?" He asks.

"The quark symmetry, it is a quark symmetry," Effion answers.

"Can you calculate the frequency of the rays and predict the maximum emissions?" Nambu asks.

"The Seth space parameters are disjoint. The angular dispersion is 260 degrees, the spectrum of the Jordan matrix is random." Effion answers.

"What is the negative value of the sum?"

"Six hundred and sixty six, six hundred and sixty six, six hundred and sixty six." Effion chants like a child with his torso moving back and forth rhythmically.

"What?" Nambu exclaims.

"Six hundred-sixty-six," Effion repeats.

"No! It cannot be, Effion. That is too high a value and there must be three distinct values, it is a cubic." Nambu states convinced he is correct.

"Six, six, six," Effion states.

"You mean each representation is six?"

"Yes, that is correct." Effion states.

The autistic man starts to convulse and he droops to the floor and starts to enter into a catatonic state.

"His medicine," Nambu cries out, "he needs his pills".

A guard rushes to Effion's aid and searches his Papal garments for the pills. He finds the pills and forces one into the mouth. A few moments later, Effion recovers and sits on the floor weeping like a child. The General is getting impatient.

"Mr. President, we must release the gas. That is the only way we know to calm the crystal's emissions," he grunts.

"No! We must wait for the Council to decide."

"But thousands are approaching the site. They will soon break the fences and start to overwhelm us." the General states.

"Get more troops!" the President orders.

"We cannot get enough goggles for the troops. They too are being affected by the rays. We must release the gas!" He insists.

"The gas will kill everyone outside of the site. We cannot do that."
The President insists.

"Then, we will all perish!" the General shouts angrily spitting on the polished floor. President Cooper approaches Elaine, disregarding the insolence of his General. He places a hand on her shoulders and speaks in a quiet voice.

"We need the codes Elaine. We must reprogram the satellite right now, or all these people and us will die."

Elaine pulls the girls close and moves to the corner of the room close to Nambu and the sitting Pontiff.

"Save us, Pope Paul, give us communion and save our souls," Elaine pleads holding the girls close to her.

Effion looks at the woman and smiles.

"Give them the codes!" he commands confidently.

Elaine believes him. Something about Effion makes her believe that there is Hope. She walks over to the President and repeats the code quietly. He removes a small sticky note pad from his jacket and writes it down.

The rays of the green crystal are becoming more intense, and they are now slowly filtering into the heavily protected room.

"The quarks, they are coming!" Effion states robotically, looking out the window into the dark skies.

Everyone turns their attention to the window. A pair of bright red and blue orbs are approaching the site from high in the already darkening sky. Nambu looks at the orbs slowly approaching the site, it is a wonderful site. All around thousands of people are roaming aimlessly about in the intense green light, while the entire Amish population fixes its gaze upon the blue and red skies.

"It is a symmetry is it not?" he asks Effion, "a quark symmetry?"

"Yes, a cubic symmetry, " Effion answers.

"But how is it a cubic symmetry and a quark symmetry at the same time?" Nambu asks.

"The u, d, and s, quarks are different from charm, top and bottom quarks. That is where six comes from. The three colors of the six quarks is three sixes, 6,6,6."

"And the three quarks u, d, s are what we see in matter, and what we can see as the red, green and blue lights of the crystals?" Nambu asks as he realizes the genius in Effion.

"Yes, the three quarks, the trinity symmetry. The box of twenty seven opens and their light spills." Effion states.

Then, suddenly as if compelled by some unknown force, the entire Amish population starts approaching the fence. They pound on the fences in a renewed frenzy of an awoken army caught in a forest fire. General Bradley gives the command to fire at the crowds. It is too late. In an instant, the building starts to vibrate with the fury of the green crystal. As its symmetry approaches, the green crystal awakens and starts to expand its field filling the entire site with a deep intense green light. The blue, and the red crystals accelerate toward the site, attracted by the unearthly power that bonds them to their brotherly symmetry.

The entire site starts to vibrate.

A guard appears in front of the door to the room.

"Mr. President, Your Holiness, it is code RGB," he states, "the time has come and the Council demands that you enter the chambers now, Sir!"

The President understands the coded message, but Effion does not. It is meant for the Pope and those allowed a final pass into the secret site in Area 51. It is time for the President to go. He quickly scurries out of the room and runs to the conference chamber. Elaine and the children follow him. As they enter into the crowded conference room, Elaine and the children are stopped by military guards. They push them out of the conference room back into the closed meeting room they had come from and the door slams shut behind them. Nambu and Effion are still in the room, Nambu aware that he could not go with the chosen ones.

"They will not let you in, Elaine, only a few have the key to enter the crystal," he explains.

Several of the Bilderberg group are now being ushered through an exit door from the conference room. The President falls in line as the procession of Kings, Presidents and other dignitaries enter the corridor to the prepared chamber below Area 51. They are ushered into a series of elevators, each with a coded key they carry in their person. The President puts his key into elevator 2's door and enters. His wife and children are waiting in the elevator together with other families that have the same coded key. The elevator rises and takes them to a waiting room. There, they each receive a pair of goggles. The elevators return empty and the one hundred members of the Bilderberg Group with their families are left alone in a secret chamber. The president enters the satellite code into a remote control and activates it. An automatic program commences a sequence, and the huge concrete wall surrounding the crystal starts lifting, exposing the fantastic green rays of the crystal.

The crystal is surrounded by a line of Rabbis all dressed in long robes in the traditional Jewish manner. In their hands they hold the reins of horses at the ends of which are bright creatures with appearance of horses of light. The Rabbis chant and sing praises to God,
"Holy, Holy, Holy,
Lord God of Hosts,
Thy glory fills all Heaven and Earth,
Hosanna in the highest,
Blessed is he that cometh in the name of the Lord,
Hosanna in the highest."

The members of the group start approaching the crystal, as if compelled to enter its sacred field. The cold rays of the crystal is slowly freezing their bodies and they now attempt to enter the crystal, but they are getting weaker as the crystal sucks the life force out of their bodies. Only children of the Bilderberg group are allowed to enter the crystal. The Rabbis remove the frozen bodies of the adults, one by one and pile them beside the crystal. With the little energy left in some of them, they try to fight their way into the field of light, but one by one, they succumb to the hypnotic rays of the freezing field and start to walk aimlessly around the field as if compelled by the unknown hypnotic force.

Nambu, Effion, Elaine, and the girls are now alone in the room. All military personnel have vanished. The power goes out leaving them in eerie darkness. The girls start to cry hysterically. Nambu pulls Effion and they come close to Elaine and the girls.

"What will they do with the codes?" Elaine asks.

"They will activate the satellite to change the colors of the crystals." Nambu answers.

"Will it work? Will it save us?" Elaine asks.

"I do not know, perhaps it will, perhaps it will destroy us all!" Nambu replies.

Elaine mulls over the ambiguous answer. The scientist wrote the programs for the satellite, but even he knows nothing about the crystals. They are doomed, abandoned by their own President.

"Stay close to me," Nambu commands them, "the emergency lights will soon come on."

They all sit down on the floor waiting for the lights to come on, but nothing happens. The darkness is becoming unbearable.

"What can we do now?" Elaine asks.

"Wait and see what happens. If we go out there, they will trample us to death. In here we are shielded from the rays of the Green crystal." Nambu explains. Effion is becoming restless. It is getting colder and the girls huddle close to their mother for warmth. Nadia is trembling, she is wearing a thin blouse that cannot keep warm. The other two girls have warm clothes on. Nambu removes his leather jacket and hands it over to Nadia. She thanks him gratefully and Elaine nods at the act of sympathy.

"My brother has a light, we must ask him for the light." Effion mumbles.

Nambu checks to make sure Effion's medication is in his pocket in case he needs them. Effion may be their salvation. The far off Red and Blue lights are becoming brighter and larger in the sky. Elaine clutches her scarab beetle and thinks of John. It starts to blaze like a green star about to burst forth in a fiery birth. She huddles close to the girls and hugs them as the green light engulfs them and brightens the room.

A faint voice comes through into Elaine's mind,

"Beetle, beetle on a rein,

my heart is bonded to a chain".

"John!" she whispers, "you are alive!"

A voice answers back from the dimension of love.

"Beetle, Beetle on a rose,

An archeologist broke my nose!"

Elaine grabs the shining scarab and shouts excitedly,

"He is alive, John is alive!"

Nambu stands up and pulls Effion up, but Effion grabs unto Michelle's arm and would not let go. Michelle starts to pull away from him, but he still will not let go. Jessica grabs Effion and starts pulling his hand away from Michelle. He still will not let go. Nadia is crying.

"Get up Effion and release the girl!" Nambu commands. Effion releases Michelle and starts rubbing his hands together vigorously.

"John Davis, did you say he is alive?" Nambu asks Elaine.

"Yes! Yes! He is in the lights!" and she points up to the approaching orbs in the sky.

"Dad, dad!" Nadia exclaims, looking around to see if John Davis would respond. Michelle and Jessica are huddled close to Elaine keeping away from Effion.

"Effion, what is the nature of the blue light?" Nambu asks,

"Three of sixes, a tertiary symmetry of quarks," he answers, repeating his previous answer.

Effion grabs Elaine's hand, and she pulls away and calls on the girls to follow her. She starts to run toward the windows. Nambu and Effion follow her. The voice is coming from the blue light. She feels the compulsion of the dimension of Love, but she cannot find a way to get to it.

"Elaine, can you hear me?" John's voice is becoming clearer.

"Yes, yes, I can!" Elaine shouts.

"Who are you talking to?" Nambu asks.

"John! He is talking to me!" She answers excitedly.

"Elaine, you and the children must go to the green light." John states in a voice only Elaine could hear.

"How?" she replies.

"Take the girls to the green light."

"How do I get into the green light?" Elaine asks Nambu.

"The only way is through the study chamber," Nambu states, "but it is closed off."

"We must find a way, John says we must go to the green light." Elaine explains.

Nambu leads the way to the study chamber. It is dark, but Elaine's scarab is lighting the way. They pass several security doors with people lying on the floor exhausted by the frenzy to get into the crystal, while some are still walking aimlessly hypnotized by the rays.

"We have to go back into the conference room, it is protected from the rays of the crystals, we will die here if we do not go back now." Nambu blurts.

"We are not going back there, we are going to the green light!" Elaine states emphatically.

"Why do we not feel the hypnotic rays?" Nambu asks.

Effion points to Elaine's scarab, it is shining with renewed intensity.

"Green, Red and Blue makes a white spectrum." Effion explains as he rhythmically rocks his torso back and forth.

"So, we are protected by the combination of lights? A resonance of a simple crystal of Egyptian emerald?" Nambu asks.

"Yes," Elaine states, "It is love. John is looking after us."

Nambu does not pursue the matter, at this point, his scientific judgement is faltering and he knows there is more to this than just science. There is a religious component that he can no longer ignore.

He recalls the Pontiff explaining about the power of the crystals and now realizes that there may some truth to all this religious business. They approach the entrance to the observation room that Nambu has been working in for the past ten years trying to understand the power of the crystal. He has not gained any further experience that could help them now, other than the discovery of the J-N theory. He might as well have discovered a quantum theology in the theory. The Pontiff must be right, his faith is increasing.

When they enter the observation room, a bright blue light fills it, as it spills through from the small opening that allows the rays to pass through to their equipment. The rays of the crystal are fluctuating as someone or something moves about on the other side of the wall.

"There are people in there!" Nambu exclaims.

"Is that the crystal that John and all of you have been studying all this time?" Elaine asks.

"Yes, this is the best kept secret of the twenty first century." Nambu explains.

"Do you know how to get into the field?" Elaine asks.

Effion is peering into the opening directly with his goggles removed.

"It is beautiful," he cries out, "Oh so beautiful".

"Effion, you must put on your goggles now", Nambu states, forcing the goggles back on Effion's face.

"How can we get to the other side?" Elaine asks again.

"If we can get the computers up and running we could open the lab panel to the crystal. I must warn you that no one has ever gone into that side of the wall and come back." Nambu states.

"We can try to squeeze through the observation hole," Jessica suggests.

"No, you are not going in there without me." Elaine shouts.

They hurry toward the bright light coming through the opening in the lab wall. Nadia reaches it first, but the intensity of the light is too great and she is temporarily blinded. She turns away from the light and removes her goggles to rub her eyes.

"I see people!" Nadia exclaims, "there are thousands of people all around us like ghosts!"

"Where do you see them? Where?" Elaine questions.

"Everywhere, they are all around us and going into the crystal," she explains.

Michelle removes her glasses and so does Jessica. Nambu rushes to the girls and shouts for them to put their goggles back on.

"I can see people too!" Jessica exclaims.

"So, can I!" Michelle states.

"They are like ghosts everywhere walking to and from the crystal!" Jessica explains.

Elaine removes her goggles to see if she can see them too. She cannot, but she does not wear the goggles again.

"All I see is the four walls and the green light. I do not see anyone," she explains.

"They are hallucinating!" Nambu explains.

"No! They are telling the truth. They are innocent." Elaine explains.

"Nadia, can you show us the way? Can you see a passage to the crystal?"

"Yes, we have to go outside, the crystal is not behind the wall, it is up in the sky!" Nadia explains.

"We cannot go outside, we will be consumed by the rays, and get trampled!" Nambu exclaims.

"No, the crystal is good, it will not harm us." Michelle says as if she is in a trance.

Nambu pulls Effion by his arm and leads them to the door that opens to the outer walls of the site. Hundreds of military guards lye dead on the floor of the hallways and rooms and there is no living soul in site. Outside, they could hear the cries of horses and their Amish masters.

"They were killed by the rays from the crystals." Nambu states, referring to the dead lying in and around the building.

"We cannot go out there, we will all die," Nambu emphasizes.

"Then, you stay here in the dark and die, I am going with the children to the light, and we have faith in the light of the scarab to guide us." Elaine states.

She hurries and grabs Nadia and Jessica's hands and commands Michelle to follow behind. They leave Nambu and Effion behind and open the door to the outside. Outside, they see an army of children entering the site, protected on all sides by an Amish army. The Amish have formed a barrier to the site and no one can pass but children. As the Amish children file into the site, thousands of other people gather like flies around the site, attracted by the light of the blue crystal. They try to force their way into the site, but as they approach, the Amish army pushes them back with their horses, bats and sticks. They are protecting the site from the invaders. Elaine and the children rush to join the children of the light. No one can stop them now. Soon they are following the children into the light. They are no longer afraid.

As they approach, they see the red and blue crystals looming large in the sky and behind them. The blue crystal is slowly rising from its burial place within Area 51 to meet its companions. As waves of children approach the light, they vanish into the crystal's brightness leaving behind the adult bodies that try to enter its field.

"It is beautiful mom, it is beautiful, " Michelle shouts.

"Where is dad?" Nadia asks.

"He is in the light" she answers.

"Go, Michelle, Nadia, Jessica, go! I cannot go, it is for children only." Elaine states.

"We can't leave you behind mom, come with us" Jessica states crying.

"No! You must go now, the light is rising, you must go!"

"Why, mommy, why can't you come?" Nadia cries.

"Dad wants me to stay here, he want you to go to the green light." She explains. They hesitate. Then, an old Rabi approaches them. He is holding a rein of light and a creature that appears like a horse made of light.

"Only the children can enter," he states. Elaine moves back and pushes the children forward, nudging them to move to the light with the old Rabi and the other children. She turns and runs back toward the building. The children look back, Elaine has vanished into the building and they vanish into the light.

A fantastic explosion is enveloping the site. The girls start to run with the other children. A warm green glow engulfs their bodies and they become as light as a feather, as the

Earth recedes from their feet. They feel that they are upside down, floating in dirt-filled air. Then, they are propelled at tremendous speed to an unknown oblivion, as if a gigantic balloon is being blown beneath them. The whole site disappears into thin air.

The children huddle together and they feel themselves being propelled through the dusty atmosphere by some gentle inexplicable force. As they follow the path of a crazy trajectory, they could see the rotating site below, heading towards space. They do not feel any gravitational or acceleration forces. This is a force that melts spacetime into a blur of higher dimensions. Then, their bodies begin a graceful descent to the ground of the site obeying the common law of gravity. They stop descending two hundred meters above the site, with debris below them gathering into a dense sheet of matter that seems to fall off the sides of the invisible surface of a huge balloon. They are inside a massive green sphere. An alien landscape spawns in midair and they suddenly feels themselves land on the new huge and soft green bubble. The bubble forms an invisible skin layered with falling debris of the site that seems flow around its surface and fall to the site below. They could see nearly two hundred meters below. A red and a blue bubble have formed beside theirs forming an eerie symmetry of a trinity of colors. Far below, they see the trinity surrounded by an army of Rabbis all holding the reins of the horses of light.

CHAPTER 44
(Children of the Light)

Nambu and Effion are alone in the darkness of the lab. Elaine arrives with the light of the scarab shining bright upon her heart. The two men feel renewed hope. They have hope. Nambu rushes to Elaine as she enters.

"What is going on out there?" he asks.

"The girls, they are in the light with the other children!" Elaine exclaims sobbing.

"What children?" Nambu asks.

"The children. Many, many children from all over. They are the only ones allowed to enter the light, no one else can enter. Many people are dead and more are dying as they try to enter. Only the children live." She explains crying and panting.

A slow rumbling is filling the building, and it starts to vibrate.

"We have to get out of here!" Elaine states.

"The building is fortified, it will not fall." Nambu tries to be encouraging, but is becoming worried.

"There must be another way!" Elaine cries out.

She is sweating even though the room is cold. Nambu is thinking. He looks at Effion huddled in a corner of the room probably frightened by the escalating vibrations of the buildings. Elaine is peering out the window, looking at the empty space filled with dead bodies. There is no living thing out there and for the first time, she faces her mortality and it frightens her deeply.

"John, please show us the way!" She cries, but no answer comes back to her. Effion is now rocking back and forth and holding on to the frame of the window. It is getting very bright outside, and Nambu recognizes that the shielded window cannot protect them.

"The roof, there is a shielded IFP on the roof!" Effion states abruptly. Nambu is surprised by the outburst. Effion might be autistic but that is a rational thought. Perhaps there is a chance with the IFP.

"Yes, yes, the IFP that brought us here, it is shielded!" Nambu exclaims.

Effion points to the window. The plastic film that screens the window is peeling off. It is freezing cold and the glass pane is being covered with condensed water crystals from the intense cold.

"It is getting too cold to go out, even if we use the shielded IFP its motors will freeze." Nambu surmises.

Effion starts to rumble some mathematical calculations.

"Minus point two degrees Kelvin, cold quark state when energy is converted to matter," he rambles.

"The energy absorption is due to energy converting to matter." Nambu explains to Elaine.

"Yes!" Effion states.

"Explain more!" Nambu commands as he runs to a table to get a sheet of paper and a pen.

"I want my brother, I want my brother, I want my brother.....!" he cries hysterically.

"He is going to flip on us!" Nambu exclaims.

"No! Look!" Elaine points to Effion.

Effion is scratching symbols on the glass with his nails. He draws the chemical symbols of hydrogen and oxygen with some quantum symbols and waves around the structure. Nambu recognizes the symbols.

"Water?" he asks.

"The resonant frequency of the water molecule," Effion states.

"The energy is being converted to water?" He asks.

"Yes! Planetary flood, Ice age!" Effion mumbles.

"Water!" The energy is being converted to water?" Nambu asks again.

"Yes, water, Holy water," Effion states lazily.

A faint voice comes through to Elaine:

"Elaine, you must have Faith and Hope, go to the Dam, go into the water, you will be safe." John Davis instructs as his voice fades out in her mind.

"I love you John" Elaine says holding the glowing emerald on her chest.

"I love you too!" John says, and his voice fades away.

"Did you communicate with him again?" Nambu asks,

"Yes, he wants us to go into the waters of the Dam." She explains.

"But how? We do not have any suits or swimming gear."

"I do not know, but I have faith in John." She says.

"We can access the bottom of the dam from here, in fact we are directly beneath the Hoover dam!" Nambu exclaims.

"Then, let us just go to the dam, perhaps there is a way." Elaine says hoping that a miracle can save them.

"What if the lake freezes over and we are trapped within it? We will all perish." Nambu states.

"We are dead anyway, and the best we can do is try!" Elaine answers.

Nambu grabs Effion and pulls him to follow them. They scurry down the corridor of the building to the Central Command chamber.

Emergency lights on the control panels are alive!

"This area has been well protected from the rays. The emergency lights survived." Nambu states.

"The shields," Effion replies, "the codes" he continues.

"What codes?" Nambu asks.

"He is talking about the satellite codes the President wanted." Elaine concludes. Nambu walks over to the main control panel. The indicators show very little activity, and there are no satellite orbits showing. He inspected the Orbiter monitor and finds nothing. Space is empty! The Resa ring is faint but the satellites are gone.

He looks at the entry panel where the codes would have been entered by Central Command and finds the panel active. The President had entered the satellite codes and the holograms must have been activated.

"Effion, the walls are ten meters thick, the concrete has a conductivity of one thousand joules per kilogram kelvin. How long before this area freezes over when it becomes zero Kelvin outside?"

Effion thinks for a moment then replies, rocking and tapping his head.

 "Forty two minutes." He answers.

"What do you mean, forty two two?" Nambu asks.

"Forty two minutes with plus or minus two minutes." Effion corrects.

"Ok, we have to leave in about twenty minutes otherwise we will freeze over!" Nambu calculates.

Nambu looks again at the satellite control station console. It is blinking. On the screen is a message showing the distance of Voyager 1, counting down.

"Damn, it is true. Voyager is coming back!" Nambu states,

"What is Voyager 1?" she asks.

"Nothing, let us go get the diving gear." He replies.

Nambu leads them to an elevator that takes them deep down to the turbine chamber of the dam. There are seven turbines still turning, but the power is gone.

"There is usually a wall map that shows were equipment is kept," Nambu states.

"Here, look, is this it?" Elaine asks.

Effion stares at Elaine and starts laughing.

"The Schrödinger equations are complete, the electron is free, no power," he states laughing out loud and pointing one hand to Elaine's crystal and another to the line of turbine generators, "no power!"

The darkness is broken only by the strange power of the light that comes from Elaine scarab emerald. Nambu squints to read the map. It shows the location of entrances and exits and where safety gear is stored for emergencies. Nambu rips the map off the wall and locates the way to a room that holds the supplies and the diving gear. He enters the room and finds a flash light. He tries to light it but it does not respond. He finds a few more and tries them, but they do not work.

"Effion, the electromagnetic field, why does it not respond?" he asks.

"The electron is intact only when shielded by the atomic field. No free electrons in superstring space." Effion replies.

"So, while atoms and other structures are held together, the free electrons are repulsed?" Nambu asks.

"Yes, the crystal field is strong." he says pointing to Elaine's scarab.

"My God!" Nambu claims, "look around you Elaine!"

Elaine strains and looks around her. A hallow of water droplets has formed a large sphere around her. The droplets are each shining like tiny spherical lamps around her.

"Emissions are the symmetry of z-particles, and Holy water is made to baptize us all." Effion states.

Nambu sees a bucket sitting on the floor with water in it. He calls out to Elaine and asks her to prepare for a wash test. He throws the bucket of water and watches as it bounces off some invisible shield around Elaine.

"It repels water. The field in your scarab repels water." Nambu exclaims surprised.

"Yes, it does, we have to enter the lake and shield ourselves." She says.

Nambu calls Elaine and Effion and hands them each a diving suit and an oxygen tank.
. Elaine looks at Nambu inquisitively.

"We do not need that!" She says.

"How do we get into the water?" Nambu asks.

"Have Faith and Hope in what John says, we will be OK." She says.

"Have faith and hope and love, aliquot sum thirteen." Effion states.

Only Effion understands what he says. At this point Nambu has no choice but to believe in Elaine and have Faith and Hope. He realizes that the oxygen tanks will be needed to keep them alive. He grabs one and assumes it will last them at least a few hours. Nambu takes out a thermometer from the instrument pack he had picked up. He measures the temperature of the room and then goes back to the control panel. The countdown to Voyager 1 leaves them with two hours to go.

"Effion, how long before the final explosion if the temperature in this area is minus thirty?" He asks.

"Three hours fifteen minutes." He answers.

They will need two tanks. Nambu grabs a second tank and hands it to Effion.

"Carry this with you, do not lose it!" he orders.

They hurry down to the main flood gate that opens to the lake. The waters have risen to midday.

"The gates were not designed to hold water at that level. They will eventually break open and flood the entire building." Nambu explains. There is a ladder leading up to the upper level of the gate. He asks the two to climb ahead of him. They arrive at the upper deck of the gate and could see the level indicator rising as the Lake fills with water formed from the crystal reactions Effion had predicted earlier.

"I am going to open the gate, and it will flood the building. You better be correct about this Elaine, or we are all going to drown."

"You must have Faith, Hope and Love," Elaine states with a serene voice that has no hint of fear. Nambu renews his faith. He looks ta Effion now huddled close to Elaine like a child. Effion knows the crystal will shield them. He knows!

Nambu opens the flood gates. The water gushes beneath them filling the chamber and starts rising fast. It is freezing cold and icicles are forming already. They must go in, or they cannot make the lake, and they will freeze to death.

He turns and looks at them.

"We either jump now, or the water will freeze and we cannot enter the lake!" Nambu commands.

They jump hugging each other in a tight circle. The water separates around them as the field of the crystal sphere pushes into the water. They descend in the spherical bubble of the crystal until they hit bottom. The water surrounds them in a huge spherical field that forms a closure around them. The water starts to freeze around them. Nambu activates the valves of the oxygen tanks. It will keep the supply going slowly for a few hours. Strangely, they do not feel too cold. The field is protecting them from the low temperatures.

Effion falls on his knees and starts to pray.

"In the beginning, there was void without form. And God said "Let there be light..", his words echo around the sphere of ice and reverberates into a continuous prayer of hope. Elaine falls on her knees, she has faith in the Pontiff, love for John and her children, and Hope that they will all make it. Nambu looks on, but the prayers hit there mark. The scientist falls on his knees, he now knows that only Faith, hope and Love can save them.

CHAPTER 45
(Desperate Guardians)

The Guardians are desperate. The unstable craft has lost faith, hope, and love, and there is no way to prevent doubt, desperation, and hate, from overwhelming them. The hatred of the Minga is boiling in the craft and their children have defected to the other side. They must rearrange the crystals to match colors and make them stable again, otherwise, the craft will be destroyed by the instability of their own emotional dimensions.

Quickly they scramble to reorder the crystals into a new combination by moving their positions to balance doubt, hate, despair, anger, disgust, deceit, inquisitiveness, depression, anxiety, boredom, envy, fear, guilt, horror, sadness, shame, worry, pride, rage and regret into a combination that can survive with four dimensions of spacetime without Love, Hope and Faith.

Lucifer is determined to build the best out of what is left.

"From Love, they can have Joy and trust, from faith, they can have anticipation and trust, and from hope they can have anticipation and joy. We must compensate for all these dimensions that have been taken," he cries out.

"They do not have Fear and Anger, so how can the Master control them?" the golden beard Guardian asks.

"Did you not read the manual for this vehicle? Love does not fear you fool. Love is patient, love is kind. It does not envy, it does not boast, it is not proud. It is not rude, it is not self-seeking, it is not easily angered, it keeps no record of wrongs. Love does not delight in evil but rejoices with the truth. It always protects, always trusts, always hopes, always perseveres. Love never fails." Lucifer answers.

"Then, how do we believe you when we have none of these qualities, how do we know you are not deceitful, how do we stabilize craft?" Another Guardian asks.

"We must change the color combinations. Awe is Fear and Anticipation, Disappointment is Anxiety and Sadness, If we add Sadness and Disgust together their colors of light will give the color of Remorse," Lucifer says.

"But why do we need Remorse?" the White Guardian asks.

"Because the Master forgives with Remorse. He will not punish the remorseful, so, we must put these crystals outside the body of the craft, so they cover the rest. " Lucifer answers.

"But that is Deceit," another guardian says, "the color of Deceit will increase and make us even more deceitful." The Golden bearded Guardian states.

"You Doubt my ability to control this craft, I have read the manual thoroughly and know its mode of operations." Lucifer answers.

"That is Pride," the Yellow Guardian states, "and that will color the craft with Shame if you fail, and, we will all perish!"

"You are Envious of me!" Lucifer states.

"We are only Inquisitive, and Anxious, and we Worry about what will happen when we fail." The Red guardian states.

"You have Rage in you Lucifer, before this, you traveled through that dimension to man's world, now, it is a road that we cannot travel through, or we will all perish." The black Guardian states.

"Take Despair and replace it with Envy. We can make the color of Anger on the far side of the craft. This will color the craft with Remorse, Anger and Deceit." Lucifer says.

"Then, we will be left with depression and boredom on the inside and we cannot survive for long" The White Guardian says.

"You are feeling fear," Lucifer says.

"No, we feel guilt, horror, and regret, Lucifer, we cannot survive in such a world, we will perish!"

"If you do not do it, I will change the colors myself!" Lucifer shouts.

"You cannot fly this craft without Love, Hope and Faith, we must separate the crystals and fly them apart!" The old Guardian states regretfully.

"Man was given the three quarks of matter, we have all the others, and if we use them right, we can make a world of exotic matter that we can survive within." Lucifer says.

The craft rumbles with a loud vibration and a flash of light burst through the desert, burning a deep hole beneath the craft. The unstable craft starts to spin out of control and the Guardians scramble to stabilize the craft by tuning their emotions.

"That is Hope, and with that you have annihilate Despair and we have lost that crystal forever, and now, we will perish!" The old Guardian states.

"The Master gave us eternal life and great wisdom, with that, we cannot perish. He cannot take away his gifts to us." Lucifer says.

Again, the craft rumbles with a loud vibration and a flash of light burst through the desert, burning a deeper hole in the desert beneath the craft. The unstable craft starts to spin out of control and the Guardians scramble to re-stabilize the craft by re-tuning their emotions.

"That is Faith, and with that you have annihilated Doubt, and we have lost that crystal forever, and now we will perish!" The old Guardian states again.

"Did you not read the manual for this vehicle, Lucifer? Hate is full of fear. Hate is impatient, and wicked. It envies, it boasts, it is proud. It is rude, it is self-seeking, it is easily angered, it keeps records of wrongs. Hate delights in evil but angers with the truth. It always harms, never trusts, never hopes, never perseveres. Hate never fails." Lucifer answers.

The old Guardian eyes Lucifer from the corner of his misty eyes. He feels love for Lucifer, but can no longer keep the secret from him.

"Lucifer, we are full of Hate and fear, we are impatient, and wicked. We envy you and you boast with pride. You are rude and self-seeking, and you are easily angered. We have kept records of your wrongs. You have never trusted us, and we are never with hope. Hate never fails." He says.

"What are you saying to me?" Lucifer asks.

"We have no trust in you. You will not save us from peril Lucifer. We have no Hope, and now, we have no Faith. I stand before you in revolt, surrounded by Abaddon the destroyer, Abezethibou the one-winged, Amduscias the Unicorn, Amon, the commander of legions, Baal, the lord of hell, Balam the serpent, Beelzebub, the prince of demons, Behemoth, of the deep, Lillith, the murderer of children, Rahab, the angel of pride and violence, Satan, the adversary, Uzza, the strong, Wormwood, the one who brings plagues upon the Earth, Xaphan, the strokes the fires of hell, and Zepar, the bearer of false love no longer have faith or hope in you. It has been taken away from us. We wish to serve you. We want the crystals divided among us so we can travel through our own dimensions." The old guardian explains.

Lucifer is angry.

"If we divide the crystals, which one will you take?" Lucifer asks.

"Abaddon, Baal, Lillith, and Uzza want the crystals of Anger. Abezethibou, Amon, Balam, Behemoth, and Satan want the crystal of Hate. Amduscias, Beelzebub, Rahab, Wormwood, Xaphan, and Zeparwant the crystal of Deceit." The old man says.

"But that leaves me with no Hate, Deceit or Anger!" Lucifer states angrily.

and and Suddenly, the cube unfolds into a massive and endless plane, expanding into a burst of fundamental particles and light. A cosmic egg unfolds at a tremendous rate in front of his spirit eyes. Then, another dimension unfolds forming a three dimensional universe of his own. A new spacetime forms around the particles, expanding at a fantastic rate, unfolding into a massive new universe of matter and force fields. A soup of uncontrolled events has not yet gelled into the solidity of laws and conditions. He observes the expanding universe until it releases itself into the dance of creation.

In the distance Jesus smiles as he heard Michael struggle with his new found role as God of a new Universe. A group of Angels giggle as Michael begins to formulate the laws of his new universe:

"...matter shall not evolve, it shall be alive at all times, obeying the will of its creator. There shall be no quantum uncertainty. Never shall the laws of my Universe be immutable!" he promises.

As the commotion in the Moroccan desert reaches a peak, millions of people start breaking into the great cavity that holds the twenty-six crystals and their strange inhabitants. They stream into the Garden of Eden and M'Goro leashes out in anger at the invading masses, killing all those who come upon the Garden and its fruits. The Alien field from the great craft has already consumed the entire Earth awaiting the final crystal, but to no avail. It is too late! The craft is becoming unstable as space and time mingle uncontrollably into twenty-six unstable dimensions. All the inhabitants of the Earth are dead, their bodies and spirits trapped forever in the field of the satanic craft that is the Unsightly Unsymmetrical Universe. Inside the craft, the demonic occupants are becoming worried, as they await the craft's linking with the elusive last crystal field! But the crystal is gone!

The field expands uncontrollably beyond Earth, spreading to the entire material Universe, capturing stars and galaxies in its hungry bid for symmetric union with the last crystal that will never come.

Jesus stands at the gates of Heaven and looks at the approaching field. In his hands is the last of the twenty-seven crystals. He replaces it with an anti-crystal. He waits until the field is only a few light years away. Then, He throws the anti-crystal into the fiery furnace that has formed within the massive cosmic cube, forever closing off the alien inhabitants into now hellish and distorted space-time of the material universe.

The angels around Jesus cheer heartily as the cosmic cube collapses into a new birth; a new Big Bang!

"The laws of the Universe I built were Immutable," He says smiling, "Perhaps that was an oversight. This new universe will be perfectly symmetrical in every way, and in another ten billion years, its inhabitants will have no need for me!"

Ten thousand years later.

he monkeys are roaming about the new Garden of Eden in Michael's new universe. The Universe evolves following the same rules of creation he had established long before. The inevitable Big Bang and the creation. A broth form that brings into being living matter. The matter evolved into a complex array of living forces. The Garden blossomed into a new earth and wondrous garden of indescribable beauty. There are no more humans. The old order of evolution was contaminated with the advent of man, woman and the devil. This time, the order of evolution is to be governed by a new hierarchy.

A monkey approaches Michael. Michael signals the monkey to come closer.

"Come here you little terror" he commands jokingly.

"You see that tree over there?" The monkey does not quite get it. Again, Michael points to a very prominent tree in the garden. "You see that tree over there? "The monkey looks in the direction of the tree and somehow seems to understand the question. It nods. "You can eat from all the fruits in this garden, but from that one tree of knowledge and evil may not eat. Do you understand?"

It is a one sided conversation. The creature nods. Its synaptic membranes are expanding at a fantastic rate by the mere presence of Michael in the garden. It whines and grunts uncomfortably. Michael smiles, he has been here before!

THE JOURNEY CONTINUES WITH
YUMUYA SCROLLS BEGINNINGS, YUMUYA SCROLLS RETRIBUTION, YUMUYA SCROLLS TRANQUILITY.